MODERN THERMODYNAMICS

Revised Second Printing

MODERN THERMODYNAMICS

Revised Second Printing

Arieh Ben-Naim

The Hebrew university of Jerualem, Israel

Diego Casadei

University of Applied Sciences and Arts,
Switzerland & University of Birmingham, UK

World Scientific

W JERSEY · LONDON · SINGAPORE · BEIJING · SHANGHAI · HONG KONG · TAIPEI · CHENNAI · TOKYO

Published by

World Scientific Publishing Co. Pte. Ltd.

5 Toh Tuck Link, Singapore 596224

USA office: 27 Warren Street, Suite 401-402, Hackensack, NJ 07601

UK office: 57 Shelton Street, Covent Garden, London WC2H 9HE

Library of Congress Cataloging-in-Publication Data

Names: Ben-Naim, Arieh, 1934– author | Casadei, Diego, 1973– author.

Title: Modern thermodynamics / Arieh Ben-Naim, The Hebrew University of Jerusalem, Israel,
   Diego Casadei, University of Applied Sciences and Arts, Switzerland &
   University of Birmingham, UK.

Description: Hackensack, NJ : World Scientific, [2016] |
   Includes bibliographical references and index.

Identifiers: LCCN 2016041558| ISBN 9789813200753 (hardcover ; alk. paper) |
   ISBN 9813200758 (hardcover ; alk. paper) | ISBN 9789813200760 (pbk. ; alk. paper) |
   ISBN 9813200766 (pbk. ; alk. paper)

Subjects: LCSH: Thermodynamics.

Classification: LCC QC311 .B395 2016 | DDC 536/.7--dc23

LC record available at https://lccn.loc.gov/2016041558

British Library Cataloguing-in-Publication Data

A catalogue record for this book is available from the British Library.

Revised Second Printing

Dedicated to all students who were exposed to the traditional
approach of Thermodynamics

Preface

This book presents Thermodynamics in a modern way, based on four fundamental physical facts: the atomic nature of matter, the indistinguishability of atoms and molecules of the same species, the uncertainty principle, and the existence of equilibrium states.

The atomic nature of matter could only be experimentally proven 2.5 millennia after the atomic hypothesis had been formulated by Democritus. Atoms and molecules are so tiny and numerous that we have the impression that matter is continuous. However, this is only an approximation: at very small scales, matter is discrete.

Furthermore, fundamental particles, atoms and molecules of the same type are indistinguishable. For example, in the scattering between two electrons there is no way of telling which particle came from the left or right direction. As there is no unique property that can be used to "label" any one of them, at best we can imagine to follow the one we choose during its evolution in time. However, strictly speaking this is not possible: the uncertainty principle establishes a lower limit to the precision of any simultaneous measurement of location and momentum of a particle.

Thus, while observing a macroscopic system, it is impossible to follow the evolution of single atoms and molecules. Even if we could, the resulting huge amount of information would be practically impossible to process in any finite time. At best, we can describe their configurations in probabilistic terms. Hence probability theory plays a fundamental role in our understanding of Nature.

Luckily enough any macroscopic system, unless perturbed, at some point reaches an equilibrium state characterized by constant macroscopic properties. The latter can be summarized by providing the values of very few physical quantities. This allows us to study macroscopic systems by focusing on some amount of information that is completely negligible, compared to a full description of the configuration at microscopic level.

Only recently it has been discovered that the concept of information is not just an incidental connection with thermodynamics. Indeed, we will show that Shannon's measure of information for an ideal gas turns out to coincide, apart from the freedom of choosing the units, with Boltzmann's entropy. The latter has the same properties as Clausius' entropy, defined in the middle of XIX century, while scientists were focusing on the work performed by thermal engines, so important in the period of the industrial revolution.

This book is divided in two parts. The first introduces all concepts that are usually covered by a first course on thermodynamics, whereas the second addresses applications ranging from phase diagrams to mixtures, chemical equilibrium and the properties of water. Depending on the curriculum chosen by the students, selected topics from the second part may be included in a first course on thermodynamics, and the others may be covered in more advanced courses.

In the first part of this book, we build up thermodynamics starting from few physical facts with the help of information theory, which is based on probability theory. This approach, in which Shannon's measure of information (SMI) plays a central role, is very different from the historical development typically followed by most textbooks. For this reason, chapter 1 provides a quick overview of our modern introduction to thermodynamics. On the other hand the history of thermodynamics is interesting by itself, hence chapter 2 summarizes the last few centuries with the goal of providing the Reader with the context in which all different pieces of the puzzle have been discovered.

The next three chapters are very unusual for a textbook on thermodynamics. Both entropy and the second law are defined and interpreted in terms of probability distributions. Therefore it is essential to be familiar with some basic ideas of probability theory. Chapter 3 introduces probability theory, a branch of applied mathematics with applications in practically all scientific disciplines. This chapter provides the building blocks for understanding the concepts of information theory, and in particular Shannon's measure of information, introduced in chapter 4. Because of the central role of SMI, chapter 5 illustrates its main properties and the three theorems which will be fundamental in building the entropy function of an ideal gas.

The bridge between information theory and thermodynamics is provided by two chapters. Chapter 6 computes the SMI of an ideal gas and identifies Shannon's measure of information for an equilibrium state with Boltzmann's entropy, apart from a multiplicative constant, corresponding to the freedom of choosing the units for the entropy. In particular, the SMI of an ideal gas at equilibrium turns out to coincide with the entropy function $S(T, V, N)$ computed by Sackur and Tetrode in 1912 in the context of Boltzmann's approach to thermodynamics, although it is derived here in a completely different way, starting from information theory and the fundamental physical facts mentioned above. The next step is to consider the fundamental entropy function $S(E, V, N)$, which can be used to derive all other thermodynamic quantities. In addition, this function achieves a maximum over all possible constrained equilibrium states for an isolated system. This is our formulation of the second law of thermodynamics for a system at constant (E, V, N).

Chapter 7 studies the properties of the fundamental entropy function $S(E, V, N)$ and shows how temperature, pressure, and chemical potential are related to its partial derivatives. Next, the entropy change is computed for few spontaneous processes, like gas expansion, heat transfer from a hot to a cold body, mixing and demixing processes. It also shows the connection with Clausius' entropy, defined in terms of (small) heat transfers at constant temperature. This chapter also clarifies some misinterpretation of entropy, like the analogy with disorder, which puzzled generations of scientists, starting from Gibbs himself.

Having shown (in chapter 6) that the SMI for an ideal gas at equilibrium coincides with the thermodynamic entropy, we postulate that for any general system at equilibrium the entropy is a particular case of SMI. All results from the historical development of thermodynamics can then be reproduced. This is done in chapter 8, where the first and second laws are illustrated in details. Thermodynamic transformations and cycles are also addressed in this chapter, including a treatment of the efficiency of thermal machines and Carnot's cycle. In particular, the *mean-*

ing of the second law is clarified, and the law itself is explained in terms of SMI properties. Chemical equilibrium is also introduced, and formulations of the second law for systems at constant (T,V,N) and (T,P,N) are provided in terms of Helmoltz and Gibbs energies, respectively. With this chapter, our introduction of thermodynamics is complete.

The second part of the book focuses on applications of thermodynamics which cover several important aspects. Chapter 9 illustrates the different states of matter and the phase transitions. Gibbs' phase rule is obtained, and the use of phase diagrams is shown in details. While addressing the coexistence of two phases of the same component, the Clausius-Clapeyron equation is obtained. Focusing on the liquid-vapor and solid-liquid equilibrium lines, one obtains the Clapeyron equations. While considering three phases of the same component, the triple point and the critical point are illustrated. Finally, allotropy is explained with examples provided by phase diagrams of sulfur, phosphorus, and carbon. Next, two-component systems are introduced, and the equilibrium conditions between different phases of both components are obtained.

Chapter 10 is about mixtures and solutions, treated in the framework of the Kirkwood-Buff theory of solutions. Theoretical results are compared to experimental results. The deviations from different kinds of ideal behavior are considered, and interpreted in the same framework. Next, global and local characterizations of mixtures are presented, and solvation thermodynamics is introduced.

Chapter 11 focuses on chemical equilibrium, starting from the simple isomerization reaction. General chemical reactions are also addressed, and the chemical equilibrium in solution is illustrated. Finally, temperature, pressure, and concentration dependences of the equilibrium constant are obtained.

Chapter 12 deals with water, of fundamental importance for all biological systems. Water has quite peculiar properties, illustrated in this chapter and interpreted at molecular level. This is the most advanced chapter, making use of several of the tools developed in the book.

Scattered through the text, there are several exercises that should help the Reader to better understand and apply the concepts presented in the book. The solutions of selected exercises are provided in appendix A.

Contents

Applications 239

PART 1

Fundamentals

Chapter 1

Introduction and Overview

Thermodynamics features in almost every process we witness in our daily lives, from simple house chores such as cleaning, cooking and refrigerating, to industrial applications like production of electricity or chemical synthesis of new materials. In addition it plays a fundamental role even in highly complex processes like our metabolism. Thus it is no surprise that learning thermodynamics is part of every curriculum in scientific education.

Teaching thermodynamics traditionally proceeds by following the historical development. However, this procedure does not lead to understanding the ultimate reasons why systems behave as described by thermodynamics. In particular, students traditionally learn about the second law of thermodynamics, which governs all spontaneous processes, and learn how to compute entropy changes, without really understanding why entropy can never decrease in a spontaneous process occurring in an isolated system, what drives its evolution, and what is the meaning of the second law. Entropy is also misused in many fields where it cannot be defined, such as life itself or the entire universe.[1]

This is the reason why we follow a different route, starting from four fundamental physical facts (the atomic nature of matter, the indistinguishability of atoms and molecules of the same species, the uncertainty principle, and the existence of equilibrium states) and analyzing the be-

---

[1]It is a fact (an astonishing one) that the cosmic microwave background radiation has the best black-body spectrum ever measured. This means that there is at least one component, which is at thermal equilibrium and permeates the entire visible universe. For this species one can define an entropy density and can apply thermodynamics. But for the *entire* universe, which today is not a system at equilibrium, entropy cannot be defined.

haviour of complex systems with the tools of information theory (in particular with Shannon's measure of information, or SMI, which can be defined on *any* probability distribution). We show that entropy is a particular type of SMI and, as such, its value is a result of the evolution of the SMI. In particular, entropy coincides (apart from the freedom of choosing the units) with the SMI of a system at equilibrium. Thus, although the SMI can change with time, entropy is not a function of time.

Any complex system spontaneously evolves toward states maximizing the SMI, and the second law of thermodynamics can be cast into the statement that, once some internal constraint is removed, an isolated system will reach a state of maximum SMI. Because it is characterized by constant values for all macroscopic (i.e. large-scale) quantities, this state is the thermodynamic equilibrium state.

Deviations from this maximum-SMI state, although possible in principle, have a probability that decreases for systems having very large number of particles. For any macroscopic system, counting 10^{23} particles or more, this probability is so low that the stability of an equilibrium state can be considered absolute: the age of the universe is way too short to witness any measurable deviation from equilibrium.

This chapter outlines few important aspects that will repeatedly pop up in the rest of the book, and offers a comparison against the traditional approach to teaching thermodynamics.

1.1 Ways of Teaching Thermodynamics

The usual approach for teaching thermodynamics starts from the concepts developed in classical mechanics. One recognizes the heat as a particular form of energy transfer and then introduces the first law of thermodynamics, which is basically the principle of energy conservation. From the fundamental observation that two systems in thermal contact only with each other eventually reach the same temperature, one defines thermodynamic equilibrium as characterized by constant values of thermodynamic variables like pressure, volume, temperature, etc. Cyclic transformations are then used to define heat engines, whose efficiency is limited by the upper bound provided by the Carnot cycle. The second law of thermodynamics is formulated in terms of the entropy, a state function that can never decrease in a spontaneous process occurring in an isolated system.

Clausius' entropy is introduced in conventional thermodynamics in terms of heat exchange and temperature of thermal engines. No explanation is provided, which tells us *what* entropy is and *why* it can only increase. Indeed, the whole thermodynamics has been developed before the atomic nature of matter was recognized and accepted. It is the very fact that matter is made of microscopic particles, such that a huge number of them randomly fills any macroscopic volume, no matter how small it is, that originates all behaviours *described* by thermodynamics.

Both the atomic nature of matter and the concept of probability are foreign to conventional thermodynamics. However, once the existence of a myriad of minuscule atoms is accepted, it becames natural to describe macroscopic quantities in terms of average properties: it is impossible to follow the exact evolution of each molecule (both in practice and in principle). This leads to statistical thermodynamics, pioneered by James Clerk Maxwell (1831–1879), Ludwig Boltzmann (1844–1906) and Josiah Willard Gibbs (1839–1903), although the atomic nature of matter was experimentally proven only after their work, at the beginning of the 20th century. It was only then that the language of probability became widely adopted. Unfortunately, not all science scholars attend a course on statistical thermodynamics.

Both classical thermodynamics and statistical mechanics are very successful theories. Because the latter provides a model for the former, it can be considered more fundamental. The deep connection between them can be better appreciated by changing point of view, and considering the behaviour of complex systems from the perspective of information theory. The key point is the measure of information introduced by Claude Shannon (1916–2001) in 1948. Recently one of the authors (Ben-Naim, 2006a) has proven that the entropy of an ideal gas at equilibrium coincides with SMI. Interactions can also be introduced as mutual information (Ben-Naim, 2008), which allows to identify also the entropy of real gases as a special case of SMI. Thus, the properties of SMI (detailed in chapter 5) *explain* both the concept of entropy and the second law of thermodynamics. This is why we opt for a non conventional approach to teaching thermodynamics.

Now let's step back for a while, and look at the basic principles of thermodynamics.

1.2 The First and Second Laws

Thermodynamics has evolved over a very long period of time. Concepts such as force, energy and work have been developed in mechanics long before thermodynamics, and the concept of *internal energy* was incorporated into thermodynamics through the first law. The first law of thermodynamics is essentially an extension of the principle of conservation of energy known in classical mechanics, to include also *thermal energy*. The latter is a form of internal energy associated with the kinetic energy of the atoms and molecules that constitute all pieces of matter.

The new element in thermodynamics, not deriving from mechanics, is the concept of entropy, intimately bound to the second law of thermodynamics. From its introduction in the 19th century, in terms of heat engines and practical problems focusing on the efficiency of cyclic processes transforming thermal energy into useful work, the concept of entropy changed significantly: the modern view presents entropy in terms of probabilities and measure of information. This is a remarkable and a profound evolution! Over a period of 150 years, the evolution of the concept of entropy and of the second law was not only conceptual, but had also many practical ramifications.

The first law states that the total energy of an isolated system is constant, a very familiar principle: it parallels the conservation of momentum and of mass.[2] On the other hand, the second law states that the entropy of an isolated system never decreases. We shall see in chapter 6 in what sense it reaches a maximum value after removing some constraints.

Clearly the second law has a non-conservative nature. In addition it features a new and unfamiliar concept, entropy. What is this quantity that must always increase? What "source" would create it? Does it "flow" from a system to another? For a long time, entropy was surrounded by a sense of mystery, which even today is still very much around us.

---

[2]Incidentally, these separate conservation laws have been unified into a single conservation law by Albert Einstein in 1905: the conservation of the so-called four-momentum.

One of the purposes of science is to understand nature. Hence it is quite unsatisfactory if some quantity that has a central role is not understood. For this reason, in this book we shall develop the concept of entropy not along the historical route, but as something spawn from the concept of Shannon's measure of information. This route provides a simple and clear meaning to the concept of entropy, and its non-conservative nature does not conjure any mystery.

There are two fundamental concepts underlying entropy and the second law of thermodynamics. The first is the atomic nature of matter, composed by a huge number of microscopic identical particles that can only be treated in some average sense. This means that one must adopt the language of probability. The second is Shannon's measure of information, which can be defined on any probability distribution. The theorems obeyed by the SMI govern the evolution of probability distributions subject to given contraints. In turn, this is connected to the spontaneous evolution of macroscopic systems toward equilibrium.

These fundamental concepts, probability and SMI, feature not only in thermodynamics, but also in almost all branches of human activities: from linguistics to communication, from mathematics to arts. It is difficult to find a subject in which none of them plays a role. Nevertheless, the connection with the second law as formulated in the 19th century is not straightforward.

1.3 Early Formulations of the Second Law

Traditionally, the birth of the second law is associated with the name Sadi Carnot (1796–1832). Engineers of the late 18th and early 19th centuries were interested in heat engines, converting thermal energy into mechanical work. Carnot was interested into the *efficiency* of a heat engine, i.e. he wanted to know how much useful work can be "extracted" from what we now call *thermal energy*. At that time the "caloric" (i.e. heat) was believed to be a kind of fluid that flows from a hot to a cold body. Carnot found that there is an upper limit to the efficiency of any heat engine, and that this limit is determined by the ratio of the two temperatures between which the heat engines operate (we shall discuss the so-called Carnot cycle in chapter 8).

The simplest way of understanding heat engines is to think by analogy with the more familiar water-fall engine. Water always flows from a higher level to a lower level. On its way down, it can rotate a turbine. The rotation of the turbine can be used to generate electricity, which in turn can be stored and later used to do useful work. Likewise, in a heat engine "the caloric falls" from a higher temperature to a lower temperature. On the way down, it can heat a gas which expands. This expansion can be used to push a piston, to lift a weight or to rotate a wheel, which in turn can produce electricity.

The Carnot cycle and other heat engines laid the cornerstone for the second law of thermodynamics. The first formulation is due to William Thomson (1824–1907), later known as Lord Kelvin. His formulation directly involves the concept of heat engine, and states that *there can be no heat engine operating in cycles, the sole effect of which is pumping energy from one heat reservoir and completely converting it into work.* Clearly this formulation does not sound like something obvious, nor like something that everyone can accept intuitively. It sounds more like a technical statement on an engineering problem.

A few years later, another formulation was proposed by Rudolf Clausius (1822–1888): *there can be no process, the sole result of which is a flow of heat from a colder to a hotter body.* Clausius' formulation rings like something obvious, something that everyone is familiar with. We all have experienced that, if a hot body is brought to contact with a cold body, heat always flows from the hot to the cold body. We accept that statement even if we do not know what "heat flow" actually means. Somehow, we associate *heating* (of the cold body by the hot body) with the noun *heat* (the "thing" that flows from the hot to the cold body). More generally, heating a body in the sense of increasing its temperature is associated with transferring of some kind of energy which we call *heat*. This statement is almost axiomatically accepted as truth. We never see the reverse of this process — flow of heat from the cold to the hot body — to occur spontaneously.

Incidentally one should notice that *heating* in the sense of increasing the temperature is usually associated with *heating* in the sense of supplying heat. However, this association is not always true. One can supply energy to a body, while its temperature remains unchanged. This happens in a phase transition, for example when heating (in the sense of supplying energy) boiling water or melting ice. In addition, the temperature of a system can be increased withouth supplying thermal energy. For example, temperature changes can occur by chemical reactions taking part inside the system without any heat exchange with the surroundings. Alternatively a system may be "heated" by performing mechanical work on it, as it was demonstrated by James Joule (1818–1889), who in 1843 determined the mechanical equivalent of the heat, opening the road to the first law of thermodynamics.[3]

The two formulations of the second law by Lord Kelvin and Clausius given above seem to be completely unrelated. One deals with a technical limitation of a heat engine, the other with a common everyday experience of everyone. Yet, the two formulations are equivalent. One implies the other, although we shall not demonstrate the equivalence of these formulations here (this is shown in most elementary textbooks on thermodynamics).

Actually, we can formulate many more statements of the second law. Here is one simple example: *there can be no process, the sole result of which is that a gas that initially occupies a region of volume V is condensed to a smaller region, say of volume V/2.* Again, this formulation does not seem to be related to the other statements. Nevertheless, one can show that it is equivalent to Clausius' statement of the second law.

We can of course reformulate the last statement in a positive language: *when a partition between two compartments is removed, a gas initially occupying one compartment will always expand to occupy the entire volume of the two compartments.* A related statement, again formulated in a positive language, is: *when a partition separating two different gases in different compartments is removed, the two gases will always mix and form a homogenous mixture.* One can show that the last two statements are in fact equivalent to each other and are equivalent to other statements of the second law of thermodynamics.

Of course, we did not exhaust all possible formulations of the second law. In fact, there are as many statements as there are processes in nature that occur spontaneously in one direction.

---

[3]The SI derived unit of energy, the joule, is named after James Joule.

It was Clausius' ingenuity that led him to find the all-encompassing formulation of the second law of thermodynamics, from which all the particular formulations may be derived. Clausius introduced a new word into the vocabulary of physics: *entropy*. With this new quantity, Clausius could proclaim the most general version of the second law:

> *In any spontaneous process occurring in an isolated system, the entropy never decreases.*

We shall further discuss in chapter 8 the meaning of the concept of *entropy*, and to what extent this term was an appropriate choice. Here, it suffices to say that the new term "entropy" was not easy to digest. Although entropy changes were originally *defined* as the ratio between the amount of transferred heat and the absolute temperature (see section 7.3.3), it was not clear what such quantity means. Furthermore, the fact that entropy is a non-conserved physical quantity was like a bitter pill to swallow. Physics was built on conservation laws. It is not surprising that the ever increasing entropy brought about an element of mystery into physics.

In spite of the mystery and in spite of the lack of understanding of the meaning of entropy, classical thermodynamics became an extremely useful tool. Although it does not offer a *meaning* on a molecular level, thermodynamics does offer a convenient and useful language with which we can understand the unifying principles underlying many processes that we encounter both at home and in our laboratories. Anyway, the history does not end here.

1.4   The Atomic Nature of Matter and Statistical Mechanics

Towards the end of the 19th century, the kinetic theory of gases was developed. In a nutshell, this theory aimed at explaining the macroscopic properties of gases in terms of the motion of individual molecules. In spite of its remarkable success in explaining in molecular terms measurable quantities such as pressure and temperature, the very assumption underlying the kinetic theory of gases — that matter consists of small units, atoms and molecules — was not universally accepted.

Not only the existence of the atoms was debated, but also the statistical methods, used to calculate the macroscopic quantities such as pressure and temperature from the molecular motion of individual particles, were foreign to physics.

Classical mechanics, electromagnetism and thermodynamics were formulated without explicit reference to the atomic nature of matter. The laws of physics were proclaimed to be exact and absolute. No provision for exceptions.

On the other hand, the kinetic theory of gases uses the methods of statistics to compute average quantities. In order to compute averages one needs probabilities, and probability theory was not yet part of the language of physics.

James Clerk Maxwell (1831–1879) was the first to use the tools of probability to express physical quantities, such as pressure and temperature, as average quantities associated with the motion of the atoms and molecules. He calculated the probability distribution for a particle to move within a certain range of velocities. From this distribution one can calculate the average kinetic energy of the particles. Pressure and temperature of the gas, two quantities that have been previously defined and measured without any reference to the atoms and molecules, are also related to average quantities.

Accepting an average quantity as a meaningful physical quantity is one thing, but accepting the *probabilities* as meaningful physical quantities is quite another. The introduction of probabilities seemed to undermine the foundations of physics.

A major stride forward was made by Ludwig Boltzmann (1844–1906). He was a staunch supporter of the atomic theory of matter. Boltzmann showed that a system of particles will tend to a specific distribution of velocities at equilibrium (chapters 5 and 6 will clarify how and why this happens), the so-called Maxwell-Boltzmann distribution. However, Boltzmann's boldest and far reaching contribution to physics was his interpretation of the entropy in terms of the number of microstates, i.e. the number of arrangements of particles in "cells" of position and momentum (we shall discuss this interpretation of entropy in chapters 5 and 6).

Boltzmann gave a molecular interpretation of entropy as well as a probability-based argument for the one-directional change of entropy.

Figure 1.1: James Clerk Maxwell (left) and Ludwig Boltzmann (right).

The direction of change of the entropy, encapsulated in the second law, is not an absolute but a probabilistic law: basically, it is a consequence of the fact that events that are more probable will occur more frequently. A surge of criticism ensued! The idea that violations of the second law are possible, although highly improbable, was untenable. The second law stated that heat will *always* flow from a hot to a cold body, never from the cold to the hot body. In the Boltzmann formulation the *impossibility* of such a process turned into a highly *improbable* occurrence.

The terms *absolutely impossible*, and *highly improbable*, are conceptually very different. However, if we recognize on one hand that we can never be sure that any law of physics will *never* be violated, and that on the other hand the "highly improbability" of violating the second law means once in many ages of the universe, then the two terms become very close to each other. Not only close: in fact one can claim that the latter is stronger than the former. In other words the admitted *non-absoluteness* of Boltzmann's formulation of the second law is more *absolute* than the proclaimed *absoluteness* of Clausius' formulation.[4]

This statement should be understood as follows. Consider the "absolute" second law, e.g. the expansion of a gas or the transfer of heat from hot to a cold body. How certain can we be, that this law (or any other law of physics for that matter) will not be violated? All we can say is that,

---

[4]A more complete discussion is presented in [Ben-Naim (2008)]

since we have observed for a few thousand years that the law was never violated, we can also project into the future and claim that this law will not be violated in the next thousand, or million or perhaps billion years. But no more than that. Yet, with this "level of certainty" we consider it as an absolute law.

On the other hand, the statistical mechanical formulation of the second law claims that the second law will be violated, but only very seldom: perhaps once in billion of billions of billions of years (the actual probability may be quantified within the statistical approach). This is the *non absolute* formulation of the second law. Obviously it is far more absolute than the "absolute" classical formulation.

Boltzmann's formulation of the second law was far from being universally accepted during his lifetime. It was not until the atomic theory of matter had gained full credibility that Boltzmann's formulation was accepted. Unfortunately, this came only after his demise in 1906. His famous formula for the entropy S as a function of the number of microstates W (in our notation $S = k_B \ln W$) was engraved on his tombstone in a Vienna cemetery (figure 1.2).

The turning point which tilted the balance towards the acceptance of the atomic theory of matter was the publication in 1905 of the theory of Brownian motion by Albert Einstein (1879–1955). The Brownian motion, named after the botanist Robert Brown (1773–1858), is the random motion of particles suspended in a liquid. Brown discovered it in 1827, while looking through a microscope at particles trapped in cavities inside pollen grains in water. He noted that the particles moved randomly through the water but could not explain the reason.

Almost 80 years later, Einstein computed the mean squared displacement of particles using the kinetic theory of fluids, by assuming that the Brownian motion is the result of random collisions with the atoms or molecules of the fluid. This explained a phenomenon known but unexplained since decades, and provided winning evidence for the reality of the atoms and for the validity of statistical mechanics. Einstein's theory of Brownian motion led to the decisive victory of the atomistic view of matter. It was also decisive for the acceptance of statistical mechanics, which was controversial at that time.

Figure 1.2: Boltzmann's tombstone at the central cemetery of Vienna.

The statistical mechanical theory of matter was developed by Josiah Willard Gibbs (1839–1903) while Boltzmann was pursuing his view of entropy and of the second law. At the heart of Gibbs' statistical approach to thermodynamics was the idea of ensemble of systems. Any measurable quantity such as pressure, volume, temperature, etc. is a time-average over many molecular events. For instance, the pressure exerted by a gas on the walls of the container is a result of many "bombardments" of microscopic particles on the wall. It takes a finite period of time Δt to measure a physical quantity such as pressure. During this period of time myriads of molecular collisions occur. The value of the pressure we read on our instrument is thus an average value over that period of time.

Unfortunately, we do not know how to calculate these time averages: such calculation would require solving some 10^{23} equations of motion. Instead, Gibbs envisaged a very large ensemble of macroscopic system, all identical in their macroscopic characterization (say, all having the same volume V, temperature T and pressure P). Instead of calculating *time averages*, Gibbs suggested to calculate *ensemble averages*. Assum-

ing a few reasonable postulates, he was able to develop a highly successful statistical thermodynamic theory. In this theory each system in the
ensemble is characterized by a few macroscopic parameters such as the
number of particles N, the temperature T, the pressure P or the total
energy E of the system. The ensemble of systems is characterized by a
probability distribution $p_1, \cdots p_n$[5], where p_i is the probability of finding
a single system in the ensemble being in the state i characterized by
some set of values for the macroscopic variables. Equivalently, p_i is the
fraction of systems in the ensemble being in state i.

One of the most astonishing results of Gibbs' approach is that the entropy of the system in any ensemble has always the same form, proportional to $-\sum p_i \log p_i$ where the sum runs over all possible states. This
was a generalization of Boltzmann's expression for the entropy, which
can be obtained as a special case of the former.

Another extremely useful quantity introduced by Gibbs is now called
the *Gibbs energy*. We shall encounter the Gibbs energy and the chemical
potential, a closely related quantity, throughout the whole book.

Thermodynamics and statistical mechanics have been developed and
used almost independently. Thermodynamics deals with the *relationships* between the macroscopic quantities that can be measured. On
the other hand, in statistical mechanics the properties of atoms and
the molecules are used to calculate the thermodynamic quantities. In
turn, they can be compared with the experimentally measured values.
Thermodynamics provides a powerful tool to understand daily phenomena. Statistical mechanics provides also an explanation of the concepts
of thermodynamics, and therefore it is a more fundamental theory.

At this point, we could have ended up the story of the entropy and the
second law and thermodynamics. During much of the 20th century, both
thermodynamics and statistical mechanics were applied successfully to
more and more systems, but no further development in the fundamental
principles was achieved.

---

[5]Here we assume a finite number of states. In general, the number of states could be infinite,
or even change continuously.

1.5   The Birth of Information Theory

In 1948 Claude Shannon (1916–2001), while working at Bell Laboratories, was engaged in developing a theory of communication. Shannon published a seminal work in this field, whose title is "A Mathematical Theory of Communication" (Shannon, 1948).

The reader might be (or should be) surprised at encountering Shannon's theory of communication after having read the story of the entropy and the second law. Indeed, Shannon's work on the theory of communication does not belong to the story of entropy and the second law. Actually, Shannon did not contribute anything to either thermodynamics or to statistical mechanics. Instead, Shannon's name was accidentally "dragged" into the story of the entropy.

Shannon developed a *measure* of *information* (or of choice, or of uncertainty). We shall refer to this measure as *Shannon's Measure of Information* (SMI in short). It turned out that the SMI had a similar *form* to the expression of the entropy found by Gibbs, namely $-\sum p_i \log_2 p_i$. Similarity in the form of a mathematical expression does not imply any relationship between two concepts. Unfortunately, this similarity led the well known mathematician John von Neumann (1903–1957) to suggest Shannon to refer to his measure as *entropy*. That was an unfortunate event: it led many scientists, and non-scientists alike, to call *entropy* the SMI applied to communication, linguistics, music and arts. All these scientists were discussing about SMI but, because an ironic accidental twist of history, referred to it as *entropy*.

SMI is a very general, useful and beautiful concept. It can be applied to *any* probability distribution. For example, it applies to the probability of the occurrence of the various outcomes of throwing a die, or the frequency of appearance of letters in the English language. All these have *nothing* to do with entropy. Yet, somehow pervertedly, these are referred to as entropy. This practice is really unfortunate.

Ironically, von Neumann's recognition of the identity between the *mathematical form* of SMI and the Gibbs' expression for the entropy can be turned around: instead of referring to SMI as entropy, one can (and should!) consider entropy as a particular case of SMI. Indeed, we

shall see that the SMI of the probability distribution of the microscopi-
cal states *at equilibrium* is proportional to the thermodynamical entropy.
They only differ by the choice of the units.[6]

Accepting this view offers a very powerful interpretation of entropy
in terms of Shannon's measure of information. We shall use this view to
build up the entropy of an ideal gas in chapter 6. Based on this foundation, and with a firm solid interpretation of entropy, we shall develop the
rest of thermodynamics.

1.6 The Basic Ideas of Information Theory

In this section we quickly present information theory, assuming some
qualitative knowledge of probability (described in details in chapter 3).
The purpose of this section is to give the reader a flavor of what information theory is, and to prepare the reader for the more quantitative and
detailed discussion in chapters 4 and 5, and its connection to the entropy
of ideal gases (and the rest of thermodynamics) in chapter 6.

"Information", like "probability", is a very general, qualitative, imprecise and possibly a very subjective concept. The same "information"
might have different meanings, effects and values to different persons.
Yet, as probability theory was developed from a subjective and imprecise concept and became a rigorous branch of mathematics, so did information theory, which was distilled and developed into a quantitative,
precise, objective and very useful theory. At the heart of the theory of
information is the concept of Shannon's measure of information (SMI).
In this section, SMI will be introduced in a qualitative way. Chapters 4
and 5 provide a more rigorous treatment.

Let us start with a familiar 20-question (20Q) game: I choose a person, and you have to find out who is that person by asking binary questions, i.e. questions which are only answerable by "yes" or "no". Suppose
I have chosen Einstein. Here are the two possible "strategies" for asking
questions:

---

[6]This was originally shown by Ben-Naim (2006a).

Dumb strategy	Smart strategy
1) Is it Nixon?	1) Is the person a male?
2) Is it Gandhi?	2) Is he alive?
3) Is it me?	3) Is he in politics?
4) Is it Marilyn Monroe?	4) Is he a scientist?
5) Is it you?	5) Is he very well known?
6) Is it Mozart?	6) Is he Einstein?
7) Is it Bohr?	
8) . . .	

In listing the two strategies above, we have qualified the two strategies as "dumb" and "smart". The reason is quite simple, although, for this particular case, it is not easy to prove that. If you use the first strategy you might, if you are lucky, hit on the right answer on the first question, while with the second strategy it is impossible to win after just one question. However, in the dumb strategy hitting on the right answer on the first guess is highly improbable. It is more likely that you will keep asking for a long time very specific questions, like those in the list, and it is possible that the right answer is never found. The reason for preferring the second strategy is that, at each answer, you get more *information*. The reason is that, at each question, you exclude a much larger number of possibilities than with the dumb strategy.

In particular, in the smart strategy the first question allows to exclude a huge number of possibilities, either all males or all females. Furthermore, if the answer to the second question is "no", then you have excluded all living persons. With each of the other answers you get, you can narrow down further the *range* of possibilities, each time excluding a large group. This is in contrast with the dumb strategy, in which each negative answer excludes just one person: practically there is no change in the *range* of unknown possibilities.

Even though we have not yet *defined* the term "information", it is intuitively clear that in the smart strategy you gain more information from each answer than in the dumb strategy.

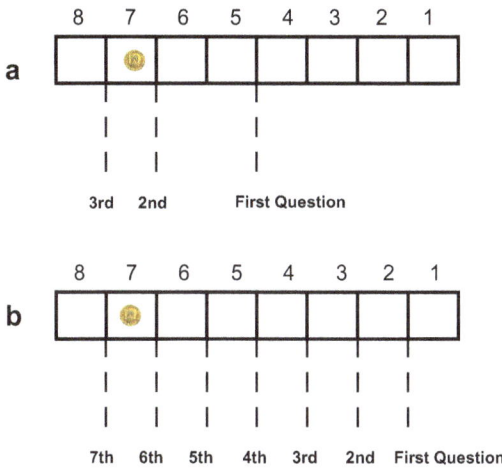

Figure 1.3: Two strategies of finding a coin hidden in one of 8 boxes: the smartest (a) and dumbest (b).

At this point we can also say something qualitative about what we mean by a *measure* of information. Suppose we decide to choose a person from a certain country, or a certain city, or from all persons who are present in the room right now. Clearly, the larger is the "pool" of persons from which I can choose from, the more difficult it is for you to find which person I chose. We can say that the *size* of the game is larger for a larger pool of persons.

All that was said above is very qualitative. The term "information" used so far is very imprecise and there are many elements of subjectivity that might enter into the game. However, it is possible to "reduce" this type of game in such a way that it becomes devoid of any trace of subjectivity. Let us now describe a new game that is essentially the same game as before, but in its distilled form: it is much simpler to describe and more amenable to a precise, quantitative and objective treatment.

Suppose we have eight equal boxes (figure 1.3). A coin is hidden in one of the boxes and you have to find where it is. All you know is that the coin *must* be in one of the boxes, and that the box was chosen at random without any preference. Equivalently, there is a chance of 1/8 to find the coin in any specific box.

Note that in this game we have completely removed any trace of subjectivity: the *information* we need is "where the coin is." The fact that I or anyone else know or do not know where the coin is, does not make this information a subjective one. The information is a concept "built in" in the game and is not dependent on the person who plays or does not play the game.

In order to acquire the *information* as to "where the coin is", you are allowed to ask only binary questions. Instead of an indefinite number of persons as in the previous game, we have only eight possibilities.

Again, there are many strategies for asking questions. Here are two extreme and well-defined strategies:

The dumbest strategy	The smartest strategy
1) Is the coin in box 1?	1) Is it in the right half (of the eight)?
2) Is the coin in box 2?	2) Is it in the right half (of the remaining four)?
3) Is the coin in box 3?	3) Is it in the right half (of the remaining two)?
4) Is the coin in box 4?	
5) Is the coin in box 5?	
6) Is the coin in box 6?	
7) Is the coin in box 7?	

With the smartest strategy, we ask questions to locate the *half* in which coin is. It takes exactly 3 questions to find the coin. With the dumbest strategy we could find the coin already with the first question, and at most it takes 7 questions to locate it. By repeating several times the game with random positions of the coin, the dumbest strategy on average takes 4 questions to locate the coin, while the smartest strategy takes exactly 3 questions to achieve the same result.

It does not look like a big advantage, but the difference between the dumbest and smartest strategies increases quickly, if we increase the number of boxes. For example, with 16 boxes the smartest strategy requires 4 questions to locate the coin, while on the average the dumbest strategy needs 8 questions. With 32 boxes one finds the coin with 5 questions with the smartest strategy, but on average one has to ask 16 questions with the dumbest one. Now it is clear why we defined them the "smartest" and "dumbest" strategies.

Please note that the *amount* of information that is required is the same, no matter which strategy you choose. The choice of strategy allows you to get the same amount of information by different number of questions. The smartest strategy guarantees that you will get it, on the average, with the minimum number of questions.

The important point to be noted at this stage is that, no matter what strategy you choose, the larger is the number of boxes, the larger is the "information" you need to locate the coin (hence, the larger the number of questions you need to obtain that information).

In information theory the amount of *missing information*, i.e. the amount of information one needs to acquire by asking questions, is *defined* in terms of the distribution of probabilities. In the game with M boxes, the distribution is $\{\frac{1}{M}, \frac{1}{M}, \ldots, \frac{1}{M}\}$. By asking the smartest question, one gains from each answer the maximum possible information (this is referred to as one *bit* of information). Maximum information is obtained in each question when you divide the space of all possible outcomes in two *equally probable* parts. The *same* information is collected by all strategies. On the *average*, the smartest strategy gathers that information with the minimal number of questions.

Intuitively, one could use the number of boxes M as a measure of the *amount* of information in this game. However, we have noticed that, if we use the smartest strategy, doubling the number of boxes requires only adding one more question. This suggests that, instead of using the number M of boxes, it would be better to use the logarithm of M as a measure of the size of information. With this choice, if we double the number of boxes from M to $2M$, the measure of the size of information changes from $\log M$ to $\log 2M = \log M + \log 2$. By choosing the base 2 for the logarithm, we obtain $\log_2 2M = \log_2 M + 1$: when doubling the number of boxes, the suggested measure of information changes only by one unit.

Clearly, the number of boxes is an "objective" quantity, belonging to the game. The fact that you or I or anyone else is ignorant of the location of the coin does not make this "information" a subjective quantity.

So far, we considered uniform probability distributions. In chapter 4 we shall extend this "measure of information" also to non-uniform distributions, and will see that the SMI is maximum for a discrete uniform distribution.

1.7 Thermodynamic Equilibrium

In order to re-connect to our discussion of the second law of thermodynamics, it suffices to say that a probability distribution with maximal SMI is much more frequent in practical systems than a probability distribution with smaller SMI. In addition, this difference becomes more and more marked when the dimensionality of the system (in our M-box game, this is the number M of boxes) increases. When one applies this sort of reasoning to macroscopic systems, composed by a huge number of atoms or molecules, the relative frequency of microscopic configurations with maximal SMI is practically 100%.

The set of microscopic configurations with maximal SMI characterizes the thermodynamic equilibrium. In addition, the probability for measurable deviations from equilibrium is so small that in practice it is impossible to observe them within the life of the universe. Furthermore, it is sufficient to allow for random fluctuations between microscopic configurations to obtain a system that *must* evolve toward equilibrium.

The reason is simply that, whatever is the initial preparation for the system, the random fluctuations will make it "explore" many different microscopic configurations. As the microscopic configurations with maximal SMI are overwhelmingly more likely to be encountered during such "exploration" than anything else, the system automatically drives toward equilibrium and will remain there practically forever. This *explains* the second law.

Thermodynamic equilibrium is a stationary state characterized by constant values of macroscopic observables, while at microscopic level there is a frenetic change across different configurations that are indistinguishable for a macroscopic observer. The latter will see an evolution from any initial state to thermodynamic equilibrium (with constant values of P, T, V, etc.) and no additional change, notwithstanding the rich dynamics at microscopic level (that is not accessible to such observer).

Thermodynamic equilibrium is also characterized by constant and maximal SMI. This means that, if the initial state has no maximal SMI, the latter will inexorably increase until attaining the maximum value. Then it will stay constant. There is no "source" of SMI, hence there is no source of entropy, which is a particular type of SMI. Maximal SMI corresponds to thermodynamic equilibrium, but this is not a single configuration of all particles in the system. Instead, it corresponds to the largest (by far!) set of microscopic configurations. Thus thermodynamic equilibrium is not a static state, but a stationary state characterized by constant values of thermodynamic (i.e. macroscopic) quantities.

Now that you know the guilty, you can start reading the romance. All that we have said so far will be clarified in the next chapters, where the connection between information theory and thermodynamics via the SMI and the second law will be explained in details. In addition to this modern approach to the second law and to the development of all important thermodynamic relationships in the first part of this book, you will also find interesting applications and will have a number of chances to switch between microscopic and macroscopic points of view in the second part.

Chapter 2

The Historical Development of Thermodynamics

The development of thermodynamics started well before the atomic nature of matter became widely accepted. The first steps were the operative definitions of the thermodynamic quantities of temperature, volume and pressure, and the study of gaseous systems. As all gasses at sufficiently low pressure behave in the same way (the so-called "ideal gas" behavior), it was recognized that it is possible to define an "absolute" (i.e. independent from the medium utilized in the construction of the thermometer) temperature, whose range is limited from below, starting from the "absolute zero" temperature.

The first bridge with classical mechanics, so successful in describing the dynamics of point-like particles, rigid bodies, and the motion of planets, was the identification of the equivalence between heat and mechanical energy. Furthermore, by assuming that the matter is composed by microscopic "molecules" obeying the laws of classical mechanics, it was possible to relate the temperature of a body with the average kinetic energy of its molecules. Hence the heat flow, which had been interpreted as some sort of fluid moving from hotter to colder bodies, could be identified as the energy transfer between objects in thermal contact. By including the heat flow in the energy budget, it was possible to extend the principle of conservation of energy, previously formulated in the framework of classical mechanics only in terms of kinetic and potential energies, also to thermodynamic systems (becoming the first law of the thermodynamics).

While measurement instrumentation was being improved, people developed the heat engines which made it possible the industrialization on large scales. Hence their efficiency was an important subject of study, which ultimately lead to the formulation of the second law of thermodynamics. But the implications of the second law extend far beyond the domain of heat engines: it deals with the evolution of isolated systems toward an equilibrium state which appears as their ineluctable destiny.

A sort of mystery surrounded the second law and the related concept of "entropy" until both the atomic nature of matter and the importance of probability in scientific models became widely accepted. Indeed, if one starts from the atomic nature and the role of probability, it is possible (and even quite natural) to *explain* the tendency of isolated systems to an equilibrium state and the concept of entropy itself. This is exactly the approach followed by this book.

However as a matter of facts, the atomic nature of matter became widely accepted only after several experiments allowed people to "see" the microscopic constituents. This required instrumentation which could be built only at the beginning of the 20th century, much later than the thermodynamics were formulated and successfully applied to a number of different systems.

Anyway, today we do know about the existence of atoms and molecules — the most important single scientific statement in the opinion of the Nobel laureate Richard Feynmann (1918–1988) (Feynman *et al.*, 2010) — and we will keep this firmly in mind in the development of a modern theory of thermodynamics. Nevertheless, it is instructive to learn about the historical development too, which is summarized in this chapter.

2.1 Thermodynamic Quantities

Measurable quantities may depend on the size of the system or not. In the first case one speaks of **extensive quantities**, otherwise they are termed **intensive quantities**. The volume is the simplest example of an extensive quantity, as it is basically a measure of the size of a system or of one portion of it, but also the total energy and the entropy belong to this category, as each portion of a system contributes to them. On the other hand, temperature and pressure are intensive quantities, as they do not depend (when measured at equilibrium) on the size or portion of the system on which one performs the measurement. This means that we can just probe a small portion of a system to measure them, which helps limiting the effects of the measurement on the system itself. A noteworthy property of these intensive quantities is that we can *feel* them by touching the system. On the other hand, we can *view* the volume and sometimes also the energy, when it is so high that the hot object radiates in the visible spectrum.

Temperature, pressure, volume, energy, entropy and chemical potential are the most commonly used thermodynamic quantities. The chemical potential is a form of potential energy that may change during a chemical reaction or a phase transition. Only the difference in chemical potential is a measurable quantity, which determines the direction of the reaction or of the flow of molecules from one state to another. In chapter 11 we shall study the role of the chemical potential in chemical reactions.

Unlike pressure and volume, which make sense also for continuous systems, temperature, entropy and chemical potential could not have existed if matter had not been constituted by an immense number of atoms and molecules. It is true that one can operatively define temperature and entropy without any reference to the atomic nature of matter — actually, this is exactly what happened in the history — but their existence (not to mention our understanding of them) is intimately dependent on it. In addition, the observed behavior of chemical reactions requires the assumption of the existence of atoms and molecules. Indeed, the very fact that in a balanced chemical reaction the proportions of reactants and products typically form ratios of positive integers (which define the so-called "reaction stoichiometry") has been a very strong indication that the substances are aggregates of a number of identical molecules.

What characterizes all thermodynamic quantities is that they provide a description of a system only from the *macroscopic* point of view. Although the system, at the microscopic level, has a very complex dynamics — at least (but not only) because of the enormous number of microscopic constituents — the latter is not directly visible: only average quantities providing a very coarse description of the system are accessible to macroscopic measurements. Remarkably, the relationships between these thermodynamic quantities are general enough that they apply to a wide variety of physical systems, making thermodynamics a very powerful tool in understanding for example chemical reactions, heat engines, seasonal effects, astrophysical processes and the entire universe as a whole!

2.2 Volume and Pressure

A ***barometer*** is an instrument for measuring the atmospheric pressure. The first barometer was created in 1643 by Evangelista Torricelli (1608–1647): a tube approximately one meter long, sealed at the top, filled with mercury, and set vertically into a basin of mercury. As mercury is about fourteen times more dense than water, Torricelli could manage to perform his experiments with reasonably short pipes. The column of mercury fell to about 76 cm, leaving a vacuum above (the "Torricellian vacuum"). As the column height fluctuates with changing atmospheric pressure, this instrument provides a measurement of it. The pressure unit of 1 torr = 1 mmHg (i.e. the pressure which corresponds to a mercury column height of one millimeter) is named after Torricelli.

In 1646, Blaise Pascal (1623–1662) along with Pierre Petit (1594–1677) reproduced Torricelli's experiment and questioned what force kept some mercury in the tube and what filled the space above the mercury in the tube. By using different liquids (including water and wine) Pascal showed that vacuum, and not some invisible substance (as the Aristotelians believed), "fills" the top of the tube. He also predicted that the column height would be smaller if the experiment were replicated in high mountains, as it was confirmed in 1648 by his brother-in-law, Florin Perier, who lived near a mountain called the Puy de Dome.

The ***pressure*** P is defined as the ratio between the force exerted perpendicularly onto a surface and the surface area. Nowadays, in the *International System of Units* or SI (from the French name Système International d'unités), the unit for pressure is the pascal: $1\,Pa = 1\,Nm^2$. The ***standard atmosphere*** (atm) is a constant, approximately equal to the typical air pressure at the mean sea level, defined as $101325\,Pa = 101.325\,kPa$.

In 1660, Robert Boyle (1627–1691) published the results of his experiments on rarified air. Critical to this work was the development of an air pump by Boyle with Robert Hooke (1635–1703) and the use of the barometer, invented by Torricelli just 16 years before, which had been introduced into the sealed chamber to measure the pressure. For the first time, it was possible to observe physical processes at both normal and reduced barometric pressures. The first experiments, among the 43 separate experiments reported in his book *New Experiments Physico-Mechanical: Touching the Spring of the Air and its Effects* (Boyle, 1682), were on the "spring of the air", that is the pressure developed by the air when its volume was changed (West, 2005).

Boyle's law (sometimes referred to as the Boyle-Mariotte law) describes how the pressure P of a gas tends to decrease as its volume V increases: at constant temperature these two quantities are inversely proportional. This means that if we measure both of them in different configurations we obtain $P_1 V_1 = P_2 V_2 = P_3 V_3$ etc. In other words, the **Boyle-Mariotte law** says that the product

$$PV = \text{constant} \tag{2.1}$$

where the "constant" can only depend on the temperature.

Half a century later, Daniel Bernoulli (1700–1782) in his book *Hydro-dynamica* (1738) (Bernoulli, 1738) laid the basis for the kinetic theory of gases, and applied the idea to explain Boyle's law. The pressure P is the orthogonal component of the force per unit area exerted by the gas on the container walls:

$$P = F_\perp / A \tag{2.2}$$

At the microscopic level, a gas is made of minuscule particles with equal mass m, randomly moving in all directions. Accordingly to Newton's second law, the force equals the variation of momentum in time. Assuming that the scattering with the container walls is perfectly elastic, only the direction changes after the interaction between the wall and the particle (whose mass is negligible compared to the wall). Hence the change in momentum per collision is $\Delta p = 2 m v_\perp$, where $v_\perp$ is the magnitude of the velocity component which is perpendicular to the wall (which is the same before and after the scattering: only the direction changes).

If we assume for simplicity that the container is a cubic box with linear dimension L, each particle impacts again on the same wall after a delay $\Delta t = 2L/v_\perp$. Hence the force which a single particle exerts on the container wall is $F = \Delta p / \Delta t = m v_\perp^2 / L$. When averaging over all molecules of the gas[1], one gets $\langle v_\perp^2 \rangle = \langle v^2 \rangle /3$, where $\langle v^2 \rangle = \langle v_x^2 \rangle + \langle v_y^2 \rangle + \langle v_z^2 \rangle$ is the average squared velocity of the molecules. As the wall area is $A = L^2$ and the container volume is $V = L^3$, it comes out that the gas pressure is

$$P = \frac{Nm \langle v^2 \rangle}{3V} = \frac{\rho m \langle v^2 \rangle}{3} = \frac{2}{3} \rho \langle E_k \rangle = \frac{2}{3} u \qquad (2.3)$$

where $\rho = N/V$ is the **number density** of the gas (with SI units of m^{-3}), $\langle E_k \rangle = \frac{1}{2} m \langle v^2 \rangle$ is the average kinetic energy of the molecules, and u is the average **energy density**. Hence *the pressure is related to the average energy density of the gas*. As we are dealing with an ideal gas, the total energy of the molecules is the sum of their kinetic energies.

2.3 Temperature, Pressure and Volume

In the XVII and XVIII centuries, various thermometers were developed by Ole Christensen Rømer (1644–1710), Guillaume Amontons (1663–1705), Daniel Gabriel Fahrenheit (1686–1736), Anders Celsius (1701–1744), and many others. They made it possible to make precise measurements of the temperature. Those thermometers made use of the fact that the volume of gases (and most liquids and solids) increases linearly with the temperature. The measurement itself is made possible by the fact that, after some time, a thermometer in contact with the system under study reaches the same temperature. Actually, the final temperature is a weighted average of the initial temperatures of the system and of the thermometer. However, the latter is build such that its **thermal capacity** (the amount of heat required to increase the temperature by one unit) is very small compared to the system under study, such that the final temperature is practically the same as the system temperature. Another way of saying this is that the thermometer is designed not to "disturb" the system in an appreciable way. The fact that two bodies in thermal contact (i.e. which can exchange heat) always reach the equilibrium at the same temperature is also known as the **zeroth law of thermodynamics**.

---

[1]Averaging implies using the concept of probability, as we shall see in the following. Bernoulli gave important contributions to the theory of probability.

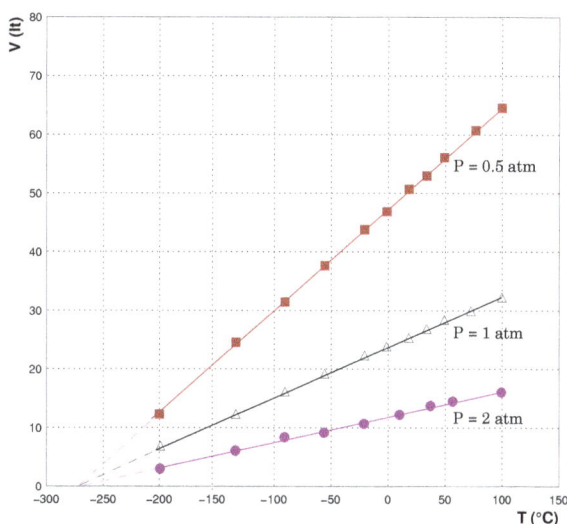

Figure 2.1: Volume as a function of temperature at different pressures.

Between 1700 and 1702, while building an "air thermometer", Amontons discovered the relationship between the pressure and temperature of a fixed mass of gas kept at a constant volume

$$P = C(t + t_0) \tag{2.4}$$

(**Amontons' law of pressure-temperature**) where C is a constant proportional to the amount of gas and t represents the temperature with arbitrary units.

In 1802 Joseph Louis Gay-Lussac (1778–1850), who credited the unpublished work by Jacques Alexandre César Charles (1746–1823), formulated the **Charles's law** (also known as the **law of volumes**), which describes how gases tend to expand when heated: at constant pressure, the volume is directly proportional to the temperature:

$$V = C'(t + t_0) \tag{2.5}$$

where C' is another constant, proportional to the amount of gas.

By comparing the volumes obtained with different gases at different temperatures and the values measured with the same gas at different pressures (figure 2.1), one realizes that there is a limiting value of the temperature, at which the *extrapolated* values of pressure and volume become zero (no pressure can be defined for arbitrarily low temperature

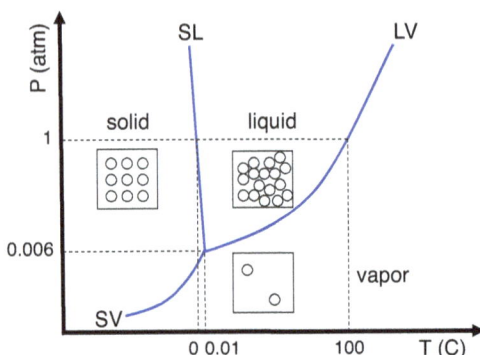

Figure 2.2: Phase space and triple point of water.

values, as any gas becomes liquid, breaking down Charles's law). This limiting temperature is called **absolute zero**, and is today the starting point of the temperature units of kelvin [K] which are used in the SI. One degree Kelvin is as big as one degree Celsius [°C] (or "centigrade"), the sole difference being the choice of the origin.

The absolute zero is *defined* to be 0 K = −273.15 °C, and the size of a degree Kelvin and Celsius is fixed by defining, at the standard atmospheric pressure, the freezing temperature of the water as 0°C = 273.15 K, and the boiling temperature as 100°C = 373.15 K. The freezing and boiling temperatures depend on the pressure, hence it is useful to look at the fixed temperature of the **triple point** of water (the point at which the solid, liquid and gaseous phases coexist; see figure 2.2), which is defined as precisely 273.16 K = 0.01 °C.

The experimental fact that the behavior of all gases at very low pressure is the same lead to the **ideal gas** concept. When expressing the temperature in kelvins (we use the symbol T in this case), the experimental gas laws above can be combined into the **equation of state of the ideal gas**:

$$PV = nRT \tag{2.6}$$

where n is the number of moles of gas and the **gas constant** is[2]

$$R = 8.3144621(75) \, \mathrm{J \, K^{-1} \, mol^{-1}}. \tag{2.7}$$

---

[2]The notation 8.3144621(75) is a shorthand form for 8.3144621 ± 0.0000075, where the second value represents the *uncertainty* on the measured value which is reported first.

The **mole** is defined in the SI as the amount of substance of a system which contains as many elementary entities as the number of atoms in 0.012 kg of carbon-12 (^{12}C). The number of molecules per mole is $N_{Av} = 6.02214 \times 10^{23}\, \text{mol}^{-1}$, where N_{Av} is the **Avogadro constant**, after Amedeo Avogadro (1776–1856). In 1811, he hypothesized that two given samples of an ideal gas, of the same volume and at the same temperature and pressure, contain the same number of molecules. Avogadro found that for a given mass of an ideal gas, the volume and amount (moles) of the gas are directly proportional if the temperature and pressure are constant (**Avogadro's law**). At the **standard conditions for temperature and pressure** of 101.325 kPa and 273.15 K, the volume occupied by one mole of an ideal gas is 22.41 liters.

In terms of the number of molecules of the gas $N = n N_{Av}$, the equation of state of the ideal gas (2.6) becomes

$$PV = N k_B T \tag{2.8}$$

where

$$k_B \equiv \frac{R}{N_{Av}} = 1.3806488(13) \times 10^{-23}\, \text{J K}^{-1} \tag{2.9}$$

is the **Boltzmann constant**, named after Ludwig Eduard Boltzmann (1844–1906), who could derive the laws of thermodynamics from the (at those times) hypothesis of the atomic nature of matter.

Dividing left and right hand sides of (2.8) by the volume V, recalling that the number density is $\rho = N/V$, and making use of (2.3), one gets

$$P = \frac{2}{3} \rho \langle E_k \rangle = \rho k_B T$$

from which one finds that

$$T = \frac{2}{3} \frac{\langle E_k \rangle}{k_B} \tag{2.10}$$

This means that *the temperature represents the average kinetic energy of the gas molecules*.

Had this be recognized earlier, we could have saved us from introducing special units for the temperature, and could have expressed the latter in energy units since the very beginning. As from the mathematical point of view the units of measure are just a burden, in theoretical physics one indeed expresses the temperature in energy units.

Another important implication of (2.10) is that *the heat*, whose flow is responsible for changes in temperature, *is a form of energy transfer*. This is the microscopic interpretation of the result obtained by James Prescott Joule (1818–1889), who discovered in 1845 the relationship between heat and mechanical work. This led to the law of conservation of energy, and to the development of the first law of thermodynamics (he also worked with Lord Kelvin to develop the absolute scale of temperature). The SI derived unit of energy, the joule (J), is named after him.

2.4 Equilibrium States and Heat Engines

Classical thermodynamics deals either with equilibrium states or with cyclical processes. An *isolated system* is a physical system which makes no interaction whatsoever with its surroundings. Strictly speaking, such a system is unobservable from outside, as any measurement implies some sort of interaction. However, the perturbations may be small enough to have no measurable effect on the system, which can then be considered isolated with very good approximation.

Our experience says that every isolated system at some point reaches an *equilibrium state* characterized by constant values of all thermodynamic quantities. This is taken as an axiom. Because no perturbation affects the system, an equilibrium state is by definition eternal, as there are no unbalanced potentials (or driving forces) within a system in thermodynamic equilibrium; there are no phase transitions and all chemical reactions are balanced.

Two systems are said to be in *thermal equilibrium* with each other if (1) the spontaneous exchanges of molecular thermal energy between them do not lead to a net exchange of energy, (2) no net flow of matter is the result of the random motion of their molecules, and (3) no difference in chemical potential exists between them.

If the system goes through very small and slow changes of the thermodynamic quantities, a so-called *quasi-static transformation*, it can be considered as passing through a series of equilibrium states in which the thermodynamic quantities are constant for a small amount of time. Quasi-static transformations are (almost) *reversible* transformations, in the sense that they could well happen in both directions (but we prefer to reserve the adjective "reversible" for the transformations in which entropy does not change). On the other hand, the majority of natural processes are instead *irreversible*, because the inverse process never happen spontaneously. Classical thermodynamics is able to describe quasi-static transformations as sequences of temporary equilibrium states, but can also deal with irreversible transformations, looking only at the initial and final states.

Systems may interact in different ways, for example by exchanging heat and/or matter. Provided that such exchanges happen on time scales much longer than the typical time it takes for a system to reach thermodynamic equilibrium, one may follow these quasi-static transformations step by step. Cyclical transformations in which a system exchanges heat with the surroundings not in a quasi-static manner, while performing (positive or negative) mechanical work, are also dealt with by classical thermodynamics. A *heat engine* is a device which performs such cycles. No assumption is made on the internal state of a heat engine, which is considered as a sort of "black box" with only known input and output quantities. In particular, it is clear that whatever is inside this black box cannot be in an equilibrium state, and in general it is not assumed that it proceeds by quasi-static transformations. The study of heat engines, including the efficiency at which they can convert heat into mechanical work (and vice versa) was at the heart of the development of thermodynamics, and lead to the formulation of the first and second laws.

The thermodynamic variables are also called "state variables" or "functions of state", because their changes depend only on the initial and final thermodynamic states, and not on the particular process which is responsible for the transition between these two states. In the mathematical idealization of a continuous process (which works well because of the tiny changes related to the molecular interactions), this means that incremental changes of state variables are exact differentials. Hence the thermodynamic description of a heat engine is made possible by the state variables specifying the thermodynamic states through which the cycle proceeds.

2.5   The Laws of Thermodynamics

The first and second laws of thermodynamics emerged in the 1850s, primarily thanks to the works of Rudolf Clausius (1822–1888) and William Thomson (also known as Lord Kelvin; 1824–1907). They describe the energy budget and the entropy changes of isolated systems, and require the axiom that they will unavoidably reach a thermodynamic equilibrium state. In addition, they assume the so-called *zeroth law of thermodynamics*[3]: if two systems are each in thermal equilibrium with a third, they are also in thermal equilibrium with each other.

The 0-th law is intended to allow for the existence of an empirical parameter, the *temperature*, as a property of a system such that systems in thermal equilibrium with each other have the same temperature. This law is necessary and tacitly assumed in every measurement of temperature. It provides an empirical definition of temperature and the justification for the construction of practical thermometers, which are based on the mechanical properties of bodies, such as their volumes, without reliance on the concepts of energy, entropy or the first, second, or third laws of thermodynamics.

The *first law of thermodynamics* states that heat and mechanical work are both forms of energy transfer, and that the total energy of an isolated system is conserved (i.e. it is constant, although it can assume different forms). In particular, perpetual motion machines of the first kind, which would do work without using the energy resources of a system, are impossible. The *internal energy* U of a closed system, defined such that it accounts for all possible contributions to its total energy, may change as heat is transferred into or out of the system or work is done on or by the system. Possible forms for the internal energy are for example vibrational and rotational states of the molecules, or chemical (potential) energy.

The first law of thermodynamics may be stated in different ways. The most common formulation is the following. *The increase in internal energy of a system is equal to the heat supplied to the system minus the work*

---

[3]When people started recognizing the importance of this principle, the other three laws had already been identified. This is why it was designated as the zeroth law.

done by the system. For a small change one can write

$$dU = \delta Q - P \, dV \qquad (2.11)$$

where the small amount δQ of heat is taken as positive when provided to the system, while the expansion work $P \, dV$ is taken positive when performed by the system. This relationship naturally emerges when considering cyclic processes like those performed by a steam engine, for example, whose expansion moves a piston.

The following statement is equivalent to the previous formulation. *For a thermodynamic cycle, the heat supplied to a closed system, minus that removed from it, equals the net work done by the system.*

The first law encompasses several principles:

- The **law of conservation of energy**. Energy can be neither created nor destroyed. However, energy can change form, and energy can flow from one place to another. The total energy of an isolated system remains the same.
- The concept of internal energy and its relationship to temperature. If a system has a definite temperature, then its total energy has three distinguishable components: **kinetic energy** (related to the motion of the system as a whole), **potential energy** (related to the presence of external force fields, like gravity), and **internal energy**, which is the sum of the kinetic energy of microscopic motions of its constituent atoms, and of the potential energy of interactions between them. If the other things do not change, the kinetic energy of microscopic motions of the constituents increases as the system temperature increases.
- *The flow of heat is a form of energy transfer.* In other words, a quantity of heat that flows from a hot body to a cold one can be expressed as an amount of energy being transferred from the hot body to the cold one.
- *Performing work is also a form of energy transfer.* For example, when a machine lifts a heavy object upwards, some energy is transferred from the machine to the object. The object acquires its energy in the form of gravitational potential energy in this example.

Combining these principles leads to one traditional statement of the first law of thermodynamics: it is not possible to construct a perpetual motion machine which will continuously do work without consuming energy.

The **second law of thermodynamics** states that an isolated system, if not already in its state of thermodynamic equilibrium, spontaneously evolves towards it. Thermodynamic equilibrium has the greatest entropy amongst all possible constrained equilibrium states. Perpetual motion machines of the second kind, which would spontaneously converts thermal energy into mechanical work, are thus impossible.

The second law of thermodynamics asserts the existence of a quantity called the **entropy** of a system. In a (theoretical) process of reversible heat transfer between the system and an external reservoir in thermal equilibrium with the system (hence at the same temperature T), an element δQ of heat transferred is the product of the temperature with the increment $\mathrm{d}S$ of the system entropy

$$\delta Q = T\,\mathrm{d}S \qquad\qquad (2.12)$$

It is important to note that the entropy of a system is defined only when the latter is at thermodynamic equilibrium.

When two initially isolated systems in separate but nearby regions of space, each in thermodynamic equilibrium in itself but not necessarily with each other, are then allowed to interact, they will eventually reach a mutual thermodynamic equilibrium. Accordingly to the second law, the sum of the entropies of the initially isolated systems is less than or equal to the total entropy of the final combination: $S_f \geq S_{i,1} + S_{i,2}$ where the equality holds only for reversible processes. In other words, the entropy of an isolated system can never decrease.

The second law refers to a wide variety of processes, reversible and irreversible, but it is expecially important when talking about irreversibility. In an irreversible process, the (total) final entropy is strictly larger than the initial one.[4]

The prime example of irreversibility is in the transfer of heat by conduction or radiation. When two bodies initially of different temperatures

---

[4]The only exception is the trivial case in which the two systems had already equal intensive variables, in which case the sum of the entropies does not change.

come into thermal connection, then heat always flows from the hotter body to the colder one. The second law tells also about kinds of irreversibility other than heat transfer, for example that of chemical reactions.

Entropy may also be viewed as a measure of the lack of physical information about the microscopic details of the motion and configuration of a system. This is indeed the preferred interpretation of the entropy in this book, thoroughly explained in the following chapters.

The **third law of thermodynamics** was stated later than the first two laws, after the formulation of statistical thermodynamics (mentioned in section 2.7). The third law states that the entropy of a system approaches a constant value as the temperature approaches the absolute zero. The entropy of a system at absolute zero is typically (but not always) zero, and in all cases it is determined only by the number of different ground states.[5] Specifically, the entropy of a pure crystalline substance at absolute zero temperature is zero, as it is assumed that the perfect crystal has only one state with minimum energy.

Boltzmann related the entropy with the number of possible microstates according to

$$S = k_B \ln W \qquad (2.13)$$

where S is the entropy of the system, k_B is the Boltzmann constant, and W is the number of microstates (e.g. possible configurations of atoms; see section 2.7). At zero temperature the system must be in a state with the minimum thermal energy. If this state is not degenerate (i.e. if only a single state has minimum energy), then at absolute zero one has $W = 1$, hence $S = 0$.

As there are materials with more than a single "ground state", a more general version of the third law is to state that the entropy of a system approaches a constant value as the temperature approaches zero. When this constant value is not zero, it is called the **residual entropy** of the system.

---

[5]Quantum mechanics is required to understand all details. Here we do not provide precise definitions for the terminology, as we are focusing only on the historical development.

2.6 The Atomic Nature of Matter

We said already that the atomic nature of matter is at the root of thermodynamics, although atoms could be observed only much later than its formulation. In the next section we will see how the statistical formulation of thermodynamics emerged from the (at those times) hypothesis of the atomic nature of matter. At the philosophical level, the **atomism** has a long history which dates back two millennia before the thermodynamics emerged as a science. Hence it is interesting to see how it influenced the development of science.

At the base of the atomism is the idea that everything is made of atoms or void, where an **atom** is something which is by definition *indivisible*. In other words, an atom has no internal structure and its size represents the limit at which one can no more subdivide a substance and observe something with the very same properties. This is in contrast with a **substance theory** wherein a prime material continuum remains qualitatively invariant under division.

In the 5-th century BC, Greek philosophers were debating about the nature of reality. Two opposite schools of though were lead by Heraclitus (circa 535–475 BC), who believed that the nature of all existence is change, and Parmenides (early 5-th century BC), who believed instead that all change is illusion.

In particular, Parmenides denied the existence of motion, change and void. He believed all existence to be a single, all-encompassing and unchanging monism, and that change and motion were mere illusions. Parmenides explicitly rejected sensory experience as a path to understanding the world, favoring pure reason. He argued against the existence of void, equating it with non-being (i.e. nothing).

Attempting to reconcile these opposite views, Democritus (circa 460–370 BC) proposed that everything is composed of fundamental and invariant atoms. The variety of shapes would be due to different configurations of unchanging atoms.

A century later, Aristotle (384–322 BC) asserted that the elements of fire, air, earth, and water were not made of atoms, but were continuous. Aristotle considered the existence of a void, which was required by atomic theories, to violate physical principles. Change took place not by the rearrangement of atoms to make new structures, but by transformation of matter from its potential form to some actual realization.

Unlike Aristotle and Plato (427–347 BC), the atomists attempted to explain the world without a scheme based on "purpose", "prime mover", or "final cause". The atomists questions should be answered with a mechanistic explanation ("What earlier circumstances caused this event?"), while their opponents searched for explanations which, in addition to the material and mechanistic, also included the formal and teleological ("What purpose did this event serve?").

While Aristotelian philosophy eclipsed the importance of the atomists, their work was still preserved and exposited through commentaries on the works of Aristotle. Epicurus (341–270 BC) said to be certain of the existence of atoms and the void, influencing the works of his follower Lucretius (99–55 BC), who wrote that the phenomena we perceive are actually composite forms. The atoms and the void are eternal and in constant motion. Atomic collisions create objects, which are still composed of the same eternal atoms whose motion for a while is incorporated into the created entity. Human sensations and meteorological phenomena are also explained by Lucretius in terms of atomic motion. However, after the work of Galen (AD 129–216) atomism remained in the shadow for several centuries.

Aristotle's views on the physical sciences profoundly shaped medieval scholarship, and their influence extended well into the Renaissance, although they were ultimately replaced by Newtonian physics. Much of the curriculum in the universities of Europe was based on Aristotle for most of the Middle Ages and his influence was still strong in the time of Isaac Newton, but in the 17-th century a renewed interest in Epicurian atomism and Corpuscularianism as a hybrid or an alternative to Aristotelian physics had begun to mount outside the classroom. The main figures in the rebirth of atomism were René Descartes (1596–1650) and Robert Boyle, among others.

Galileo Galilei (1564–1642) identified some basic problems with Aristotelian physics through his experiments, and utilized a theory of atomism in his 1612 *Discourse on Floating Bodies* (Galilei, 1612). Later in *The Assayer* (Galilei, 1623) Galileo offered a more complete physical system based on a corpuscular theory of matter, in which all phenomena — with the exception of sound — are produced by "matter in motion".

Galileo created the basis for the modern scientific method, based on the experimental study of natural phenomena and the expression of quantitative statements with mathematical laguage, in contrast with the Aristotelian tradition and its qualitative analysis. The scientific method is at the heart of the incredible success of modern science, but at that time it could not yet be used to decide between the atomistic and continuous hypotheses.

The first experimental evidences of the atomic nature of matter came at the end of the 18th century, when two laws about chemical reactions emerged without referring to the notion of an atomic theory. The first was the ***law of conservation of mass***, formulated in 1789 by Antoine Lavoisier (1743–1794), which states that the total mass in a chemical reaction remains constant: the reactants have the same mass as the products. The second was the ***law of definite proportions***, first proven in 1799 by Joseph Louis Proust (1754–1826), stating that if a compound is broken down into its constituent elements, then the masses of the constituents will always have the same proportions, regardless of the quantity or source of the original substance.

John Dalton (1766–1844) summarized the empirical evidence on the composition of matter with his ***law of multiple proportions*** (1803): if two elements came together to form more than one compound, then the ratios of the masses of the second element which combine with a fixed mass of the first element will be ratios of small integers. Dalton also stated that the implication of his law of proportions is that atoms of the same species are all identical, and that chemical reactions happen when particles separate and aggregate again forming different compounds. He also provided examples, like the case of water (hydrogen and oxygen whose relative weights are in the proportion of 1:7) and ammonia (hydrogen and nitrogen whose relative weights are in the proportion of 1:5) (Dalton, 1808).

In 1811 Amedeo Avogadro proposed that equal volumes of any two gases, at equal temperature and pressure, contain equal numbers of molecules. This implies that the mass of the gas particles does not affect the gas volume, and that the latter is much larger than the volume of the molecule itself. In turn, this allowed him to deduce the diatomic nature of numerous gases by studying the volumes at which they reacted. For example, the fact that two liters of hydrogen will react with just one liter of oxygen to produce two liters of water vapor (at constant pressure and temperature) implies that a single oxygen molecule splits in two in order to form two particles of water. Thus, Avogadro was able to offer more accurate estimates of the atomic mass of oxygen and various other elements, and made a clear distinction between molecules and atoms. However, it took about half a century for Avogadro's ideas to be widely accepted.

In 1827, the botanist Robert Brown (1773–1858) observed that dust particles inside pollen grains floating in water constantly jiggled about for no apparent reason. In 1905, Albert Einstein (1879–1955) theorized that this **Brownian motion** was caused by the water molecules continuously hitting the grains, and developed a mathematical model to describe it. This model was validated experimentally in 1908 by Jean Perrin (1870–1942), thus providing additional validation for the particle theory and, by extension, for the atomic theory. So, it took until the beginning of the XX century to prove that matter is indeed made of atoms, roughly 25 centuries later than Democritus proposed their existence. But at the same time, people also started to realize that such atoms, which for chemistry (and hence for all substances which humans could play with until then) are indeed indivisible, do have some internal structure.

In 1897 J.J. Thomson (1856–1940) discovered the electron through his work on cathode rays. Electrons are the electric charge carriers in most conductors and surround the atomic nucleus, discovered in 1909 by Ernest Rutherford (1871 1937), which carries the vast majority of the atom mass, although it occupies a volume which is negligible compared to the atomic volume. Rutherford's model of the atom (1911) could explain the results of the experiment performed with Hans Geiger (1882–1945) and Ernest Marsden (1889–1970) in 1909, in which alpha particles were seen to be (very seldom) deflected at very high angles by a thin gold foil. The nucleus is positively charged, balancing the total electronic charge (which is negative) such that an atom is electrically neutral in normal conditions. The electrons form a sort of "cloud" around the nucleus, and the configuration of this cloud is responsible for all chemical properties.

Although we currently believe that electrons are indeed elementary particles without any internal structure, we also know that the nucleus is made of two different types of particles, the protons (carrying a positive unit of electric charge) and the neutrons (without electric charge). The fact that neutrons were also present in the nucleus was discovered in 1932 by James Chadwick (1891–1974). They were also the first example of unstable particles, i.e. particles which spontaneously *decay*, transforming into other particles. A neutron outside a nucleus lives for about 15 minutes — which is the average *life time* of an exponential distribution of decay times — and spontaneously decays into a proton, an electron (to conserve the electric charge) and a neutrino (to conserve energy and momentum). This process is responsible of the so-called "beta decay" of some atomic nucleus. Neutrinos were postulated in 1931 by Wolfgang Pauli (1900–1958) and named by Enrico Fermi (1901–1954) to be produced in beta decays of neutrons, but were not discovered until 1956, because of their extremely weak tendency to interact with other particles.

The development of new particle accelerators and particle detectors in the 1950s led to the discovery of a huge variety of new particles, the most recent being the Higgs boson, detected in 2012 at CERN (ATLAS Collaboration, 2012; CMS Collaboration, 2012). Today, we know many subatomic particles, most of which are unstable. Only very few of them appear to have no internal structure, like the neutrinos and the muons, discovered in 1936 by Carl D. Anderson (1905–1991), whereas the others are indeed made of other particles. In the so-called "standard model of particle physics", there are elementary particles which are responsible for the interactions (like the photons), elementary particles which are directly observable (like the electrons), and elementary particles called "quarks" which are never directly observable but are thought to be the constituents of the "hadrons", the class of particles that includes protons and neutrons.

2.7 Statistical Thermodynamics

The foundations of statistical thermodynamics were set out by James Clerk Maxwell (1831–1879), Ludwig Boltzmann (1844–1906), Josiah

Willard Gibbs (1839–1903), and Max Planck (1858–1947). In the rest of the book we will investigate many details, hence here we restrict ourselves only to the historical development.

In section 2.2 we have seen that in 1738 Bernoulli laid the basis for the kinetic theory of gases. By postulating that a gas consists of a large number of molecules moving in all directions, he could relate the macroscopic pressure with their impact on a surface. In the same picture, one also finds that what we experience as heat is simply the (flow of) kinetic energy of the molecular motion.

In 1859 Maxwell found the distribution of molecular velocities, which gives the fraction of particles having velocity in a specific range (presently known as the Maxwell distribution; more on this on chapter 5). This was the first statistical law in physics. Five years later Boltzmann, so inspired by Maxwell's work, started studying similar subjects. Boltzmann can be considered the "father" of statistical thermodynamics, with his derivation in 1875 of the relationship between entropy S and the **multiplicity** W, that is the number of microscopic arrangements (**microstates**) producing the same macroscopic state (**macrostate**) for a particular system. Much of his work was collectively published in Boltzmann's 1896 *Lectures on Gas Theory* (Boltzmann, 1995). The term "statistical thermodynamics" was proposed for use by Gibbs in 1902.

The central concept of statistical thermodynamics is that the thermodynamic quantities provide a very coarse description of the system, at the macroscopic level in which the constituents and their interactions are not directly observable. This means that there can be many microstates which are indistiguishable for a macroscopic observer, simply because they happen to be characterized by the same values of volume, pressure, temperature, etc.

If one accepts the ideas that (1) a system at equilibrium is actually not in a constant and unique state, but is actually randomly passing through many microscopic configurations, and that (2) all accessible microstates are equally probable (the so-called **ergodic hypothesis)**, then it follows that the system spends most time in the macrostate which represents the largest number of microstates. In other words, the system wanders among different microstates all the time. Its macroscopic evolution toward equilibrium is simply the result of its continuous sampling of its *phase space*, a point of which represents one specific microstate. While wandering through the phase space, the thermodynamic quantities at some point reach the values which characterize the macroscopic state comprising the largest number of microstates. While it is true that the system will leave this macrostate at some point in its future evolution, it comes out that the probability that this happens is so low that, in practice, the system will remain forever in this state, which is its thermodynamic equilibrium state.

The modern formulation of statistical mechanics is due to Gibbs, who based the description of the physical system on an *ensemble* that represents all possible configurations of the system. Each ensemble is associated with a *partition function* that, with mathematical manipulation, can be used to extract the values of all thermodynamic quantities of the system. The partition function is used to find the average values of microscopic properties, which can then be related to macroscopic observables.

Gibbs classified the ensembles into three general types, in order of increasing complexity of its interactions with the surroundings:

- The *microcanonical ensemble* represents an isolated system, which does not exchange energy or mass with the rest of the universe. All thermodynamic quantities are constant for such a system, and the entropy

$$S = k_B \ln W \qquad (2.14)$$

  can be used to derive all other quantities. The independent variables are the energy E, the number of particles N, and the volume V;

- The *canonical ensemble* describes a system in thermal equilibrium with its environment, which can only exchange energy in the form of heat with the surroundings. The independent variables are the temperature T, the number of particles N, and the volume V.

- The *gran-canonical ensemble* is an open system which can exchange energy and mass with the surroundings. The independent variables are the temperature T, the chemical potential μ, and the volume V.

After having introduced the basic notions of probability and information theory, we will examine the connection between the latter and the statistical formulation of thermodynamics. Hence, in the rest of this book we will develop all thermodynamics results starting from the modern knowledge of the atomic nature of matter.

Chapter 3

Elements of Probability Theory

Probability theory is a branch of mathematics, and it is used in all fields of science, from physics and chemistry, to biology and sociology, to economics and psychology. When we take decisions in presence of partial information, which happens very often in everydays life, we also (consciously or unconsciously) make a sort of probabilitstic reasoning. In short, the concept of probability is used everywhere and anytime in our lives.

3.1 The Axiomatic Approach to Probability

The axiomatic approach to probability was developed mainly by Andrey Kolmogorov (1903–1987) in the thirties of the last century. It consists of the three elements denoted as $\{\Omega, F, P\}$ which, together, define the **probability space**, and it is based on the theory of sets.

The **sample space** Ω is the set of all possible outcomes of a specific experiment (sometimes referred to as a "trial"). For example, the sample space of all possible outcomes of tossing a coin consists of two elements $\Omega = \{H, T\}$ corresponding to "head" and "tail", whereas the sample space of throwing a die consists of six possible outcomes $\Omega = \{1, 2, 3, 4, 5, 6\}$. Each possible outcome is called **simple event** or **elementary event**, and is a "point" of the sample space.

A **compound event**, or simply an **event**, is defined as a union (or a sum) of elementary events. For example, when throwing a die one may define the (compound) event "even" consisting of the elementary events $\{2, 4, 6\}$, and the (compound) event "odd" consisting of the elementary events $\{1, 3, 5\}$. Whenever the result is 2, 4, or 6, the "even" event occurs; whenever the result is 1, 3, or 5, the "odd" event occurs. Another example is the event "larger than or equal to 5", which consists of the elementary events $\{5, 6\}$ and occurs whenever the outcome is 5 or 6.

The **field of events** F consists of all partial sets of the sample space Ω. Note that Ω itself belongs to F and is called the **certain event**, as well as the empty set $\emptyset$ which represents the **impossible event**. For a finite sample space Ω with n elementary events, the total number of partial sets is 2^n, which is then the cardinality of the field of events.

The proof is obtained by straightforward counting, with the help of the **binomial coefficient**

$$\binom{n}{k} \equiv \frac{n!}{(n-k)!\,k!} \tag{3.1}$$

(few examples are shown in figure 3.1) and of Newton's **binomial theorem**

$$(x+y)^n = \sum_{k=0}^{n} \binom{n}{k} x^k\, y^{n-k} \tag{3.2}$$

Proof — Let us list all possible events in F:

- $\binom{n}{0} = 1$ event denoted by $\emptyset$ and representing the impossible (or "empty") event (by definition $0! = 1$)
- $\binom{n}{1} = n$ simple (or elementary) events
- $\binom{n}{2}$ events consisting of two simple events
- $\binom{n}{3}$ events consisting of three simple events

$$\vdots$$

- $\binom{n}{n} = 1$ event consisting of the entire sample space Ω and representing the certain event

Altogether, we have

$$\binom{n}{0} + \binom{n}{1} + \binom{n}{2} + \cdots + \binom{n}{n} = (1+1)^n = 2^n$$

where we have made use of Newton's binomial theorem with $x = y = 1$. ∎

Now, we introduce some notations regarding operations between events, illustrated in figure 3.2 with the help of Venn diagrams.

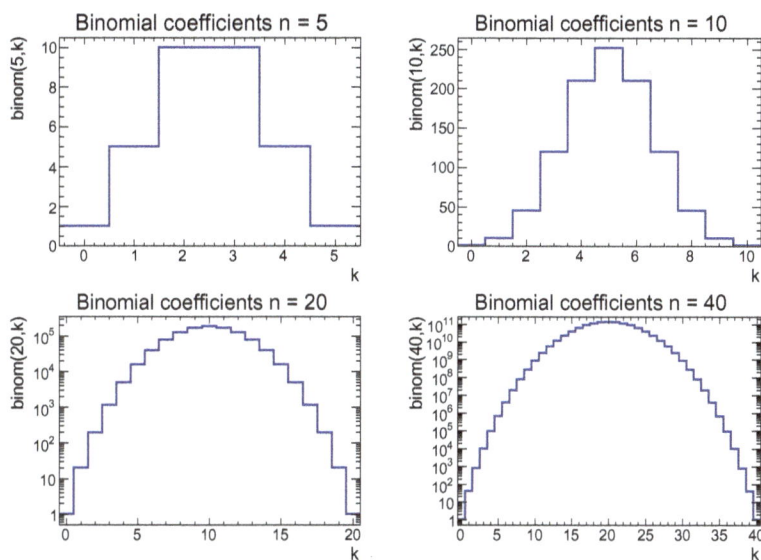

Figure 3.1: Binomial coefficients, for $n = 5, 10, 20, 40$ and k ranging from 0 to n. The vertical scale is linear in the top plots and logarithmic in the bottom plots.

- The event $A \cup B$ (or $A + B$) is the **union** (or sum) of the two events A and B. It represents the compound event "either A or B has occurred".

- The event $A \cap B$ (or $A \cdot B$) is the **intersection** (or product) of the two events A and B. It represents the compound event "both A and B have occurred".

- When $A \cap B = \emptyset$ the two events are **mutually exclusive**, in the sense that it is impossible that both A and B occur. In the language of set theory, we also say that A and B are "disjoint": there is no simple event that is common to both. When the intersection is not the impossible event, A and B are "overlapping" events.

- The **complementary event** of A is denoted as $\overline{A}$ or as $\Omega - A$, and represents the event "A did not occur". Obviously, Ω and $\emptyset$ are complementary to each other.

- When the occurrence of A implies that B has occurred, we write $A \subset B$ and say that A is **included** in the event B.

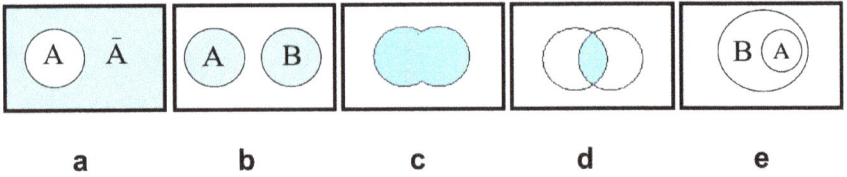

Figure 3.2: Some relations between events (and sets): (a) the event A and its complementary event $\bar{A}$; (b) A and B are disjoint events; (c) union of overlapping events; (d) intersection of overlapping events; (e) the event A is included in the event B.

The last element of the probability space is the **probability function** P, which is defined on each element $A \in F$ and has the following properties:

$$P(\Omega) = 1 \tag{3.3}$$

$$0 \leq P(A) \leq 1 \tag{3.4}$$

$$\text{For } A \text{ and } B \text{ disjoint}, P(A \cup B) = P(A) + P(B) \tag{3.5}$$

The first two conditions define the range of numbers for the probability function[1]. The first condition simply means that the event Ω has the largest value of the probability, which is set to 1 by definition. The second property fixes the range of probability values to the unit interval $[0, 1]$. The minimum probability value is clearly assigned to the impossible event, which is given 0 probability.

The third condition is intuitively clear. When two events A and B are mutually exclusive, the occurrence of one event excludes the occurrence of the other. In this case, the probability that any of them occurs is simply the sum of their individual probabilities. For example, the two events A = "the outcome of throwing a die is even" and B = "the outcome of throwing a die is odd" are clearly disjoint: the occurence of one *excludes* the occurrence of the other. If a third event is defined as $C = \{5, 6\}$, then C has non empty overlap with both A and B. The event $D = \{1, 2\}$ is clearly disjoint with C, although it overlaps with both A and B.

In the general case of non mutually exclusive events, one has

$$P(A \cup B) = P(A) + P(B) - P(A \cap B) \tag{3.6}$$

---

[1]We shall use either P or Pr for probability. When we have a distribution we also use lowercase letters $p_1, \cdots, p_n$.

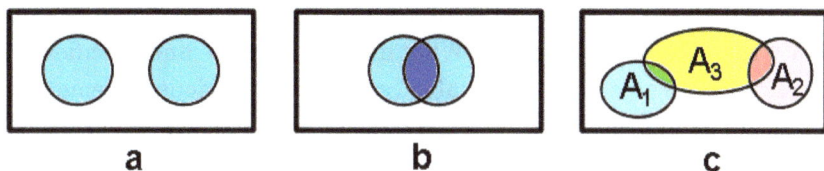

Figure 3.3: Venn diagrams showing (a) two mutually exclusive events, (b) two overlapping events, and (c) an event which overlaps with two mutually exclusive events.

These properties can be represented graphically with the help of Venn diagrams, like those in figure 3.3, because P can be interpreted as a measure on F. Suppose we throw a dart on a board. If we assume that the dart always hits some point on the board (i.e. any region in this board is considered to be an event), then the larger is the area representing an event, the higher is the probability to hit it. In the case of overlapping events, summing their areas would give a too large probability, as the overall area covered by them is equal to the sum of their areas minus the area of the overlap region, which is also what equation (3.6) says.

On this relatively simple axiomatic foundation, the whole edifice of the mathematical theory of probability can be erected. In the axiomatic structure of the theory of probability, the probabilities are said to be *assigned* to each event. Although they must obey the four properties (3.3), (3.4), (3.5) and (3.6), nothing is said on the *interpretation* of these values.

The theory does not *define* the meaning of probability, nor provides a method of calculating or measuring these probabilities. In facts, one may choose different possible ways of defining the probability. For simple experiments, like tossing a coin or throwing a die, it comes out that from the knowledge (or assumption) of some symmetry between the possible outcomes one has a very simple method for assigning probabilities. However, this kind of definition is not general enough to cover all possible applications of probability theory. A couple of different choices are illustrated below, which are good in many practical problems although they cannot be considered genuine definitions. A definition which has no internal inconsistency is based on the concept of subjective degree of belief, which is general (and loose) enough to adapt iself to all cases. However a degree of belief is by itself very subjective. In what follows, we will not need to use this definition, as relying on some property of the physical systems under investigation will be sufficient to assign "reasonable" probabilities to all events. What matters for us — and specifies the meaning of the adjective "reasonable" used in the previous statement — is that the results obtained with the help of the probability theory make physical sense, i.e. they match the experimental results.

It is important to note that it is not possible to *measure* probabilities directly. Instead, one typically measures the **relative frequency** of some result, by taking the ratio between the number of its occurrences and the total number of trials (assumed to be all identical repetitions of the same experiment). Whatever definition of probability we adopt (provided that it is "reasonable" in the sense precised above), the **weak law of large numbers** (or **Bernoulli's theorem**) ensures that *in the limit of an infinite number of measurements the relative frequency converges in probability to the probability*. Hence we can always estimate the probability of an event by looking at its relative frequency on the long run.

Note that in this theorem the probability is used twice, in the type of the convergence and in the limit. Convergence in probability implies that "unusual" outcomes become less and less likely as the sequence of trials progresses. As such, it is weaker than the usual mathematical limit.[2]

3.1.1 *The classical definition*

The classical definition of probability is sometimes referred to as the *a priori* definition. Let us denote as N(total) the total number of outcomes of a specified experiment. For example, when throwing a die N(total) = 6. Next, we denote as N(event) the number of elementary events which are included in the (composite) event of interest. For a die, the "odd" event

---

[2] Formally, one may consider the sequence of relative frequencies $\{f_n\}$, where $f_n \equiv k/n$ is the ratio between the number k of occurrences of the event under consideration and the total number n of trials. If p is the (unknown) value of the true probability, Bernoulli's theorem states that $\forall \varepsilon > 0$ the probability that the distance between f_n and p exceeds ε tends to zero:

$$\lim_{n \to \infty} P(|f_n - p| \geq \varepsilon) = 0$$

This is *not* the same behavior of a mathematical limit, for which it never happens that, after some point, the distance to the limit exceeds a given small quantity: *convergence in probability is weaker*. In other words, it is always possible to find (even for large n) a frequency that is not very close to the probability p, although the *probability* for this to happen decreases with increasing number of trials.

consists of the three elementary events $\{1,3,5\}$, hence $N(\text{odd}) = 3$. Then one may *define* the probability of this event as

$$P(\text{event}) = \frac{N(\text{event})}{N(\text{total})} \qquad (3.7)$$

For example, the probability of obtaining an odd result when throwing a die is $P(\text{odd}) = 3/6 = 1/2$.

Some care must be used in applying (3.7) as a *definition* of probability. First, not every event can be "decomposed" into elementary events. For instance, the probability of the event "tomorrow it will start raining at ten o'clock", cannot be calculated from (3.7). More importantly, the above formula presumes that each of the elementary events has the same likelihood of occurrence. For instance, in throwing a die, each elementary event is presumed to have the same *probability*, 1/6. The motivation for this is the assumption of a perfect physical symmetry between all die faces. But this means that the "definition" (3.7) of the probability of an event is based on the knowledge of the probabilities of each of the elementary events. As the concept of probability is already present in the "definition" of probability (3.7), the latter is ill-posed and cannot be used *bona fide*: it is a circular definition. Nevertheless, this method of assigning probabilities is extremely useful in a number of practical problems, in particular when symmetry considerations support the choice of equal probability for all elementary events.

3.1.2 *The frequentist definition*

The frequentist definition is referred to as the *a posteriori* or the "experimental" definition, since it is based on the actual relative frequency of occurrence of events.

The simplest example is tossing a coin. There are two possible outcomes, head (H) or tail (T). We exclude the rare events that the coin will fall exactly perpendicular to the floor, or break into pieces during the experiment, or even disappear from sight so that the outcome is indeterminable. On the other hand, we do not assume perfect symmetry between H and T (otherwise we would just apply the classical assignment of probabilities $P(H) = P(T) = 1/2$). We now toss the die N times, and the number $N(H)$ of occurrence of heads is recorded. If $N(\text{total})$ is the total number of trials, then the frequency of occurrence of head is $N(H)/N(\text{total})$. One may "define" the probability of occurrence of the event H as the limiting relative frequency when $N(\text{total})$ goes to infinity:

$$P(H) = \lim_{N(\text{total}) \to \infty} N(H)/N(\text{total}) \tag{3.8}$$

This is a well-defined and feasible experiment, and the limiting result is guaranteed by Bernoulli's theorem. However, this is not a proper definition of probability, because the concept of probability is also used in the type of convergence. Nevertheless, if we perform a long enough sequence of trials, the fluctuations about the true (but unknown) value of the probability will be small enough that in practice we can take the relative frequency as a very good estimate of the true probability.

3.1.3 *Probability as degree of belief*

When we need to take a decision (e.g. whether to carry an umbrella with us) in presence of partial information (e.g. a cloudy sky during the Fall season), we implicitly perform some sort of probabilistic reasoning. We need to evaluate the *possible* benefits or disadvantages that a correct or incorrect guess implies, so we need to quantify the vague concept of "possible", which may range from "impossible" to "certain". In this case, the probability is the value which quantifies "possible" for each event. This value is what is usually called "degree of belief", as it is inherently subjective.

Different people may assign different probabilities to the same event. However, a rigorous use of probabilities requires that, in presence of the very same information, two honest and coherent persons (or artificial intelligences) assign the same probability to the same event. A few prescriptions exist, which make this requirement possible in practice. One example is the "coherent bet" (de Finetti, 1974), in which you set the odds of two mutually exclusive events, but your opponent decides which side of the bet will be yours. The price you set is the "operational subjective probability" that you assign to the proposition on which you are betting.

When a practical prescription is followed, which leads to a honest and coherent assignment of probabilities, the concept (and use of) probability extends the Boolean logics, in which only values of 0 = impossible and 1 = certain are considered (Jaynes, 2003). This definition of probability is crucially connected to the information which is available to the individual who assigns its value. It is at the heart of the Bayesian approach to statistics, and contains as particular cases the classical and frequentist definitions, when cast in terms of information about the experiment and of degree of belief about each possible result.

When considering statistical inference, for example when learning about a physical model from the data obtained with a particular experiment, this definition of probability is very powerful. However, in this book we will not use probability to learn from data, but only to construct models that are meant to represent our "understanding" of macroscopic systems when viewed from a microscopic point of view. In this context the classical definition, which assumes a large sample of indentical replicas of the same system, will be sufficient. The "goodness" of this choice is proven by the agreement between the theoretical results which follow from it and the experimental results.

3.2 Independent Events and Conditional Probability

The concepts of dependence between events and conditional probability, are central to probability theory and have many uses in sciences.

Two events are said to be ***independent***, if the occurrence of one event has no effect on the probability of occurrence of the other. Mathematically, two events A and B are said to be independent, if and only if

$$P(A \cdot B) = P(A)P(B) \tag{3.9}$$

For example, if two persons who are far apart from each other throw a fair die each, the outcomes of the two dice are independent, in the sense that the occurrence of, e.g. "5", on one die, does not have any effect on the probability of occurrence of a result, say "3", on the other. On the other hand, if the two dice are connected by an inflexible wire, the outcomes of the two results cannot be considered independent. Similarly, if we throw a single die consecutively, and at each throw the die is deformed or damaged, the outcomes would not be independent. Intuitively, it is clear that whenever two events are independent, the probability of the occurrence of both events (for example "5" on one die and "3" on the other) is the *product* of the two probabilities. The reason is quite simple. By tossing two dice simultaneously, we have altogether 36 possible elementary events. Because (1) the outcomes of a single die are independent events and (2) the result obtained with one die does not influence the outcome of the other die, the 36 possible elementary events have equal probability. This means that each of them has probability 1/36, equal to the product $1/6 \times 1/6$ of the probabilities of any two single-die events.

A fundamental concept in the theory of probability is that of **conditional probability**. This is the probability of the occurrence of an event *A given that an event B has occurred*. We use the notation $P(A|B)$ for the conditional probability. It can be read as "probability of A given (the occurrence of) B". By *definition*, we write

$$P(A \cdot B) = P(A|B)P(B) \tag{3.10}$$

which, in case $P(B) \neq 0$, gives

$$P(A|B) = P(A \cdot B)/P(B) \tag{3.11}$$

As $P(A \cdot B) = P(B \cdot A)$, eq. (3.10) implies that

$$P(A|B)P(B) = P(B|A)P(A)$$

By comparing (3.10) with (3.9), one immediately gets the important result that

$$A \text{ and } B \text{ independent} \iff \begin{cases} P(A|B) = P(A) \\ P(B|A) = P(B) \end{cases} \tag{3.12}$$

We can define the **correlation** between the two events as

$$g(A,B) \equiv \frac{P(A \cdot B)}{P(A)P(B)} \tag{3.13}$$

With the help of (3.10), we obtain

$$g(A,B) = \frac{P(A|B)}{P(A)} = \frac{P(B|A)}{P(B)} \tag{3.14}$$

We say that the two events are **positively correlated** when $g(A,B) > 1$, i.e. when the occurrence of one event enhances or increases the probability of the second event. We say that the two events are **negatively correlated** when $g(A,B) < 1$, i.e. when the occurrence of one event reduces the probability of the other. Finally, the events are said to be **uncorrelated** or **indifferent** when $g(A,B) = 1$.

As an example, consider the following events:

A = {The outcome of throwing a die is "4"}

B = {The outcome of throwing a die is even}

C = {The outcome of throwing a die is odd}

We can calculate the following two conditional probabilities:

$$P(A|B) = 1/3 > P(A) = 1/6$$
$$P(A|C) = 0 < P(A) = 1/6$$

The knowledge that B has occurred *increases* the probability of the occurrence of A. Without that knowledge, the probability of A is only 1/6 (half of 1/3). Given the occurrence of B, the probability of A becomes larger, but if instead C has occurred the probability of A becomes zero, i.e., *smaller* than the probability of A without that knowledge.

It is important to distinguish between *disjoint* (i.e. mutually exclusive events) and *independent* events. Disjoint events are events that can never occur together: the occurrence of one excludes the occurrence of the second. For example, the dice events "even" and "5" are disjoint.

That two events are disjoint is a property of the events themselves: the two events have no common elementary event. In terms of Venn diagrams, two regions that are non-overlapping are disjoint.

On the other hand, the independence between events is not defined in terms of the elementary events comprising the two events. Instead, it is defined in terms of their probabilities.

If the two events are disjoint, then they are strongly dependent. Indeed, the occurrence of one makes the other impossible (hence they are negatively correlated). On the other hand, if the two regions representing the A and B events do overlap (i.e. A and B are not disjoint), then the two events could be either dependent or independent. In fact, the two events could either be positively or negatively correlated.

Exercise 3.1. Show that if three events are disjoint $A \cap B \cap C = \emptyset$, it does not follow that these events must be disjoint in pairs. On the other hand, show that disjoint in pairs imply disjoint in triplets.

Exercise 3.2. Show that independence between pairs does not imply independence between triplets, and that independence between the triplets does not imply independence in pairs.

3.3   Random Variables, Average, Variance and Correlation

A *random variable* (abbreviated as RV) is a real-valued function defined on a sample space. If $\Omega = \{w_1, \ldots, w_n\}$ is the sample space, then for any $w_i \in \Omega$, a RV is a function $X(w_i)$ of the elements of the sample space whose values are real numbers.

We shall use capital letters like X, Y for the random variables and lowercase letters like x, y for their values.

Note that the outcomes of the experiment, i.e. the elements w_i of the sample space Ω, are not necessarily numbers. The outcomes can be colors, different objects, different figures, etc. For instance, the outcomes of tossing a coin are $\{H, T\}$. Another example is a die, whose six faces have different colors, say red, white, blue, yellow, green and black. In such a case, we cannot plot the function $X(w)$, but we can still write it down as a table, for example like table 3.1.

Table 3.1: **Example of Random Variable Defined on Sample Space of Colors**

w =	red	white	blue	yellow	green	black
$X(w)$ =	2	2	2	1	0	1

Let's now define the compound event $\{w : X(w) = x_i\}$ as the event "all w for which $X(w) = x_i$", and write $P(X(w) = x_i)$ as a shorthand notation for $P(\{w : X(w) = x_i\})$. We always have the identity

$$P(\Omega) = \sum_{w \in \Omega} P(w) = \sum_i P(X(w) = x_i) = 1 \qquad (3.15)$$

The first sum is over all the elements $w \in \Omega$ and the second sum is over all values x_i of the RV $X(w)$.

In general, if the outcomes w_i are numbers, and their corresponding probabilities are $P(w_i) = p_i$, then we can define the *average outcome* of the experiment as:

$$\langle w \rangle \equiv \sum_i w_i p_i \qquad (3.16)$$

For instance, for the ordinary die we have

$$\langle w \rangle = \sum_{i=1}^{6} w_i p_i = \frac{1}{6} \sum_{i=1}^{6} i = 3.5 \qquad (3.17)$$

However, it is meaningless to talk about the average of the outcomes of an experiment when they are not numbers. For instance, for the six-colored die, there is no average outcome that can be defined for this experiment. On the other hand, if a RV is defined on the sample space, then one can *always* define the **average value** of the RV as[3]

$$\overline{X} \equiv \langle X \rangle \equiv \sum_w X(w)P(w) = \sum_x x P\{w : X(w) = x\} \qquad (3.18)$$

The first sum in (3.16) is over all possible outcomes $w \in \Omega$, whereas the second sum is over al possible values x of the RV X.

With $\{w : X(w) = x_i\}$ we indicate the (compound) event consisting of all possible outcomes w such that the value of the RV $X(w)$ is x. In the example of table 3.1, these events are

$\{w : X(w) = 0\} = \{\text{green}\}$	with	$P\{w : X(w) = 0\} = 1/6$
$\{w : X(w) = 1\} = \{\text{yellow, black}\}$	with	$P\{w : X(w) = 1\} = 2/6$
$\{w : X(w) = 2\} = \{\text{red, white, blue}\}$	with	$P\{w : X(w) = 2\} = 3/6$

Hence, the average of the RV is $\overline{X} = 0 \times \frac{1}{6} + 1 \times \frac{2}{6} + 2 \times \frac{3}{6} = \frac{4}{3}$.

The extension of the concept of random variable to two or more dimensions is quite straightforward. Let X and Y be two RV, which can be defined on the same or on different sample spaces. The **joint distribution** of X and Y is defined as the probability that X attains the value x and Y attains the value y, and is denoted by $P(X = x, Y = y)$. It must satisfy the normalization condition

$$\sum_{x,y} P(X = x, Y = y) = 1 \qquad (3.19)$$

The **marginal distribution** of X is defined as

$$P(X = x) \equiv \sum_y P(X = x, Y = y) \qquad (3.20)$$

where the sum runs over all possible values y of Y. Similarly the marginal distribution of Y is defined as

$$P(Y = y) \equiv \sum_x P(X = x, Y = y)$$

where the sum runs over all possible values x of X.

Now, let us consider a real-valued function $f(X)$ of the RV X, which associates a number $f(x)$ to each possible value x of X. As X is a random

---

[3]We shall use either $\overline{X}$ or $\langle X \rangle$ to denote the average of the RV X. Another possible notation is $E[X]$, which reads "expectation of X".

variable which depends on the result w of some experiment, $f(X)$ also depends on w and is a random variable. The **average of the function** $f(X)$ is defined as

$$\langle f(X) \rangle \equiv \sum_x f(x) P\{w : X(w) = x\} \qquad (3.21)$$

A function which has special interest is the **variance**, which quantifies how much the distribution is spread out or dispersed. The variance is defined as

$$\mathrm{Var}(X) \equiv V[X] \equiv \sigma^2 \equiv \langle (X - \langle X \rangle)^2 \rangle = \langle X^2 \rangle - \langle X \rangle^2 \qquad (3.22)$$

Exercise 3.3. Prove by calculation the last equality in eq. (3.22).

The positive square root σ of $\mathrm{Var}(X)$ is called the **standard deviation**. Note that, from its definition (3.22), it follows that $\mathrm{Var}(X)$ is always positive. For two random variables X and Y one defines the **covariance** of X and Y as

$$\mathrm{Cov}(X, Y) \equiv \langle (X - \langle X \rangle)(Y - \langle Y \rangle) \rangle = \langle XY \rangle - \langle X \rangle \langle Y \rangle \qquad (3.23)$$

and the **correlation coefficient** as

$$\mathrm{Cor}(X, Y) \equiv R(X, Y) \equiv \frac{\mathrm{Cov}(X, Y)}{\mathrm{Var}(X)\mathrm{Var}(Y)} \qquad (3.24)$$

where the denominator is introduced such that the range of possible values of $\mathrm{Cor}(X, Y)$ is $[-1, +1]$. Two variables such that $\mathrm{Cor}(X, Y) = 0$ are said to be **uncorrelated**.

Two random variables are said to be **independent** if and only if

$$P(X = x, Y = y) = P(X = x) P(Y = y) \qquad (3.25)$$

for any values x and y.

If two RV X and Y are independent, then they are also uncorrelated (but not vice versa).

Proof — If X and Y are independent, the first term in the r.h.s. of (3.23) is

$$\langle XY \rangle = \sum_{x,y} xy P(X = x, Y = y) = \sum_{x,y} xy P(X = x) P(Y = y)$$

$$= \left[\sum_x x P(X = x) \right] \cdot \left[\sum_y y P(Y = y) \right] = \langle X \rangle \langle Y \rangle$$

hence $\mathrm{Cov}(X, Y) = 0$.   ∎

Note that if $\text{Cov}(X,Y) = 0$ it does not necessarily follow that X and Y are independent. The reason is that independence of X and Y applies for *any* value of x and y, but the uncorrelation applies only to the *average* value of the RV XY.

In summary, independence implies uncorrelation, but nor vice versa.

3.4   Continuous Random Variables

One might consider random variables which can assume any value in $\mathbb{R}$ or any other continuous interval $\mathbb{X} \subseteq \mathbb{R}$. In this case, the probability that $x \in \mathbb{X}$ assumes any given specific value is zero, because there are infinite possible values. However, the probability $P(x \in \mathbb{A} \subseteq \mathbb{X})$ that x falls in some finite interval $\mathbb{A}$ is finite. This allows to associate with x a probability distribution, in the following way.

A ***probability density function*** (abbreviated as PDF) is any function $f(x) \geq 0$ with non-negative values satisfying the ***normalization*** condition

$$\int_{\mathbb{X}} f(x)\,\mathrm{d}x = 1 \qquad (3.26)$$

where $\mathbb{X}$ is the domain of x. Provided that we define $f(x)$ such that it gives zero outside $\mathbb{X}$, we can also extend the integral above to run from $-\infty$ to $+\infty$.

The ***cumulative distribution function*** $F(x)$ is then defined as the probability distribution which corresponds to all intervals starting from $-\infty$:

$$F(x) \equiv \int_{-\infty}^{x} f(x')\,\mathrm{d}x' \qquad (3.27)$$

Clearly, $F(x)$ is a monotonically increasing function of x which maps the domain of x onto the interval $[0,1]$. Hence, $F(x)$ can be interpreted as a probability measure, and $P(a < x < b) = F(b) - F(a)$. On the other hand, $f(x)$ is *not* a probability: rather, its integral over some interval is a probability (this is why it is called a probability "density").

The ***mean value*** (or simply the ***mean***) of a probability density function $f(x)$ is defined as

$$E[x] = \langle x \rangle = \overline{x} \equiv \int_{\mathbb{X}} x f(x)\,\mathrm{d}x \qquad (3.28)$$

The three notations are all commonly used, with $E[x]$ often being read as the **expected value** (or simply the **expectation**) of x. Similarly, the expectation of any function $g(x)$ is defined as:

$$E[g(x)] \equiv \int_{\mathbb{X}} g(x) f(x) \, dx \qquad (3.29)$$

which means that $E[\cdot]$ acts as a linear operator over a function, and $dm = f(x)dx$ is a measure on $\mathbb{X}$.

The n-th (raw) **moment** of the PDF $f(x)$ is defined as $\mu'_n \equiv E[x^n]$, so that $\mu'_1 = E[x]$ is the mean value of x. The n-th **central moment** is defined as $\mu_n \equiv E[(x - \mu'_1)^n]$, the raw moment about the mean.

The second central moment has special importance, and is called the **variance** of the distribution, usually denoted as $V[x]$ or σ^2. The (positive) square root σ of the variance is called the **standard deviation** (or std.dev.) of the distribution. The variance is a measure of the dispersion of the probability density function about its mean. Higher moments may be used too, although they will not be necessary in most cases addressed in this book.

3.5 The Binomial Distribution

Suppose we have a series of experiments, each with only two possible outcomes, for example H and T in tossing a coin, or a particle being in the left (L) or the right (R) compartment of a container. Let p be the probability of occurrence of one of the outcomes, say H, which implies that the other outcome has probability $P(T) = q = 1 - p = 1 - P(H)$. If all the trials in the series are independent, then the probability of any *specific* sequence of outcomes, say $HTHHTTH$, is simply the product of the probabilities of each outcomes, that is $pqppqqp = p^4 q^3 = p^4 (1 - p)^3$.

Similarly, if we have seven particles distributed in two compartments such that $P(L) = p$ and $P(R) = q$, the probability of finding exactly particles (1), (2), (4) and (7) in R and particles (2), (5), (6) in L is also $p^4 q^3 = p^4 (1 - p)^3$.

In most cases, when dealing with particles in different compartments, we are not interested in the *specific* configuration, i.e. which particle is in which compartment, but only in the *number* of particles in each compartment.

Clearly, we have many possible *specific* sequences of outcomes for which the number of H's (or of R's) is constant. A **specific configuration** is one particular sequence which indicates the actual outcome of each experiment in the series. For example, in the case of four particles and two compartments we have all possible specific configurations listed in table 3.2.

Table 3.2: **All Possible Specific Configurations of 4 Particles in 2 Compartments.** N_L is the number of particles in the left comparment, M is the multiplicity of the state identified by N_L alone (i.e. of the "compound event"), and P is the probability of each specific configuration, assuming that p = probability for a single particle to be in the left compartment and $q = 1 - p$ = probability to be in the right compartment.

sp. cnf.	N_L	M	P	sp. cnf.	N_L	M	P	sp. cnf.	N_L	M	P
$RRRR$	0	1	q^4	$LLRR$	2			$LLLR$	3		
————				$LRLR$	2			$LLRL$	3		
$LRRR$	1			$LRRL$	2			$LRLL$	3	4	$p^3 q$
$RLRR$	1			$RLLR$	2	6	$p^2 q^2$	$RLLL$	3		
$RRLR$	1	4	pq^3	$RLRL$	2						
$RRRL$	1			$RRLL$	2			————			
								LLLL	4	1	p^4

There is only one configuration for which all particles are in R, in other words the (compound) event "all four particles are in R" has multiplicity one. There are 4 specific configurations in which one particle is in L (i.e. the multiplicity is 4), 6 specific configurations with two particles in each side, etc.

Clearly, any two *specific* configurations are disjoint, or mutually exclusive events. Therefore, the probability of occurrence of all particles in R is q^4, the probability of finding 3 particles in R and 1 particle in L is pq^3, etc. In general, given N particles which can be found in either of two compartments, the probability of occurrence of the event "n particles in L" (or "n coins showing T") is

$$P(n;N,p) = \binom{N}{n} p^n (1-p)^{N-n} \tag{3.30}$$

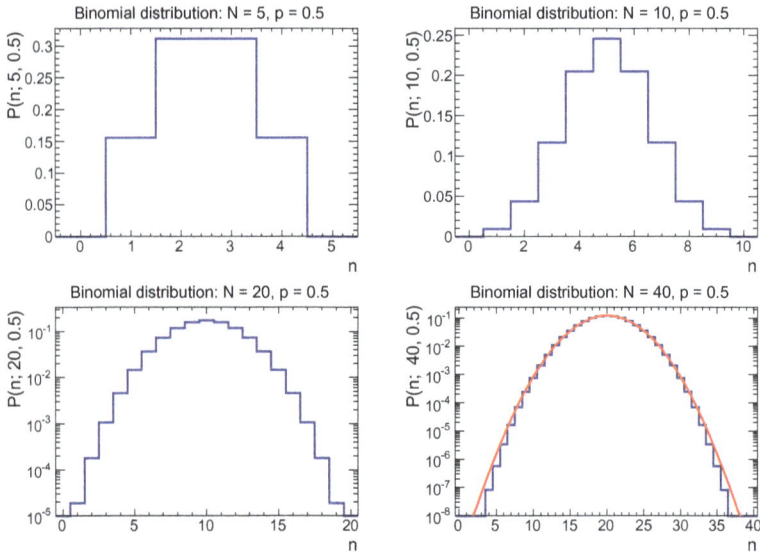

Figure 3.4: Binomial distribution, for $n = 5, 10, 20, 40$ and $p = 0.5$. The last plot also shows the Gaussian approximation (see section 3.6). Note that the vertical scale is linear in the top plots, and logarithmic in the bottom plots.

which is the ***binomial distribution***. The mean and variance of n are

$$E[n] \equiv \langle n \rangle = \sum_{n=0}^{N} n\, P(n; N, p) = Np \tag{3.31}$$

$$V[n] \equiv \sigma^2 = \langle n^2 \rangle - \langle n \rangle^2 = Npq \tag{3.32}$$

Note that in constructing the expression (3.30), we used the *product rule* for the probabilities of the independent events "R" and "L" and the *sum rule* for the disjoint events "specific sequence" having n L-events and $N-n$ R-events. The number of specific disjoint events is simply the number of ways of selecting a group of n particles out of N identical particles. The binomial coefficients $\binom{N}{n}$ are the coefficients in the binomial expansion which we have introduced to formulate Newton's binomial theorem (3.2) on page 50.

Figure 3.4 shows few examples of $P(n; N, p)$ with $p = q = 1/2$ (a fair coin, for example), in which case it is also known as ***Bernoulli's distribution***. Note that, as N increases, the form of the distribution becomes more and more similar to the (continuous) normal distribution explained in the next section (the red curve in the last plot in figure 3.4).

3.6 The Normal Distribution

As we have seen in figure 3.4, when N is very large, the form of the binomial distribution becomes very similar to the **normal distribution**, also known as **Gaussian distribution**:

$$\mathcal{N}(x;\mu,\sigma) = \frac{1}{2\pi\sigma^2}\exp\left[-\frac{(x-\mu)^2}{2\sigma^2}\right] \tag{3.33}$$

This is the most important probability density function, as it represents the limiting form of a number of different distributions.

The function (3.33) does not represent a probability. In order to get a probability, we must integrate $\mathcal{N}(x;\mu,\sigma)$ over some range of $x \in \mathbb{R}$:

$$P(a < x < b;\mu,\sigma) = \int_a^b \mathcal{N}(x;\mu,\sigma)\,\mathrm{d}x \tag{3.34}$$

and this is the reason why we call $\mathcal{N}(x;\mu,\sigma)$ a probability *density* (function).

Mean and variance of the normal distribution (3.33) are $E[x] = \mu$ and $V[x] = \sigma^2$.

The normal **cumulative distribution function** gives the probability to obtain a value smaller than x:

$$F(x;\mu,\sigma) \equiv \int_\infty^x \mathcal{N}(x';\mu,\sigma)\,\mathrm{d}x' \tag{3.35}$$

such that $\mathcal{N}(x;\mu,\sigma) = \frac{\mathrm{d}}{\mathrm{d}x}F(x;\mu,\sigma)$. The normal distribution $\mathcal{N}(x;0,1)$ and its cumulative distribution function are shown in figure 3.5.

We shall now show that for large N, we obtain the normal distribution as a limiting form of the binomial distribution. In the proof, we make use of the **Stirling approximation** in the form

$$\ln N! \approx N\ln N - N \tag{3.36}$$

which is often used in statistical thermodynamics.

> **Proof** — We start with the binomial distribution (3.30) and treat n as a continuous variable.
>
> The binomial mean and the variance were given by equations (3.31) and (3.32)
>
> $$\overline{n} = \sum_{n=0}^N nP(n;N,p) = Np$$
>
> $$\sigma^2 = \left\langle n^2 \right\rangle - \langle n \rangle^2 = Npq$$

N(x;0,1)

Normal c.d.f.

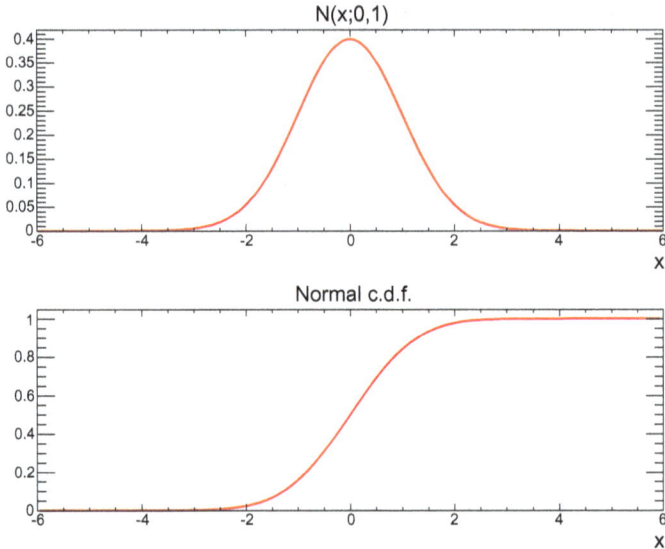

Figure 3.5: Normal distribution centered on zero and with unit standard deviation (top) and the corresponding cumulative distribution function (bottom).

As shown in figure 3.4, the binomial distribution $P(n;N,p)$ has a sharp maximum at $\overline{n}$, hence also its logarithm $\ln P(n;N,p)$ has a maximum there. Treating n as a continuous variable (which is a good approximation when n is very very large[4]) and writing the Taylor expansion of $\ln P(n;N,p)$ with only the first few terms, we get

$$\ln P(n;N,p) = \ln P(\overline{n}|N) + \left.\frac{\partial \ln P(n;N,p)}{\partial n}\right|_{n=\overline{n}} (n - \overline{n})$$
$$+ \frac{1}{2} \left.\frac{\partial^2 \ln P(n;N,p)}{\partial n^2}\right|_{n=\overline{n}} (n - \overline{n})^2 + \dots \tag{3.37}$$

At the maximum, the first derivative is zero. Therefore, we need to consider the expansion (3.37) from the second term. Taking the second derivative of $\ln P(n;N,p)$ and using the Stirling approximation (3.36) we get

$$\left.\frac{\partial^2 \ln P(n;N,p)}{\partial n^2}\right|_{n=\overline{n}} \simeq -\frac{1}{Npq} \tag{3.38}$$

---

[4]We are considering the limit $N \to \infty$ and values of n near $N/2$, hence this is the case.

Note that second derivative is always negative which means that the function $\ln P(n;N,p)$, hence also $P(n;N,p)$, has maximum at $n = \overline{n}$.

Next take the exponential of left and right side terms in equation (3.37), neglecting all higher order terms in the expansion, which can be shown to be negligible for large N, and obtain

$$P(n;N,p) = C \exp\left[-\frac{(n-\overline{n})^2}{2Npq} \right] \qquad (3.39)$$

where the constant C is determined by imposing the normalization condition

$$\int_{-\infty}^{\infty} P(n;N,p)\, dn = 1 \qquad (3.40)$$

from which we obtain $C = (2\pi Npq)^{-1/2}$. When inserted into (3.39), we finally obtain

$$P(n;N,p) = \frac{1}{\sqrt{2\pi Npq}} \exp\left[-\frac{(n-\overline{n})^2}{2Npq} \right] \qquad (3.41)$$

which is identified with the normal distribution (3.33), with mean $\mu = \overline{n} = Np$ and variance $\sigma^2 = Npq$. This is the curve drawn on the last plot of figure 3.4. ∎

3.7 Multidimensional Distributions

Let's suppose that X, Y, Z are 3 random variables. If they are discrete, we define their *joint probability distribution* as

$$P(x, y, z) \equiv \Pr(X = x, Y = y, Z = z) \qquad (3.42)$$

i.e. $P(x, y, z)$ specifies the probability that the three results $X = x, Y = y, Z = z$ occur at the same time. On the other hand, if they are continuous we define their *joint probability density function* $f(x, y, z)$ as the density which satisfies the following relationship:

$$\Pr(X \in [x, x+dx], Y \in [y, y+dy], Z \in [z, z+dz]) = f(x, y, z)\, dx\, dy\, dz \qquad (3.43)$$

In both cases, the new functions must satisfy the normalization condition: when summing (or integrating) over all possible values one must obtain one. The generalization to an arbitrary number of random variables is straightforward.

One example of multidimensional discrete distribution is the multinomial distribution, which extends the binomial distribution to the case of k possible independent values. For example, for a dice $k = 6$. Let $\{p_1, p_2, \ldots, p_k\}$ be the probability distribution of obtaining each value, and imagine to play dice N times. The probability of obtaining n_1 times the value 1, n_2 times the value 2, $\ldots$, and n_k times the value k is given by the **multinomial distribution**

$$P(n_1, \ldots, n_k | N, p_1, \ldots, p_k) = \frac{N!}{n_1! \cdots n_k!} p_1^{n_1} \cdots p_k^{n_k} \tag{3.44}$$

subject to the conditions

$$\sum_{i=1}^{k} n_i = N \quad \text{and} \quad \sum_{i=1}^{k} p_i = 1 \tag{3.45}$$

such that only $k - 1$ among the n_i's are independent.

A **marginal distribution** for the discrete case is obtained by summing over all values of one or more random variables, starting with a n-dimensional distribution and ending with a m-dimensional one, with $m < n$. Similarly, for the continuous case one obtains a **marginal density** by integrating over the domain of one or more random variables, again reducing the dimensionality of the function. The operation of reducing the dimensionality by summing or integrating over all possible values is called **marginalization**.

In the case of a multinomial distribution, summing over all possible values of all but one variable (say the first one) leaves with a binomial distribution. In other words, the marginal distribution of each of the n_i's is binomial, with probabilities $p = p_i$ and $q = 1 - p_i$. Another interesting property is that grouping together some variables again givs a multinomial distribution. For example, by merging the first two categories and defining the random variables $n_1 + n_2, n_3, \ldots, n_k$ one gets a multinomial with probabilities $p_1 + p_2, p_3, \ldots, p_k$.

The **expected value** or **mean of a random vector X** is the vector $E[X]$ whose elements are the expected values of the respective random variables. For the continuous case with joint density $f(x_1, x_2, \ldots, x_k)$, the **expectation operator** acts on a generic function $g(X_1, X_2, \ldots, X_k)$ as follows:

$$E[g(x_1, \ldots, x_k)] = \int \cdots \int g(x_1, \ldots, x_k) f(x_1, \ldots, x_k) \, dx_1 \cdots dx_k \tag{3.46}$$

(for a discrete distribution, the integral is replaced by a sum over all variables). By means of this operator, one can also define the **covariance**

matrix of a random vector as

$$\text{Var}[\mathbf{X}] = E[(\mathbf{X} - E[\mathbf{X}])(\mathbf{X} - E[\mathbf{X}])^T] \tag{3.47}$$

The diagonal elements of the covariance matrix are the variances of each random variable and the off-diagonal ones are the covariances of each possible pair.

In addition, the $n \times m$ **cross-covariance matrix** between two random vectors $\mathbf{X}$ (of dimension n) and $\mathbf{Y}$ (of dimension m) is

$$\text{Cov}[\mathbf{X}, \mathbf{Y}] \equiv E[(\mathbf{X} - E[\mathbf{X}])(\mathbf{Y} - E[\mathbf{Y}])^T]$$
$$= E\left[\mathbf{X}\mathbf{Y}^T\right] - E[\mathbf{X}]E[\mathbf{Y}]^T, \tag{3.48}$$

One example of multidimensional continuous density is provided by the **multivariate normal distribution**

$$f(x_1, \ldots, x_k) = \frac{1}{\sqrt{(2\pi)^k |\Sigma|}} \exp\left[-\frac{1}{2}(\mathbf{x} - \boldsymbol{\mu})^T \Sigma^{-1}(\mathbf{x} - \boldsymbol{\mu})\right] \tag{3.49}$$

where $\mathbf{x} \equiv (x_1, \ldots, x_k)$, $\boldsymbol{\mu}$ is the mean vector, Σ is the **covariance matrix**, and $|\Sigma|$ is its determinant. The marginal distribution of each random variable is a normal distribution, but the reverse is not automatically true: the joint distribution of two Gaussian variables is a bivariate normal distribution only if they are statistically independent. In the bivariate case, the covariance matrix has the form

$$\Sigma = \begin{pmatrix} \sigma_X^2 & \rho \sigma_X \sigma_Y \\ \rho \sigma_X \sigma_Y & \sigma_Y^2 \end{pmatrix} \tag{3.50}$$

where σ_i is the standard deviation and ρ is the correlation between X and Y.

In the continuous case, sometimes it is interesting to look at the shape of the joint PDF when one or more variables is assumed to be fixed at some value. For example, if we fix $Z = z_0$ in the 3-dimensional case we reduce the dimensionality of the joint density defined in (3.43), obtaining a 2-dimensional function $f(x, y, z_0)$, which is sometimes called **profile function**. Please note that $f(x, y, z_0)$ is not normalized to one in the (X, Y) space, hence it is not a probability density function. Thus, this operation results in a quite different object than the "proper" (i.e. normalized to one) PDF that one obtains by marginalization. This kind of dimensionality reduction (called "profiling") can also be achieved in the discrete case, again ending up into a non-normalized result, which is not a probability distribution in the lower-dimensional space.

Chapter 4

Shannon's Measure of Information (SMI)

4.1 Introduction

In 1948 Shannon published a landmark paper entitled "A Mathematical Theory of Communication" (Shannon, 1948). In section 6 of this paper Shannon writes:

> Suppose we have a set of possible events whose probabilities of occurrence are $p_1, p_2, \ldots, p_n$. These probabilities are known but that is all we know concerning which event will occur. Can we find a measure of how much "choice" is involved in the selection of the event or how uncertain we are of the outcome?
>
> If there is such a measure, say $H(p_1, p_2, \ldots, p_n)$, it is reasonable to require of it the following properties:
>
> (i) H should be continuous in the p_i.
> (ii) If all the p_i are equal, $p_i = 1/n$, then H should be a monotonic increasing function of n. With equally likely events there is more choice, or uncertainty, when there are more possible events.
> (iii) If a choice be broken down into two successive choices, the original H should be the weighted sum of the individual values of H.

Then Shannon proved that the only function H satisfying the three assumption above has the form:

$$H = -K \sum_{i=1}^{n} p_i \log p_i \tag{4.1}$$

where K is a positive constant which fixes the units of measure of H.[1] We note here that, if the sum in (4.1) is finite, one may equivalently say

---

[1]The base of the logarithm is also a free parameter. We will always use a base > 1, which makes $H > 0$: the most common choices are binary and natural logarithms.

that H is proportional to the expectation of the function $-\log p_i$: $H = KE[-\log p_i]$.

In this book we shall not be interested in the derivation[2] of the quantity H, to which we shall refer to as **Shannon's measure of information** (SMI). In this chapter, we shall only be interested in the *properties* of the SMI, and its *meaning* as a measure of information. The SMI is used in many branches of science: it is a very general concept. At the moment we shall study the quantity H defined in (4.1) without any reference to thermodynamics. We shall show in chapters 5 and 6 that the entropy interpretation in terms of information spawns from the SMI. Let us quote another paragraph from Shannon's paper.

> This theorem, and the assumptions required for its proof, are in no way necessary for the present theory. It is given chiefly to lend a certain plausibility to some of our later definitions. The real justification of these definitions however, will reside in their implications.
>
> Quantities of the form $H = -K \sum_i p_i \log p_i$ (the constant K merely amounts to a choice of a unit of measure) play a central role in information theory as measures of information, choice and uncertainty. The form of H will be recognized as that of entropy as defined in certain formulations of statistical mechanics where p_i is the probability of a system being in cell i of its phase space. H is then, for example, the H in Boltzmann's famous H theorem. We shall call $H = -K \sum_i p_i \log p_i$ the entropy of the set of probabilities $p_1, p_2, \ldots, p_n$. If x is a chance variable we will write $H(x)$ for its entropy; thus x is not an argument of a function but a label for a number, to differentiate it from $H(y)$ say, the entropy of the chance variable y.

Note carefully that Shannon describes H as a "measure of information, choice and uncertainty". All these are valid interpretations of the quantity H as defined in (2.1). Shannon goes on to say that "the *form* of H will be recognized as that of entropy as defined in certain formulation of statistical mechanics where p_i is the probability..."

The reader is urged to read carefully the above quotation. There is no need to understand the details of the various statements. However, three points will be crucial for understanding the relevance of these words to thermodynamics.

---

[2]The interested reader may consult the original paper by Shannon (1948) or Ben-Naim (2008).

First, note that Shannon presents his problem in terms of a probability distribution: $p_1, \ldots p_n$. This distribution must be such that $0 \leq p_i, \leq 1$ and $\sum p_i = 1$. In other words, we assume that there are n possible outcomes of an experiment, with probabilities p_i of occurrence, and that one of the outcomes must occur. Shannon seeks a measure of how much "choice" or "uncertainty" is in the outcome, and later he refers to the quantity H as a measure of "information, choice and uncertainty" (we shall further discuss these interpretations of the quantity H in section 4.8). He did not seek a measure of the general concept of information, only a measure of *information contained in* (or carried out by) *a probability distribution*.

Second, Shannon proposed three plausible properties of such a measure, presuming that such a measure exists. We shall discuss these properties and their plausibility in the following sections. Here, we draw the attention of the reader to the "methodology" of seeking and finding a quantity that is not even clear *a priori* whether it exists.

Finally, note carefully that Shannon was not interested in thermodynamics in general, nor in the *entropy* in particular. However, he noted that "the form of H will be recognized as that of entropy as defined in certain formulations of statistical mechanics..." Therefore, he suggested to call H "the entropy of the set of probabilities $p_1, \ldots, p_n$". Indeed, the *form* of the function H is the same as the *form* of the entropy as used in statistical mechanics. However, this does *not* imply that H is the entropy. We shall further discuss this point in section 6.6, after we learn some of the properties of the SMI and the entropy. For the moment we shall study the SMI without any reference to entropy.

Shannon's measure of information is a very general concept. It can be defined on *any distribution function*. For example, it can be defined for the outcomes of throwing a die or the frequencies of the appearance of alphabet letters in certain languages. There is a vast range of fields in which the quantity H is definable, and as such the SMI became a very useful tool in so many fields of research.

As we shall see in the next chapters, the entropy is defined only on very small sets of probability distributions. When H is applied to those distributions it becomes identical with the statistical mechanical entropy, hence one can safely say that *entropy is a particular case of SMI*.

In other words, the statistical mechanical entropy is a particular case of a SMI, but in general the SMI is not the entropy. Unfortunately, confusion of the two concepts abounds. The source of this confusion is probably due to von Neumann's suggestion to Shannon to name the quantity H as "entropy" (we will come back on this detail in section 8.11.1).

In this book, we shall refer to the quantity defined in (4.1) as the Shannon's measure of information (or SMI, in short). We shall not be interested in the formal proof of the uniqueness of this function. Instead we shall survey the properties, the applications and the meanings of the quantity as defined in equation (4.1).

4.2   Shannon's Measure of Information (SMI)

In the mathematical theory of information as developed by Shannon, one starts with a random variable X and the corresponding probability distribution $\{p_1, \ldots, p_n\}$. Here, we shall use a simpler language. We shall denote by X an experiment, say throwing a die, or tossing a coin. The possible outcomes of this experiment are denoted $A_1, A_2, \ldots, A_n$, and the corresponding probabilities are $p_1, \ldots, p_n$. The set of events $A_1, \ldots A_n$ is assumed to be **complete**, and the evens are assumed to be **mutually exclusive**. These two requirements mean that when we perform an experiment, one and only one of these events must occur.

An example is useful to clarify what is a complete set of mutually exclusive events. Suppose we have a board of unit area divided into n non-overlapping regions denoted $A_1, \ldots, A_n$ such that their union covers the entire board. If we throw a dart on the board it must hit one and only one of the regions denoted $A_1, \ldots, A_n$. The probabilities of these events are supposed to be known and fulfill the *completeness condition*

$$\sum_{i=1}^{n} P(A_i) = 1 \qquad (4.2)$$

i.e., the occurrence of either A_1, or A_2 or $\ldots A_n$ is the certain event. The *condition of mutual exclusiveness* of each pair of events is

$$\forall i \neq j, \, A_i \cap A_j = \emptyset \quad \text{or equivalently} \quad P(A_i \cdot A_j) = 0 \qquad (4.3)$$

where $\emptyset$ is the empty event, the probability of which is zero.

If the experiment consists of throwing a single die, clearly the six possible outcomes $1, 2, 3, 4, 5, 6$ form a complete set of mutually exclusive events. If the die is fair, or well balanced, the probabilities are all equal: $p_1 = p_2 = p_3 = p_4 = p_5 = p_6 = \frac{1}{6}$. We say that in this case the probability distribution is ***uniform***.

Clearly, *knowing* the distribution is equivalent to having some *information* about the experiment. The question posed by Shannon is: how can we measure the "size" of this information?

Before we elaborate on Shannon's measure of information, consider the following simple example. I tossed a coin, looked at the outcome, and you have to *guess* which outcome has occurred. If you guess correctly you win a prize, if not, you get nothing. Consider the following two cases:

A) I tell you that the coin is fair. This means that the probability distribution is uniform, and the probability of "head" (H) and "tail" (T) are $\frac{1}{2}, \frac{1}{2}$.

B) I tell you that the coin is "unfair", it is heavier on one side, and that the probability distribution of the two outcomes is $\frac{1}{100}$ and $\frac{99}{100}$ for the H and the T, respectively.

Recall the 20Q game of section 1.6, in which I have n possible choices and you have to find out which one of the possibilities was chosen. The *information* required there was "which person I chose" or "in which box the coin is hidden". We have stressed that we were not interested in the *information* itself but in some measure of the *size of the problem* of retrieving this information. Likewise here we have to guess which one of the two outcomes has occurred. Obviously, since there are only two possible outcomes one might be tempted to believe that the size of the problem in case A) is the same as in case B), that is guessing one of two possibilities. However, you also feel that being given different *distributions* you are actually given also different *information* in each case. Intuitively you feel that in case B) you are given more information than in case A). More in what sense? We shall soon discuss the quantitative measure of information as introduced by Shannon. For now we only want to have a qualitative feeling for this kind of measure.

Suppose that we play the game: I toss a coin and you have to guess the outcome. Clearly, in case A), the chances of guessing the outcome correctly is about 50%. There is no way you can use the information contained in the uniform distribution to enhance your chances of gaining the prize. To put it differently, the uniform distribution does not suggest any "preferred" outcome; being given the uniform distribution is equivalent to being given *no information* that we can use. If you play game A) many times, you cannot do any better than guessing the outcomes H or T at random. In this case, you will win the prize on the average about 50% of the time. On the other hand, if we play game B), you can *use* the *information contained* in the distribution to your advantage. In this particular case, it will be wiser to guess that the outcome was T, rather than H. This does not guarantee that you will guess correctly the outcome in playing the game once. However, if you play the same game B) many times, and if you always bet on T, you will win on the average about 99% of the time.

Let us make one generalization of this game. Suppose you are given the distribution $\{p,(1-p)\}$ for the outcomes H and T, respectively: how would you make the guess? Clearly, if $p = 1$, $1 - p = 0$, then you know *with certainty* that the result is H. We say that in this case we have the *maximum information* on the experiment. Suppose now that $p = \frac{9}{10}$, $1 - p = \frac{1}{10}$. In this case you do not know for sure the outcome, but it is more likely that the result is H rather than T. When $p = \frac{7}{10}$, $1 - p = \frac{3}{10}$ again you know that H is more likely to occur than T, but the degree of confidence in H is lower (or the degree of uncertainty is higher than in the previous case). When $p = \frac{1}{2}$, $1 - p = \frac{1}{2}$ your degree of confidence is the lowest (or the degree of uncertainty is the highest). If we increase p furthermore the situation with H and T gets reversed, but what matters is that we get again more information than in the uniform case. For example, if $p = \frac{3}{10}$, $1 - p = \frac{7}{10}$ the degree of confidence is larger than the uniform case, and you should guess that T is the preferred outcome, winning 70% of the time. In the case $p = \frac{1}{10}$, $1 - p = \frac{9}{10}$ if your guess is T you will win 90% of the time. When $p = 0$, $1 - p = 1$ you know that the result is T, and you will win 100% of the time if you choose T, hence you have again maximum information.

Thus, when p changes from zero to half (and $1 - p$ changes from one to half), the amount of confidence about the outcome of the experiment changes from certainty to a minimum (or the uncertainty changes to a maximum). The same is true when p changes from 1 to $\frac{1}{2}$. Qualitatively, we draw in figure 4.1 how the degree of uncertainty changes with p. In

SMI of an experiment with two possible outcomes

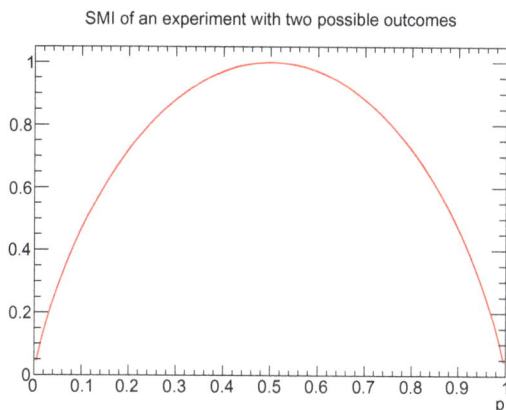

Figure 4.1: Degree of uncertainty in the outcome of tossing a coin, as a function of the probability of "head" (base 2 is used for the logarithms).

section 4.4, we shall discuss the quantitative SMI for this case.

In the previous example, we considered only two possibilities. Shannon sought a measure of the information contained in a more general distribution, say $\{p_1,\ldots,p_n\}$. For instance, a coin could be hidden in one of n boxes, with probability p_i that the coin is in box number i. The measure sought out by Shannon can be formulated in qualitative terms as follows. Given the distribution $\{p_1,\ldots,p_n\}$, you have to find out where the coin is by asking binary questions, i.e. questions that are answerable by either YES or NO. How easy or how difficult is the task of finding the coin?

It turns out that *the Shannon's measure of information is related to the minimum average number of binary questions one needs to ask* in order to find where the coin is hidden.

As we can see from the quotation reported at the beginning of this chapter, Shannon started with the general concept of information. Information can be subjective or objective; it can be interesting or dull, or even meaningless; it can be important or totally irrelevant. Then he restricted himself to one kind of information, the one *contained* in a probability distribution. Next, Shannon imagined to have an experiment (or a random variable, or a game) with a *known* probability distribution for all the possible outcomes. This is a well defined situation where information is involved. The question is how to *measure* such information. In order to answer this question, Shannon argued that if such a measure exists then it must have some properties — those listed in section 4.1 — that are very reasonable to expect from such a measure. With these plausible requirements of the expected function, Shannon then proved that the only function that satisfies these three requirements must be of the form

$$H = -K \sum_{i=1}^{n} p_i \log p_i$$

where K is some positive constant.

In this chapter, we shall take $K = 1$ and we shall take the base 2 for the logarithm. In the rest of the book we will instead choose K to be the Boltzmann constant k_B and use the natural logarithm to conform with the units of entropy customarily used in the literature.

Clearly, the choice of K and the base of the logarithm do not affect the properties or the meaning of the quantity H. They only affect the units we choose to measure H.

Sometimes we may use a simpler notation $H(X)$ instead of $H(p_1, \ldots, p_n)$. But one should be careful not to interpret $H(X)$ as implying that H is a function of X: H is a function of the *distribution* $\{p_1, \ldots p_n\}$ pertaining to the random variable X, not a function of X itself.

Exercise 4.1. In tossing a coin let the distribution of the two outcomes be $\{p, (1-p)\}$. Taking $K = 1$ and the base 2 for the logarithm, calculate the quantity H. Draw the function $H(p)$ and compare the result with our qualitative discussion.

Exercise 4.2. Two fair, identical and independent dice are thrown. Calculate the probability distribution of the sum of the outcomes and the associated quantity H.

Exercise 4.3. Generalize exercise 4.2 for N dice. Find the probability distribution for the sum of the outcomes of N dice and calculate the corresponding SMI, as a function of N.

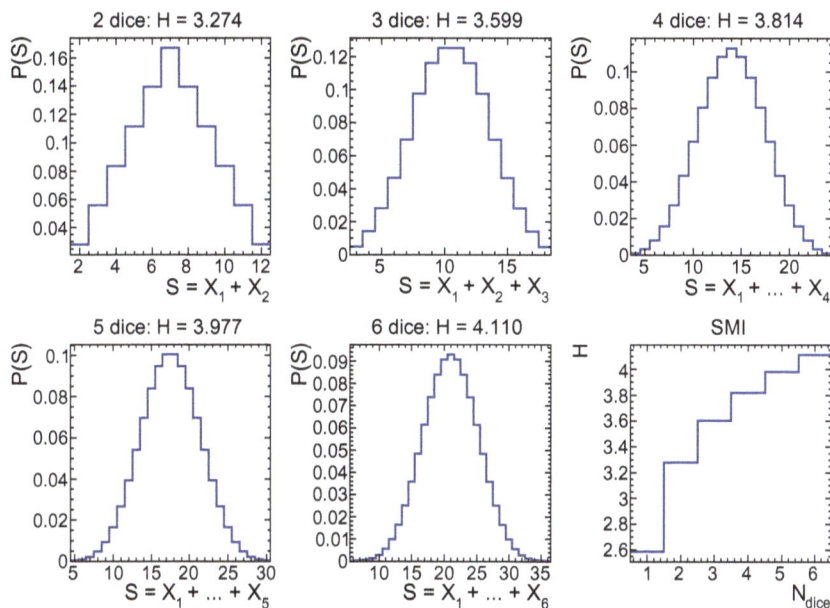

Figure 4.2: Probability distribution of the sum of 2, 3, 4, 5, and 6 dice, with the corresponding SMI values.

4.3 The 20-Question Game Interpretation of H

Consider the 20-question game (20Q in short). I hid a coin in one of n boxes. I also informed you that I have chosen the box in which to place the coin at random, i.e. each box has a probability $p_i = \frac{1}{n}$ to contain the coin.

Your task is to find where the coin is, by asking binary questions, i.e. questions that are answerable by YES or NO only. To make the game a bit more dramatic, suppose you have to pay \$1 for each answer you get. Once you find where the coin is, you get a prize, say \$5. Clearly, it is in your interest to ask as few questions as possible to maximize your earnings.

Intuitively, it is clear that *information* is involved in this game. However, the information "where the coin is", in itself, is of no interest to us. We are interested only in the *size* of the problem, how many questions you need to ask (or how many dollars you have to invest) to obtain the required information (the prize of \$5).

Average number of questions

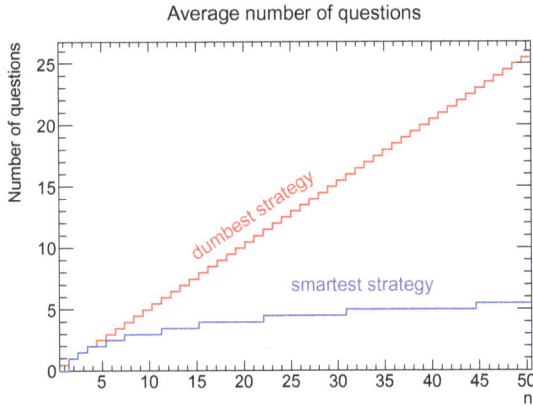

Figure 4.3: Average number of questions needed to find the coin hidden in one of n boxes using the dumbest and smartest strategies.

If you ever played the 20Q game you know that there are different ways (or strategies) of asking questions. Thus, even for a fixed value of n, choosing different *strategies* of asking questions entails different number of questions. One example was already shown in figure 1.3 on page 19, where $n = 8$. Two extreme strategies are to ask sequential questions ("Is the coin in box 1? Is it in box 2?" etc.) or to repeatedly divide in two halves the remaining boxes.

We can call the first strategy the "dumbest" strategy, because on the average (assuming to repeat the game many times with the coin randomly placed every time) one has to ask $n/2$ questions to find the coin. The second is instead the "smartest" strategy, because it takes only $\log_2 n$ questions to find the solution. When n is large, the difference between $n/2$ and $\log_2 n$ is also large, hence the bigger is the range of possibilities, the "smarter" the second strategy is (see figure 4.3).

If $n = 8$, we need to ask exactly 3 ($= \log_2 8$) questions in order to know where the coin is. If we double the number of boxes to $n = 16$, the number of possibilities also doubles. If we ask the dumbest way, we shall also double the average number of questions. However, if we adopt the smartest strategy we shall ask only 4 questions, i.e. only *one* more question than in the previous case of $n = 8$. When n is not a power of 2, one can not divide the possibilities into equal parts. In this case, in the "smartest" strategy one has to balance the two subsets as much as possible, getting the solution into a number of questions which is of the order of the smallest integer larger than $\log_2 n$ (one example is provided by the exercise 4.4 on page 90).

What have all these to do with SMI? Although we shall not prove it here, *the SMI is equal to the average number of questions one needs to ask using the "smartest" strategy.* Thus, the 20Q game interpretation is the simplest and most convenient one for the SMI.

Suppose for simplicity that we have n equally probable events, i.e. $p_1 = p_2 = \ldots = p_n = 1/n$. In this case if we use the base 2 for the logarithm we have

$$H\left(\frac{1}{n},\ldots,\frac{1}{n}\right) = -\sum_{i=1}^{n} p_i \log_2 p_i = \log_2 n$$

Now, suppose that n is a number of the form 2^m, such that $H = \log_2 2^m = m$. For $n = 2$ we need one question, as $m = 1$. For $n = 4$ ($m = 2$) we need two questions, and in general if we double the number of boxes we know that we have to add one question only, because $\log_2(2n) = \log_2 2 + \log_2 n = 1 + \log_2 n$. Hence the SMI increases only by one unit when doubling the number of possibilities.

If n is not in the form 2^m, then we can always find an integer m such that $2^m \leq n \leq 2^{m+1}$. The change in SMI when replacing n with either 2^m or 2^{m+1} is ± 1, which becomes negligible when n is large.

Actually, whenever n is not a power of 2, assuming that half of the times m questions are sufficient and $m + 1$ questions are needed for the other half, the average number of questions will be $m + 1/2$, with $m = \lfloor \log_2 n \rfloor$, the largest integer smaller than $\log_2 n$. In any case, the number of questions is never larger than $m + 1$.

There remains the question of the relationship between the SMI and the number of questions for non-uniform distributions. As we shall prove in the next section, the uniform distribution maximizes H. However we can intuitively understand that any non-uniformity in the distribution always decreases the average number of questions in the 20Q game, hence also the SMI.

Thus, what we have achieved in this section is transforming a formal and somewhat abstract quantity into a quantity that is simple and familiar to anyone who has ever played the 20Q game. We hope that the reader will feel comfortable with the concept of the SMI discussed in this and the next chapters, as well as with the concept of entropy which will be spawned from the SMI for the specific distributions featuring in a thermodynamic system (see chapter 6).

4.4   Some Properties of the Function H

We discuss here the properties assumed by Shannon to obtain the measure H, defined in equation (4.1), as reported in the first quotation at the beginning of this chapter. We can omit to specify the base of the logarithm in this section, because it is irrelevant. Indeed, $\log_b(x) = \log_a(x)/\log_a(b)$ and the denominator is a constant value, not affecting the behavior of H.

We start with a concrete and simple example, the game of tossing a coin with two possible outcomes. The SMI in this case is

$$H = -p \log_2 p - (1-p) \log_2 (1-p)$$

This is a continuous function of the single variable p, with a maximum at $p = \frac{1}{2}$ where $H = 1$.

> **Proof** — As we are free to choose the units of H, we perform analytic computations using $K = 1$ and the natural logarithms, in order to simplify the notation while preserving the properties of the SMI. The derivative of H with respect to p is
>
> $$\frac{\mathrm{d}H}{\mathrm{d}p} = \ln \frac{1-p}{p}$$
>
> At the maximum $\frac{\mathrm{d}H}{\mathrm{d}p} = 0$, which implies $p = 1-p$, that is $p = \frac{1}{2}$.
> We can further check that this is indeed a maximum by computing the second derivative of H with respect to p:
>
> $$\frac{\mathrm{d}^2 H}{\mathrm{d}p^2} = \frac{-1}{p(1-p)} < 0$$
>
> Thus, the second derivative is always negative, which means that the function is concave downwards.   ∎

Note that the quantity $p \log_2 p$ tends to zero when p tends to zero.

> **Proof** — By using the theorem by de l'Hôpital:
>
> $$\lim_{x \to 0} (x \log_2 x) = \lim_{x \to 0} \frac{\log_2 x}{1/x} = \lim_{x \to 0} \frac{\frac{\mathrm{d}}{\mathrm{d}x} \log_2 x}{\frac{\mathrm{d}}{\mathrm{d}x} 1/x} = \lim_{x \to 0} \frac{1/(x \ln 2)}{-1/x^2} = 0$$
>
>   ∎

If we choose $K = 1$ and use binary logarithms, the SMI is measured in **bits**. In section 4.3 we have seen that, with these units, the SMI of a uniform distribution with n possible events is

$$H = \log_2 n \text{ bits} \tag{4.4}$$

such that when $n = 2^m$ the SMI assumes integer values. When $n = 1$ there is no uncertainty and $H(1) = 0$: there is no missing information in this case. When $n = 2$ we have $H(\frac{1}{2}, \frac{1}{2}) = 1$ bit. Thus, one bit measures the amount of information one gets from a binary question, when the two alternatives are equally probable. When $n = 4$ we have $H(\frac{1}{4}, \frac{1}{4}, \frac{1}{4}, \frac{1}{4}) = 2$ bits, and so on.

We now turn to the general case of a random variable (or an experiment or a game) having a probability distribution $\{p_1, \ldots, p_n\}$, and illustrate the few important properties requested by Shannon for the function H defined by (4.1). The examples in section 4.5 should help the reader to understand all details.

We shall not discuss here the proof of the uniqueness of the function H satisfying all such properties. In addition, as the units do not matter we choose again to simplify things by adopting the natural logarithms and $K = 1$.

(i) Continuity of the function H

The SMI is a continuous function of all the p_i's.

> **Proof** — Since the logarithm function is continuous, the SMI is also a continuous function of all the variables $p_1, \ldots, p_n$. For the same reason, H is also a differentiable function of its variables. ∎

(ii) H has a unique maximum at the uniform distribution, whose value monotonically increases with n

We have already seen that the SMI is maximum for a uniform distribution when $n = 2$. Here we consider any $n > 1$.

> **Proof** — We seek the condition for maximum of the function

$$H = -\sum_{i=1}^{n} p_i \ln p_i$$

subject to the condition

$$\sum_{i=1}^{n} p_i = 1 \tag{4.5}$$

We use the Lagrange method of undetermined multipliers by defining the auxiliary function

$$F(p_1,\ldots,p_n) \equiv H(p_1,\ldots,p_n) + \lambda\left(\sum_{i=1}^{n} p_i - 1\right)$$

The condition for maximum is

$$\frac{\partial F}{\partial p_i} = -\ln p_i - 1 + \lambda = 0 \qquad \forall i = 1,\ldots,n$$

which implies that at the maximum one has $p_i = p_i^* = \exp(\lambda - 1)$ $\forall i = 1,\ldots,n$, i.e. all probabilities are equal. This is the uniform distribution with $p_1 = \cdots = p_n = 1/n$.

Indeed, substituting p_i^* in (4.5) we may also find

$$1 = \sum_{i=1}^{n} p_i^* = n\exp(\lambda - 1)$$

which means that $\exp(\lambda - 1) = \frac{1}{n} = p_i^*$ (we are not interested into the particular value assumed by the Lagrange multiplier λ, which is only a way of accounting for the constraint (4.5) while looking for the maximum of H).

Note that the concavity of H is downwards, since

$$\frac{\mathrm{d}^2 H}{\mathrm{d}p_i^2} = \frac{-1}{p_i} < 0$$

for all values of $p_i \in [0,1]$. This also implies that the point at which all $p_i = \frac{1}{n}$ is a maximum (indeed, $\mathrm{d}^2 H/\mathrm{d}p_i^2|_{p_i^*} = -n < 0$).

At the maximum, the value of H is

$$H_{\max} = -p_i^* \sum_{i=1}^{n} \ln p_i^* = -\frac{1}{n} n \ln\frac{1}{n} = \ln n \tag{4.6}$$

while eq. (4.4) provides its value in bits.

Thus, we have also found that the maximum of H is a monotonically increasing function of n, the second property required by Shannon. ∎

(iii) The consistency property of H

The third condition posed by Shannon is referred to as the *consistency property* of the function H, and is equivalent with the request of independence on the grouping of the events. This condition essentially states that the amount of missing information in a given distribution $\{p_1,\ldots,p_n\}$ is independent of the strategy we choose to acquire this information. In other words, the same amount of information is obtained regardless of the way one follows in order to acquire this information.

For example, if we have to locate a coin hidden in one among n boxes with probabilities $p_1,\ldots,p_n$, we could decide to ask about each of them (the "dumbest" strategy) or to make sure that, at each answer, we can discard about half of them (the "smartest" strategy). In the latter approach, one deals with groups of possible choices, rather than with individual possibilities. The consistency property requires that the amount of information contained in the initial system is the same as the amount of information in the new set of two groups, plus the average amount of information in each group.

Suppose that we have a complete set of n mutually exclusive outcomes $\{A_1,\ldots,A_n\}$ of a given experiment and that the corresponding probabilities are $p_1,\ldots,p_n$. We can regroup the outcomes to form a new set of compound events. For example,

$$
\begin{array}{lcccc}
\textit{Regrouping:} & \{A_1,A_2,A_3\}, & \{A_4,A_5\}, & \ldots & \{A_{n-1},A_n\} \\
\textit{New set of events:} & \{A'_1, & A'_2, & \ldots & A'_r\}
\end{array}
$$

As A'_1 is a compound event consisting of the original events A_1,A_2,A_3 and the A_i's are mutually exclusive, the probability of the new event A'_1 is $p'_1 = p_1 + p_2 + p_3$. In addition A'_2 is a compound event comprising A_4 and A_5, whose probability is $p'_2 = p_4 + p_5$. And so on.

With a more general notation, we can say that A'_1 is the union of the first m_1 original events, A'_2 is the union of the next m_2 events, etc. such that $m_1 + m_2 + \ldots + m_r = n$. In our example, $m_1 = 3$, $m_2 = 2$, $\ldots$, and $m_r = 2$. Thus, from the initial set of n outcomes, we constructed a new set of r new events $\{A'_1,A'_2,\ldots,A'_r\}$ with probabilities $p'_1,p'_2,\ldots,p'_r$ given by

$$p'_1 = \sum_{i=1}^{m_1} p_i , \quad p'_2 = \sum_{i=m_1+1}^{m_1+m_2} p_i , \quad \text{etc.} \quad \text{with} \quad \sum_{k=1}^{r} m_k = n$$

The **consistency requirement** is now written as (Ben-Naim, 2008)

$$H(p_1,\ldots,p_n) = H(p'_1,\ldots,p'_r) + \sum_{k=1}^{r} p'_k H_k \tag{4.7}$$

where H_k is the SMI associated with the k-th compound event. For example, $H_1 = H(p_1, p_2, p_3)$ and $H_2 = H(p_4, p_5)$. The last term in (4.7) is the weighted sum of the SMI of each group, with the weight being the probability of that group. Thus, it represents the average SMI over the r groups.

4.5 Examples

Consider the case of four boxes, where one of which a coin is hidden. We assume that the probabilities of finding the coin in any one of the boxes are equal: $p_1 = p_2 = p_3 = p_4 = \frac{1}{4}$. Hence the amount of missing information is $H\left(\frac{1}{4}, \frac{1}{4}, \frac{1}{4}, \frac{1}{4}\right) = \log_2 4 = 2$ bits. In other words, we need two bits of information to locate the coin. We can obtain this information along different routes, but the consistency property of H implies that the amount of information is independent of the strategy, hence we are free to choose a route that is easier to compute.

The quickest route is to divide the total number of boxes into two halves, each half having probability $\frac{1}{2}$ (figure 4.4a). For this case we have from eq. (4.7)

$$H\left(\tfrac{1}{4}, \tfrac{1}{4}, \tfrac{1}{4}, \tfrac{1}{4}\right) = H\left(\tfrac{1}{2}, \tfrac{1}{2}\right) + \left[\tfrac{1}{2}H\left(\tfrac{1}{2}, \tfrac{1}{2}\right) + \tfrac{1}{2}H\left(\tfrac{1}{2}, \tfrac{1}{2}\right)\right] \tag{4.8}$$

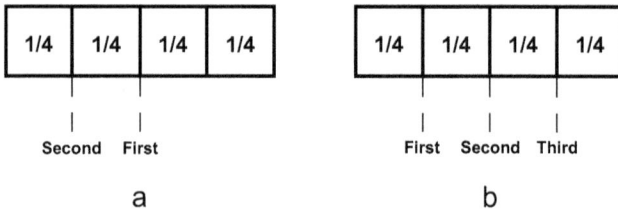

Figure 4.4: Smartest (a) and dumbest (b) strategies of finding a coin hidden in one of 4 boxes.

which simply means that the value of H in the original set of events is the sum of the value of H associated with the two groups (left and right halves) and the average value of H *within* the newly formed groups.

After the first question, we do not know yet where the coin is. However, we know which half contains it. So, we know something more than before. The reduction of SMI at this stage is $H\left(\frac{1}{2},\frac{1}{2}\right) = 1$ bit: *reducing the possibilities by half amounts to gaining 1 bit of information*. In the next step, the average reduction in SMI is $\frac{1}{2}H\left(\frac{1}{2},\frac{1}{2}\right) + \frac{1}{2}H\left(\frac{1}{2},\frac{1}{2}\right) = 1$ bit, as there are only two answers with equal probability, and each of them consists of a set of two boxes, for which the reduction in SMI is again 1 bit. Thus, the sum in (4.8), i.e. the total SMI, is 2 bits.

Now, suppose we take a different route, by asking sequentially about each box as in figure 4.4b. At the first step, instead of dividing by two halves, we divide in two groups: one box and three boxes. For this case, by applying eq. (4.7), we have

$$H\left(\frac{1}{4},\frac{1}{4},\frac{1}{4},\frac{1}{4}\right) = H\left(\frac{1}{4},\frac{3}{4}\right) + \left[\frac{1}{4}H(1) + \frac{3}{4}H\left(\frac{1}{3},\frac{1}{3},\frac{1}{3}\right)\right]$$
$$= H\left(\frac{1}{4},\frac{3}{4}\right) + \frac{3}{4}H\left(\frac{1}{3},\frac{1}{3},\frac{1}{3}\right)$$

We have then to consider the group of three boxes, which is addressed the same way, by splitting it into one box and two boxes:

$$H\left(\frac{1}{4},\frac{1}{4},\frac{1}{4},\frac{1}{4}\right) = H\left(\frac{1}{4},\frac{3}{4}\right) + \frac{3}{4}\left[H\left(\frac{1}{3},\frac{2}{3}\right) + \frac{2}{3}H\left(\frac{1}{2},\frac{1}{2}\right)\right]$$
$$= H\left(\frac{1}{4},\frac{3}{4}\right) + \frac{3}{4}H\left(\frac{1}{3},\frac{2}{3}\right) + \frac{1}{2}H\left(\frac{1}{2},\frac{1}{2}\right)$$
$$= -\left[\frac{1}{4}\log_2\frac{1}{4} + \frac{3}{4}\log_2\frac{3}{4}\right] - \frac{3}{4}\left[\frac{1}{3}\log_2\frac{1}{3} + \frac{2}{3}\log_2\frac{2}{3}\right] + \frac{1}{2}$$
$$= 0.8113 + 0.6887 + 0.5 = 2$$

Note that with the first strategy we *gained* one bit of information at each step. Therefore, we needed exactly two steps to gain the required 2 bits of information. On the other hand, in the second route, the average amount of information gained at the first step is only 0.8113, less than the maximum (one bit). In the second step is 0.6887 bits, and in the third step is 0.5 bits. Clearly, the total amount of information required is 2 bits. However, in this route we shall need on the average more steps or more questions to gain the same amount of information.

Table 4.1: Frequency of Each Letter in the English Language

i	a_i	p_i	$-\log_2 p_i$	i	a_i	p_i	$-\log_2 p_i$
1	a	0.0575	4.1	14	n	0.0596	4.1
2	b	0.0128	6.3	15	o	0.0689	3.9
3	c	0.0263	5.2	16	p	0.0192	5.7
4	d	0.0285	5.1	17	q	0.0008	10.3
5	e	0.0913	3.5	18	r	0.0508	4.3
6	f	0.0173	5.9	19	s	0.0567	4.1
7	g	0.0133	6.2	20	t	0.0706	3.8
8	h	0.0313	5.0	21	u	0.0334	4.9
9	i	0.0599	4.1	22	v	0.0069	7.2
10	j	0.0006	10.7	23	w	0.0119	6.4
11	k	0.0084	6.9	24	x	0.0073	7.1
12	l	0.0335	4.9	25	y	0.0164	5.9
13	m	0.0235	5.4	26	z	0.0007	10.4
				27	(space)	0.1928	2.4

Exercise 4.4. Suppose you are given the 26 letters (27 if we include the space between words) from the English languageas in the table 4.1. One picks up a letter at random (e.g. pick up a book and choose a page, and within the page point out at a letter while the eyes are blindfolded). What is the best strategy for you to ask binary questions in order to find out what the letter is? How many binary questions you need to ask on the average if you use the best strategy? What is the SMI in this game (calculate with logarithm to the base 2)? Why the difference between the SMI and the number of questions?

4.6 The Case of Infinite Number of Outcomes

The case of discrete infinite possibilities is straightforward. First, we recall that for a finite and uniform distribution we found $H = \log n$, where n is the number of possibilities. Taking the limit $n \to \infty$ we obviously get $H \to \infty$. This is clear: if we have to find out one outcome from infinite possibilities, we need an infinite number of questions.

For non-uniform distributions defined on countably infinite sets of possible outcomes the SMI H may be finite or infinite, depending on whether the series $\sum_{i=1}^{\infty} p_i \log p_i$ converges or diverges.

The case of a continuous distribution is problematic.[3] If we start from the discrete case and proceed to the continuous limit we get into some difficulties. We shall defer discussing this problem to section 5.2. Here we shall only follow Shannon's treatment for a continuous distribution for which a probability density function $f(x)$ exists.

Let $f(x)$ be a probability density function, i.e. $f(x) dx$ is the probability to find the corresponding random variable having values between x and $x + dx$. In analogy with the definition of the H function for discrete probability distribution, and omitting the overall constant and the base for the logarithms (in other words, leaving the units unspecified), we *define* the SMI for a continuous distribution as

$$H[f(x)] \equiv E[-\log f(x)] = - \int_{-\infty}^{\infty} f(x) \log f(x) \, dx \qquad (4.9)$$

In other words, H is defined as the expectation value of the random variable $-\log f(X)$, where X is the random variable having probability density $f(x)$.

Note that H defined in (4.9) is clearly a *functional*, not a function, as it is defined in terms of the expectation operator $E[\cdot]$, which has a function (not a number or variable) as argument.

Similarly, H may be defined for an n-dimensional density function $f(x_1, \ldots, x_n)$ as the expectation of the function $-\log f(x_1, \ldots, x_n)$.

In chapter 5 we shall use the definition (4.9) to derive the three theorems proved by Shannon, which are useful for the application of the SMI to real systems of particles in a box.

---

[3]Shannon himself was apparently not worried about the mathematical difficulties involved in the generalization of the measure of information to the continuous case. See also Khinchin (1957), and Ben-Naim (2008).

4.7 Conditional and Mutual Information

In this section we define two important quantities that are derived from the SMI: conditional information and mutual information. These two quantities are very useful in interpreting the SMI of two or more random variables which are not independent. Again, in our notation we will leave the units and the base for the logarithms unspecified.

Consider first the case of two random variables X and Y with distributions $p_i \equiv p_X(i) = P(X = x_i)$ and $q_j \equiv p_Y(j) = P(Y = y_j)$, with $i = 1,\ldots,n$ and $j = 1,\ldots,m$. Let $p_{ij} \equiv p(i,j) = P(X = x_i, Y = y_j)$ be the joint probability of occurrence of the events $X = x_i$ and $Y = y_j$. This means that the marginal probabilities of X and Y are

$$p_i = \sum_{j=1}^{m} p(i,j) \qquad \text{and} \qquad q_j = \sum_{i=1}^{n} p(i,j) \tag{4.10}$$

The SMI of the probability distribution $\{p(i,j)|i = 1,\ldots,n, j = 1,\ldots,m\}$ is defined as

$$H(X,Y) \equiv -\sum_{i,j} p(i,j)\log p(i,j) = E[-\log p(i,j)] \tag{4.11}$$

The SMI associated to the two random variables are

$$H(X) = -\sum_{i=1}^{n} p_i \log p_i = -\sum_{i=1}^{n}\sum_{j=1}^{m} p(i,j)\log\left[\sum_{j=1}^{m} p(i,j)\right] \tag{4.12}$$

$$H(Y) = -\sum_{j=1}^{m} q_j \log q_j = -\sum_{j=1}^{m}\sum_{i=1}^{n} p(i,j)\log\left[\sum_{i=1}^{n} p(i,j)\right] \tag{4.13}$$

For simplicity, we now consider the case $m = n$.[4] For any two distributions $\{p_i\}$ and $\{q_i\}$ such that $\sum_{i=1}^{n} p_i = 1$ and $\sum_{i=1}^{n} q_i = 1$ the following inequality holds

$$H(q_1,\ldots,q_n) = -\sum_{i=1}^{n} q_j \log q_i \leq -\sum_{i=1}^{n} q_i \log p_i \tag{4.14}$$

Proof — This can be proven directly from the elementary inequality (see appendix B.1) $\log x \leq x - 1$. Choosing $x = \frac{p_i}{q_i}$ we get $\log\frac{p_i}{q_i} \leq \frac{p_i}{q_i} - 1$. Multiplying by q_i and summing over i, we get $\sum q_i \log\frac{p_i}{q_i} \leq \sum p_i - \sum q_i = 0$, from which (4.14) follows. ■

---

[4] What if $m \neq n$? If $m < n$ one adds $n - m$ null q_j's; otherwise one adds $m - n$ null p_i's.

Now we have all ingredients that are needed to show the important inequality:

$$H(X,Y) \leq H(X) + H(Y) \tag{4.15}$$

Proof — From (4.11), (4.12) and (4.13), we obtain

$$\begin{aligned}
H(X) + H(Y) &= -\sum_{i=1}^{n} p_i \log p_i - \sum_{j=1}^{m} q_j \log q_j \\
&= -\sum_{i,j} p(i,j) \log p_i - \sum_{i,j} p(i,j) \log q_j \tag{4.16} \\
&= -\sum_{i,J} p(i,j) \log(p_i q_j)
\end{aligned}$$

Now, applying the inequality (4.14) to the two distributions $p(i,j)$ and $p_i q_j$ and using (4.10), we obtain the inequality (4.15):

$$\begin{aligned}
H(X,Y) &= -\sum_{i,j} p(i,j) \log p(i,j) \\
&\leq -\sum_{i,J} p(i,j) \log(p_i q_j) = H(X) + H(Y)
\end{aligned}$$

∎

The equality in (4.15) holds if and only if the two RV are independent, as immediate consequence of (4.16):

$$p(i,j) = p_i q_j \ \forall i,j \iff H(X,Y) = H(X) + H(Y) \tag{4.17}$$

This simply mean that if we have two experiments (or two games), the outcomes of which are independent, then the SMI associated with the outcomes of the two experiments is the sum of the SMI associated with the outcomes of each one of the experiment: knowing the result of one experiment tells us nothing about the other. A simple example will clarify this even further. Suppose we have two dice, one red and one blue. If the two games consist of guessing the outcome of the red die and whether the sum of red and blue dice is even or odd, then knowing the result of one game will not help guessing about the other. In this case, (4.15) says that the SMI associated to the compound experiment is equal to the sum of the SMI of each individual experiment.

On the other hand, if there is a dependence between the two experiments, then the SMI associated with the compound experiment will never be larger than the SMI associated with the two experiments separately. In this case, knowing the result of one experiment provides some information about the other. For example, if the two games consist of guessing the outcome of the red die and the sum of the two dice, then knowing the result of one of them does affect our guess about the other. For example, if the outcome of the red dice is 4 we know that the sum can not be lower than 5. Alternatively, knowing that the sum is 4 implies that the outcome of the red die can not be larger than 3. Hence, in this case the SMI associated to the compound experiment is lower than the sum of the individual SMIs.

For dependent experiments, we write the conditional probability that $Y = y_j$ given that $X = x_i$ (which cannot be impossible) as

$$P(y_j|x_i) = \frac{P(x_i, y_j)}{P(x_i)} \tag{4.18}$$

The corresponding ***conditional SMI*** is defined as

$$H(Y|x_i) \equiv -\sum_j P(y_j|x_i) \log P(y_j|x_i) \tag{4.19}$$

This is simply the SMI associated with the conditional probability defined in (4.18).

The conditional SMI of Y given X is defined as the expectation of $H(Y|x_i)$, i.e.[5]

$$
\begin{aligned}
H(Y|X) &\equiv E[H(Y|x_i)] = \sum_i P(x_i) H(Y|x_i) \\
&= -\sum_i P(x_i) \sum_j P(y_j|x_i) \log P(y_j|x_i) \\
&= -\sum_{i,j} P(x_i, y_j) \log \frac{P(x_i, y_j)}{P(x_i)} \\
&= -\sum_{i,j} P(x_i, y_j) \log P(x_i, y_j) + \sum_{i,j} P(x_i, y_j) \log P(x_i) \\
&= H(X, Y) - H(X)
\end{aligned}
\tag{4.20}
$$

Thus, $H(Y|X)$ measures the difference between the SMI of X *and* Y and the SMI of X alone. This can be rewritten as

$$
\begin{aligned}
H(X, Y) &= H(X) + H(Y|X) \\
&= H(Y) + H(X|Y)
\end{aligned}
\tag{4.21}
$$

---

[5]The quantity $H(Y|X)$ is sometimes denoted by $H_X(Y)$.

From (4.15) and (4.21) we also obtain the inequality

$$H(Y|X) \le H(Y) \qquad (4.22)$$

which means that the SMI of Y can never increase by knowing X. Alternatively, $H(Y|X)$ is the average uncertainty that remains about Y when X is known. This uncertainty is always smaller than the uncertainty about Y. If X and Y are independent, then the equality sign in (4.22) holds. This is another reasonable property that we can expect from a quantity that measures information or uncertainty. The quantity $H(Y|X)$ has been referred to as **conditional information**, or **equivocation** (Shannon, 1948).

Qualitatively the meaning of (4.22) is intuitively clear. If we have two independent experiments, performing one does not provide any information on the other. On the other hand, if the two experiments are not independent, then performing one can only *add* information on the other (which means decreasing the Shannon *missing* information).

Another useful quantity is the **mutual information** defined by

$$I(X;Y) \equiv H(X) + H(Y) - H(X,Y) \qquad (4.23)$$

which is of course a symmetric function: $I(X;Y) = I(Y;X)$.[6] Sometimes, $I(X;Y)$ is referred to as the average amount of information conveyed by the RV X on the RV Y and vice versa. An immediate consequence of the inequality (4.15) is that $I(X;Y) \ge 0$ always.

From (4.23) and (4.21), we can also write

$$
\begin{aligned}
I(X;Y) &= H(Y) - H(Y|X) \\
&= H(X) - H(X|Y)
\end{aligned} \qquad (4.24)
$$

From the definition (4.11) of $H(X,Y)$ and from (4.12) and (4.13), we obtain

$$
\begin{aligned}
I(X;Y) &= -\sum_{i=1}^{n} p_i \log p_i - \sum_{j=1}^{m} q_j \log q_j + \sum_{i,j} p(i,j) \log p(i,j) \\
&= -\sum_{i,j} p(i,j) \log(p_i q_j) + \sum_{i,j} p(i,j) \log p(i,j) \\
&= \sum_{i,j} p(i,j) \log \frac{p(i,j)}{p_i q_j} = E[\log g(i,j)] \ge 0
\end{aligned} \qquad (4.25)
$$

---

[6]Note that sometimes the notation $H(X;Y)$ is used instead of $I(X;Y)$. This is potentially a confusing notation because of the similarities between $H(X,Y)$ and $H(X;Y)$.

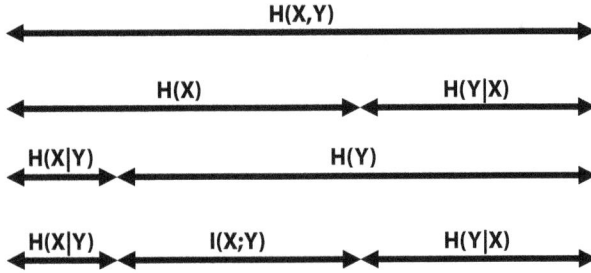

Figure 4.5: The relations between $H(X), H(Y), H(X,Y)$ and $I(X;Y)$.

where $g(i,j) \equiv p(i,j)/(p_i q_j)$ is the correlation between the two events $\{X = x_i\}$ and $\{Y = y_j\}$. Hence $I(X;Y)$ quantifies the *average logarithmic correlation* between two random variables. Thus, we see that $I(X;Y)$ is a measure of the extent of dependence between X and Y.

The inequality in (4.25) is a result of the general inequality (4.15). The equality holds if X and Y are independent. It should be noted that the correlation between two specific events $g(i,j)$ could be larger or smaller than 1, i.e. $\log g(i,j)$ could be negative or positive. However, the average of $\log g(i,j)$ is always positive.

The mutual information is always positive. It is zero when the two experiments are independent. Thus, $I(X;Y)$ measures the average reduction in SMI associated with the experiment X resulting from knowing Y, and vice versa. The relations between $H(X)$, $H(Y)$, $H(Y,Y)$ and $I(X;Y)$ is shown in diagram of figure 4.5.

4.8 The Various Interpretations of H

After having defined the quantity H and seen some of its property, let us discuss a few possible interpretations. Originally, Shannon referred to the quantity he was seeking to define as "choice," "uncertainty," "information" and "entropy." Except for the last term, which does not have any intuitive meaning (and should not belong to this section), the first three terms do have. Let us discuss these for a simple case.

Suppose we are given n boxes and we are told that a coin was hidden in one and only one box. We are also told that the events "the coin is in box k" are mutually exclusive (i.e. the coin cannot be in more than one box), and that the n events form a complete set of events (i.e. the coin is certainly in one of the boxes). In addition, the box in which the coin was placed was chosen at random, with probability $1/n$.

The term "choice" is easily understood, in the sense that in this particular game we have to *choose* between n boxes to place the coin. Clearly, for $n = 1$, there is only one box to choose and the amount of "choice" we have is zero: we must place the coin in that box. It is also clear that as n increases, the larger n, the larger the "choice" we have to select the box in which the coin is to be placed. The interpretation of H as the amount of "choice" for the case of unequal probabilities is less straightforward. For instance, if the probabilities of say, 10 boxes are $\frac{9}{10}, \frac{1}{10}, 0, \ldots, 0$, it is clear that we have less choice than in the case of a uniform distribution, but in the general case of unequal probabilities, the "choice" interpretation is not satisfactory. In addition, this interpretation is not easy to apply for the conditional SMI and the mutual information. For these reasons, we shall not use this interpretation of H.

The term "measure of information" is clearly and intuitively more appealing. If we are asked to find out where the coin is hidden, it is clear even to the lay person that we *lack information* on "where the coin is hidden." It is also clear that if $n = 1$ we need no additional information: we know that the coin is in that box. When n increases, so does the amount of the information we lack, or the missing information.[7] This interpretation can be easily extended to the case of unequal probabilities. Clearly, any non-uniformity in the distribution only increases our information on the problem, or decreases the missing information. As we have seen, this reduction in the missing information can be translated in terms of fewer binary questions to be asked. In facts, all the properties of H listed in the previous section are consistent with the interpretation of H as the amount of missing information. For instance, for two sets of independent experiments (or games), the amount of the missing information is the sum of the missing information of the two experiments. When the two sets are dependent, the occurrence of an event in one experiment affects the probability of an outcome of the second experiment. Hence,

---

[7]We can say either that the problem contains more information, or that our missing information is larger. Sometimes, this interpretation is expressed as our ignorance, but this could be potentially misleading if by ignorance one understands our *subjective ignorance*.

having information on one experiment can only reduce the missing information on the second experiment.

The "information" interpretation of H is also intuitively appealing, since it is also equal to the average number of questions we need to ask in order to *acquire* the missing information (see section 4.3). For instance, increasing n will always require more questions to be asked. Furthermore, any deviation from uniform distribution will reduce the average number of questions.

Note that the informational interpretation of the SMI applies to the *experiment* as a whole, not to individual outcomes.[8]

There are two more interpretations that can be assigned to H which are useful. First, the interpretation of H as the amount of *uncertainty*, which is derived from the meaning of probability. When either $p_i = 1$ or $p_i = 0$, we are *certain* that the event i did occur or did not occur, respectively. If on the other hand $p_1 = \frac{9}{10}, p_2 = \frac{1}{10}, p_3 = 0, \ldots, p_n = 0$, then we have more uncertainty of the outcome compared with the previous example where our uncertainty was zero. It is also intuitively clear that the more uniform the distribution is, the more uncertainty we have about the outcome, and the maximum uncertainty is reached when the distribution of outcomes is uniform. One can also say that H measures the average uncertainty that is removed once we know the outcome of the RV X. The "uncertainty" interpretation can be applied to all the properties discussed in the previous section.

A slightly different but still useful interpretation of H is in terms of "likelihood" or "expectedness", also derived from the meaning of probability. When p_i is small, the event i is less likely to occur, or its occurrence is less expected. When p_i approaches one, the occurrence of i becomes more likely or more expected. Since $\log p_i$ is a monotonically increasing function of p_i, we can say that the larger $\log p_i$, the larger the likelihood or the larger the expectedness of the event i. Only the range of numbers is different: instead of $0 \le p_i \le 1$, we now have $-\infty \le \log p_i \le 0$. The quantity $-\log p_i$ is thus a measure of the "unlikelihood" or the "unexpectedness" of the event i. Therefore, the quantity $H = -\sum p_i \log p_i$ is a

---

[8]Some authors interpret $-\log p_i$ as self-information, but this is not related to the SMI: the two concepts are not equivalent.

measure of the *average* unlikelihood (or unexpectedness) of the entire set of the events.

There are several other interpretations of H. We shall specifically refrain from using the meaning of H as a measure of "entropy" and of "disorder". The first is an outright misleading term because, as we will see in the next chapters, the entropy is a special case of the SMI, which has a broader applicability. The second, although very commonly used, is very problematic. First because the term "order" and "disorder" are fuzzy concepts. Like "beauty", they are very subjective terms and lie in the eyes of the beholder. Second, many examples can be given showing that the amount of information does not correlate with what we perceive as order or disorder (we will come back to this point in section 6.7).

Perhaps the most important objection to the usage of order and disorder interpretation of the entropy, or of H, is that concepts of order or disorder (unlike information or uncertainty) do not have the properties that we require. When Shannon searched for a measure of information, or uncertainty, he posed several plausible requirements that such a measure should fulfill. None of these can be said to be a plausible requirement for the concepts of order or disorder. Certainly, the additivity and the consistency properties cannot be assigned to the concept of disorder. Also, the concepts of conditional information and mutual information cannot be claimed to be *plausible* properties of disorder.

As we have seen in section 4.1, Shannon recognized the similarity between the expression for the entropy as it appeared in statistical mechanics and the quantity he discovered. We shall further discuss the relation between the thermodynamic entropy and SMI in chapter 6. Here, we stress again that SMI is defined for any probability distribution, like the outcomes of throwing a die, or the frequencies of appearance of the various letters in a certain language. On the other hand, the statistical mechanical entropy is defined on a very limited number of probability distributions. Thus, one can safely say that entropy is defined on a small subset of probability distributions. Therefore, the entropy is a special case of SMI, which is defined on *any* probability distribution. It is not appropriate to refer to the general concept of SMI as entropy.

4.9   Summary of What We Have Learned in this Chapter

We have defined Shannon's measure of information as the quantity H from eq. (4.1). The SMI is a property of the probability distribution, in other words it characterizes the experiment under consideration. As it can be defined for every probability distribution $\{p_i\}$, H is a very general concept.

The SMI is related to the minimum average number of binary questions needed to find which is the outcome of the experiment. The natural units for the SMI are the bits, where one bit measures the amount of information one gets from a binary question, when the two alternatives are equally probable.

When considered as a function defined over all possible discrete distributions, the SMI is continuous and has a unique maximum, achieved when $\{p_i\}$ is the uniform distribution. Its consistency property ensures that it can be computed following different routes, always obtaining the same result: H is independent of the strategy adopted in asking questions.

The SMI defined in (4.9) for continuous distributions is a functional, defined in terms of the expectation operator, whose argument is the logarithm of a probability density function. The expectation operator can also be used to define the SMI for discrete and multidimensional distributions.

When conditional probabilities are taken into account, one may introduce the conditional information $H(Y|X)$ between the probability distributions characterizing two random variables X and Y. When the latter are statistically independent, and only in this case, it comes out that $H(Y|X) = H(Y)$, i.e. the conditional information becomes identical to the SMI. Otherwise the conditional information is smaller than the SMI.

Finally, we have defined the mutual information $I(X;Y)$ showing that it equals $H(Y) - H(Y|X)$ and $H(X) - H(X|Y)$. The mutual information is zero when X and Y are independent random variables, and characterizes the amount of information that one variable conveys on the other. More precisely, $I(X;Y)$ quantifies the average logarithmic correlation between two random variables and is a measure of the extent of dependence between X and Y.

Most of the content of this chapter is not directly relevant to thermodynamics. However, we urge the reader to get some degree of familiarity with the concept of SMI. There are two reasons for doing so.

First, the SMI is a very general, useful and interesting quantity, used in many fields of science. Therefore, there is a good chance that one day you will encounter a problem where the SMI may be useful to measure or to interpret certain quantities associated with an experiment.

Second, the SMI is indispensable for understanding the entropy. This is more than just one application of SMI: it makes the concept of entropy simple, clear and mystery-free. Without this interpretation, the entropy remains to be a highly mysterious and a hard concept to be understood. We shall define entropy as a special case of a SMI in chapter 6.

Chapter 5

Three Theorems on Shannon's Measure of Information

In 1948 Shannon published a seminal article on the theory of communication (Shannon, 1948). In this article Shannon proved several theorems on the function H, which we call the Shannon's measure of information, or Shannon's measure of information (SMI in short). We discuss in this chapter three of his theorems that are relevant to statistical mechanics. They appeared in section 20 of Shannon's paper, although their numbering here is different from the original article.

These theorems are of interest for three reasons. First, their results are used in chapter 6 to construct the entropy function of an ideal gas. Second, these theorems spawn three fundamental probability distributions that are important in statistics in general, and in statistical mechanics in particular. Finally, their results shed a new light on the meaning of the entropy and the probabilistic interpretation of the second law of thermodynamics.

In this chapter, to avoid cluttering equations we will choose $K = 1$ and natural logarithms. For the continuous case, we choose the same convention for the H function defined in (4.9), but using natural logarithms. In addition, we will ignore all subtleties related to the continuous case and perform all computation in this limit, using (4.9) in place of (4.1).[1]

---

[1] As we will be interested in macroscopic systems, the number of possible configurations is enormously large. In this case there is no appreciable difference between a discrete and continuous representation of a probability distribution. For additional details, see Ben-Naim (2008).

5.1 The First Theorem: The Uniform Distribution

Consider a particle that is confined in a one-dimensional box of length L. We are interested in the probability density function $f(x)$ that maximizes the quantity H, as defined in (4.9), with support on the interval $(0,L)$ and satisfying the normalization condition

$$\int_0^L f(x)\,dx = 1 \qquad (5.1)$$

This condition simply means that the probability of finding the particle anywhere within the interval $(0,L)$ (i.e. inside the box) is one. Note also that, if the particles were not confined to be within the limits of a box, there would be no equilibrium state: the particles would expand their "volume" indefinitely.

The procedure for obtaining the maximum of H subject to the constraint (5.1) involves the Lagrange's method of undetermined multipliers. We define the auxiliary functional $A[f(x)]$ by

$$A[f(x)] \equiv H[f(x)] + \lambda\left(\int_0^L f(x)\,dx - 1\right)$$

Recall that the quantity defined in (4.9) is a function of the probability distribution, hence H is a *functional* of the continuous PDF $f(x)$.

A functional is a function of a function. In other words, its argument is an entire function, not just a number. As $H[f(x)]$ is a functional, so is also $A[f(x)]$. In this case, both are defined over all probability densities with support in $(0,L)$, i.e. over all non-negative functions satisfying the normalization condition (5.1).

In the discrete case (addressed in section 4.4), where the method of Lagrange multipliers is based on a normal function, one takes the partial derivatives of the auxiliary function A with respect to all the components p_i of the distribution. On the other hand, in the continuous case one takes the *variation* of a functional $A[f(x)]$. The extremum of A is obtained by setting the functional derivative of A with respect to the function $f(x)$ to zero. We will not enter into all details here. It suffices to say that this leads to the condition:

$$-1 - \ln f^*(x) + \lambda = 0$$

or equivalently to

$$f^*(x) = \exp(\lambda - 1) \qquad (5.2)$$

The mathematical details are not important.[2] The important thing is that this procedure provides us with the function denoted $f^*(x)$ for which the quantity H defined in (4.9) has a maximum value.

We shall later refer to $f^*(x)$ as the *equilibrium probability density*. The reason will become clear when we will discuss the evolution of the function $f^*(x)$ towards the equilibrium state. For now, $f^*(x)$ is the density function that maximizes the SMI as defined in (4.9).

We have found that the PDF $f^*(x)$ maximizing H, eq. (5.2), does not depend on the variable x. Hence the SMI is maximized by a constant function. Substituting $f^*(x)$ into eq. (5.1), we get

$$1 = f^*(x) \int_0^L \mathrm{d}x = f^*(x)L \qquad (5.3)$$

Hence the function which maximizes H subject to the normalization condition (5.1) is the *uniform distribution*

$$f^*(x) = \frac{1}{L} \qquad (5.4)$$

and the corresponding value of the SMI is[3]

$$H[f^*(x)] = -\int_0^L f^*(x)\ln f^*(x)\,\mathrm{d}x = \ln L \qquad (5.5)$$

The result (5.4) sounds quite natural. We know that the particle is somewhere in the box of length L, and the uncertainty on its location is maximum when the probability of finding the particle in a specific interval between x_1 and $x_1 + \mathrm{d}x$ does not depend on the location of the interval, i.e. when

$$P(x_1 \le x \le x_1 + \mathrm{d}x) = \frac{\mathrm{d}x}{L} \qquad (5.6)$$

This is the uniform distribution (figure 5.1): the probability of finding the particle within an interval $\mathrm{d}x$ at any point x_1 is *independent* of x_1.

---

[2]The interested reader may check Ben-Naim (2008).
[3]Here L is taken as a dimensionless number. When physical units are used, one must remember to divide L by the chosen unit of length. Also $f(x)$ has the dimensions of $1/\mathrm{d}x$, thus expressions like $\ln f(x)$ are meaningless if x is not a pure number. However, in real applications one always considers differences between logarithms, which in practice solves the problem of the physical units.

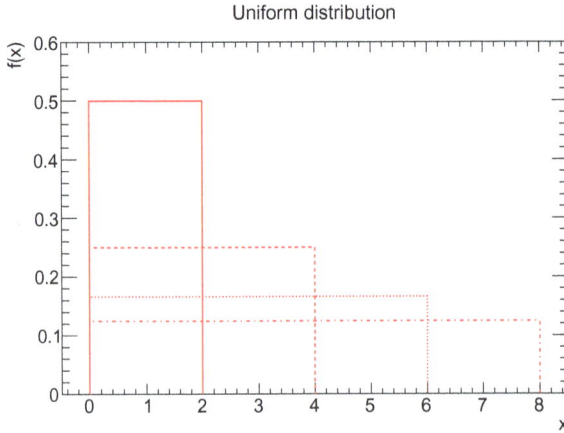

Figure 5.1: Uniform distribution with domain on $[0,2]$, $[0,4]$, $[0,6]$, and $[0,8]$.

5.2   Comparison of Discrete and Continuous SMI

We have seen in section 4.4 that also in the discrete case the SMI is maximized by the uniform distribution. The maximum SMI value in that case, eq. (4.6), is $\ln n$. Here we want to understand the differences and similarities between the continuous and discrete cases.

First, suppose that we have, instead of one box of length L, n cells each of length $h = L/n$. We ask ourselves which is the cell occupied by the particle. If f_i is the probability of finding the particle in cell i, then the SMI for this case is written as

$$H(f_1,\ldots,f_n) = -\sum_{i=1}^{n} f_i \ln f_i \qquad (5.7)$$

Here, H is a *function* of the vector with components $(f_1,\ldots,f_n)$ representing the probability distribution of the discrete case. In (5.5) H is instead a *functional* of the probability density function $f(x)$, which we could view as an "infinite vector" indexed by the continuous variable x.

As we assume that the particle must be in one of the cells, we know that

$$\sum_{i=1}^{n} f_i = 1 \qquad (5.8)$$

We recall from section 4.4 that the distribution that maximize the SMI H defined in (5.7), subject to the condition (5.8) is the discrete uniform distribution

$$f_i^* = \frac{1}{n} \qquad \forall i = 1,\ldots,n \tag{5.9}$$

The corresponding maximal value of H is

$$H_{\max} = -\sum_{i=1}^{n} f_i^* \ln f_i^* = \ln n \tag{5.10}$$

We now proceed in a slightly different way to obtain the result (5.10). The new method applies the consistency property of H discussed in section 4.4.

We start again with the interval $(0,L)$, which contains an infinite number of points (instead of the finite number of possible outcomes $A_1,\ldots,A_n$ that we had in section 4.4). Let us denote them as $A(x)$ with $0 \le x \le L$. The next step is to partition these events into n groups (figure 5.2). The first group A_1 is the set of all *infinite* outcomes $A(x)$ with $0 < x \le h$ with $h \equiv L/n$, corresponding to the particle confined inside the first box. The second group A_2 is the set of all infinite outcomes with $h < x \le 2h$ which correspond to the particle contained by the second box, etc. By applying the consistency property of H we obtain

$$
\begin{aligned}
H_L &= H\left(\tfrac{1}{n},\ldots,\tfrac{1}{n}\right) + \sum_{i=1}^{n} \frac{1}{n} H_h \\
&= H\left(\tfrac{1}{n},\ldots,\tfrac{1}{n}\right) + H_h \\
&= \ln n + \ln h = \ln L
\end{aligned} \tag{5.11}
$$

as in eq. (5.5). Hence $\ln n = \ln \frac{L}{h}$.

Here, the SMI (or the uncertainty) associated with the entire box of length L is written as the sum of the SMI associated with the discrete uniform distribution over the n cells in which the box is subdivided, and the average SMI associated with the location of the particle *within* a single cell of length $h = L/n$. The first term was computed in the discrete case in eq. (4.6), and is $\ln n$. The second term is equal to H_h, which is the SMI for the continuous case which corresponds to a single cell. This was computed above for the interval $(0,L)$. Replacing L with h in eq. (5.5) we obtain the result for the interval $(0,h)$ which corresponds to a single cell, which is $H_h = \ln h = \ln L - \ln n$.

From eq. (5.11) we immediately obtain

$$H\left(\tfrac{1}{n},\ldots,\tfrac{1}{n}\right) = H_L - H_h = \ln n \tag{5.12}$$

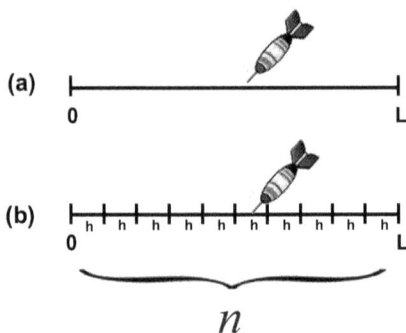

Figure 5.2: Partitioning an interval into n boxes.

Now, suppose that h is so small that we do not care (or cannot care) about the exact location of the particle *within* a cell of length h. All we care of is in which, among the n cells, the particle is located. For example, this is the case if h is much smaller than our experimental accuracy in the measurement of the particle position. In this case we retain only the first term on the r.h.s. of (5.11) and ignore the second term. Indeed, once we know which cell contains the particle, there is no additional missing information: we know the position of the particle, hence $H_h = 0$.

In practice, what we did in (5.12) is subtracting two quantities that are calculated from the continuous analog of the SMI, to obtain a discrete SMI associated with the finite number n of cells. Since we *always* have a limit on the accuracy of determination the location of the particle, we can always choose h smaller than our accuracy, in which case (5.12) is the only meaningful quantity.

Of course, choosing h very small implies that n is very large, so that (5.11) will diverge if we let $h \to 0$. On the other hand this limit is never reached, because quantum mechanics tells us that there is a lower bound to the intrinsic precision of any position measurement, related to the Heisenberg's uncertainty principle. Although this is so small that it has no practical impact on any macroscopic measurement, such that taking the continuous limit provides an excellent approximation, it does set an upper limit to (5.11) that prevents it from diverging.

As H_h can be taken as zero when the size h is chosen smaller than our experimental resolution, (5.12) implies that we can replace H_L by

$H\left(\frac{1}{n}, \ldots, \frac{1}{n}\right)$. In practice, this removes the divergent part of H_L. In actual applications we shall always be interested in differences in SMI values (or in entropy values in the next two chapters). Such differences will be dependent of the number of cells we choose to divide the continuous interval $(0, L)$.

5.3 Reinterpretation of $f^*(x)$ as an Equilibrium Density

In this section we shall show that the distribution that maximizes the SMI, also maximizes the probability of occurrence of that distribution. In other words, starting from any distribution of N particles in c cells, we shall find that if we "shake" the system (in the sense discussed below) the system will reach a uniform distribution, which is the most probable distribution. For a real system of particles in a box, the uniform distribution is also the equilibrium distribution for the locations of the particles.

5.3.1 *Specific versus generic configurations*

Consider a system of N marbles distributed in c cells. Denote by N_i the number of marbles in the cell i, and assume that there is no limit on the number of marbles in each cell.

A ***specific configuration*** of the N marbles in the c cells is a list of the locations of all the specific marble. For instance, if the marbles are numbered $1, 2, 3, \ldots, N$, then a specific configuration would be "marble 1 in cell 7, marble 2 in cell 3, marble 3 in cell 5, ... marble N in cell 2". Since we can place marble 1 in any of the c cells, and marble 2 in any one of the c cells, and so on, we have altogether c^N possible *specific configurations* of the N marbles in the c cells. An example of possible specific configurations of two marbles in four cells is shown in figure 5.3.

We assume that each of the c^N specific configurations has the same probability of occurrence: $P(\text{specific configuration}) = c^{-N}$.

Next, suppose that we are not interested into *specific* configurations, i.e. into knowing which particle is in which cell, but we only care about knowing a ***generic configuration***, that is how many particles are in each cell.

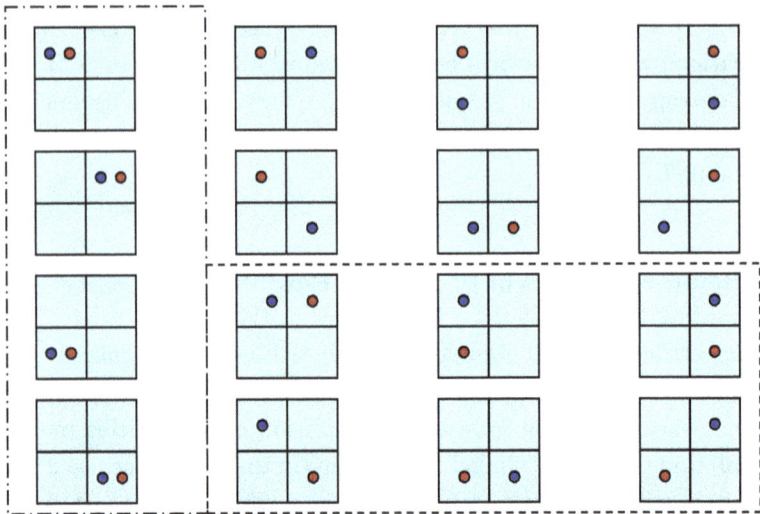

Figure 5.3: All possible specific configurations of 2 marbles in 4 boxes. There are 4 possible ways to put both marbles in the same box and 12 configurations with a single marble in each box (6 configurations plus 6 obtained from them by swapping the marbles.

We shall denote by $\boldsymbol{N} \equiv (N_1, N_2, \ldots N_c)$ a generic configuration, with $\sum_{i=1}^{c} N_i = N$. This simply expresses a distribution of the N marbles in the c cells. We shall also refer to $\boldsymbol{N}$ as describing the state of the system, or better the **state-distribution**. When considering a state-distribution of the marbles, we are interested only into the *number* of marbles in each cell, and not into which specific marble is in which cell.

Note that from the *state-distribution* $\boldsymbol{N}$ we can construct a probability distribution associated with this state, simply by dividing each component by the total number N of marbles. Define $f_i \equiv N_i/N$ and define the new vector $\boldsymbol{f} \equiv (f_1, f_2, \ldots f_c)$, where f_i is the fraction of marbles in cell i in the given generic configuration. We can then define each f_i as the probability of finding any chosen marble in cell i.

Exercise 5.1. Consider the specific configuration (B,R,0,0) shown in figure 5.3, where B indicates the blue marble and R indicates the red one. What is its probability? Write the state-distribution and the probability distribution pertinent to this configuration. If we know only the generic configuration, what is the probability of finding a specific particle in cell 1, in cell 2, in cell 3, and in cell 4? What is the probability of obtaining this generic configuration?

5.3.2 *Probability of a generic configuration*

Now suppose that we "shake" the system in such a way that the marbles can move from a cell to any another one in some random manner. What is the probability of occurrence of the state-distribution $N = (N_1, N_2, \ldots, N_c)$?

Since we know the total number of *specific* configurations, all we need to calculate is the number of specific configurations that "belong" to a given state-distribution. In other words, in how many ways we can change the specific arrangements in such a way that the state-distribution N is unchanged.

The number of possible configurations is

$$W(N) = \frac{N!}{\prod_{i=1}^{c} N_i!} \tag{5.13}$$

This number is the number of ways we can choose groups of N_1, N_2, ..., N_c marbles from a total of N marbles. We shall refer to $W(N)$ as the **multiplicity** of the distribution N, or of the corresponding state of the system. Using the classical definition of the probability (see section 3.1.1) of an event, we can calculate the probability of the event N as the ratio between the multiplicity and the total number of possibilities

$$P(N) = \frac{W(N)}{\sum_N W(N)} = \frac{1}{c^N} \frac{N!}{\prod_{i=1}^{c} N_i!} \tag{5.14}$$

Note that one gets the same result from the multinomial distribution (3.44) by setting $p_1 = p_2 = \cdots = p_c = 1/c$.

5.3.3 *Different levels of detail*

The reader should be aware that we are speaking about different levels of detail and probability. So far, we met 2 levels of detail. At the lower level, we have all specific configurations, which can only be specified by providing a fully detailed description of the system. For example, at this level we specify the location of each individual particle. At the higher level, we have the generic configurations, which correspond to descriptions of the system with (typically much) less details. For example, we report only the number of particles in each cell. We can anticipate here that, for macroscopic systems, one further level exists, at which only thermodynamic quantities are available. This is the least detailed level, characterized by few measurable values.

As a consequence, when speaking in terms of probabilities, we also have different levels. One starts with the distribution of N particles in cells: $\boldsymbol{N} \equiv (N_1, N_2, \ldots, N_c)$, with $N = \sum_i N_i$. Next one defines the probability distribution $\boldsymbol{f} \equiv (f_1, \ldots, f_c)$ by taking the ratio $f_i \equiv N_i/N$ between the cell **occupancy** N_i and the total number N of particles.

At the higher level, we are interested into the probability $p(\boldsymbol{N})$ that a given state-distribution $\boldsymbol{N}$ happens to be the result of our experiment. Clearly $p(\boldsymbol{N}) = p(\boldsymbol{f})$ and, on the long term, this probability is also the fraction of time spent by the system in such generic configuration during its evolution.

As we have assumed that all specific configurations are equally probable, the probability associated with the given state of the system is proportional to the number of specific configurations whose union represents that generic configuration. This probability is given by eq. (5.14).

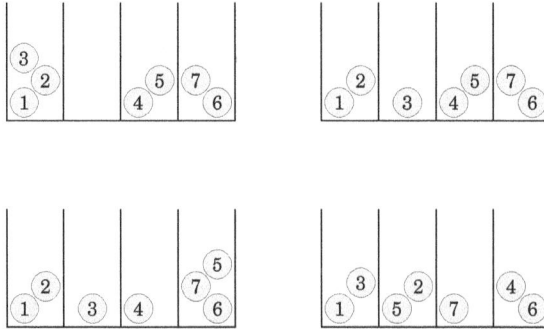

Figure 5.4: Example with 7 marbles in 4 boxes. The state-distribution evolves from (3,0,2,2) to (2,1,2,2), then moves to (2,1,1,3) and (2,2,1,2).

For example, figure 5.4 shows the evolution of a set of 7 marbles in 4 compartment. We start initially with a state-distribution $\boldsymbol{N} = (3,0,2,2)$, whose specific configuration is $\{(1,2,3), \emptyset, (4,5), (6,7)\}$. Then we start shaking the system, and the generic configuration $\boldsymbol{N}$ moves to $(2,1,2,2)$, then to $(2,1,1,3)$, next to $(2,2,1,2)$, and so on. If we follow the evolution at the lowest level, each specific configuration is equally probable, hence nothing special is noticed. At the next level, the probability distribution

f of any single marble is initially defined as $\{\frac{3}{7}, 0, \frac{2}{7}, \frac{2}{7}\}$, then evolves to $\{\frac{2}{7}, \frac{1}{7}, \frac{2}{7}, \frac{2}{7}\}$, and so on. At the higest level, the state-distributions have quite different probabilities $P(N)$, because more uniform distributions can be realized in many more ways than less uniform ones.

5.3.4 *Probability of state distributions*

Now we show that the distribution that maximizes the SMI is also the most probable state.

> **Proof** — Taking the natural logarithm of (5.14) and using the Stirling approximation (3.36) in the form $\ln N! \approx N \ln N - N$, we get
>
> $$
> \begin{aligned}
> \ln P(N) &= -N \ln c + \ln N! - \sum_{i=1}^{c} \ln N_i! \\
> &= -N \ln c + N \ln N - N - \sum (N_i \ln N_i - N_i) \\
> &= -N \ln c + N \ln N - \sum N_i \ln N_i \\
> &= -N \ln c + N \ln N - \sum (N_i \ln f_i + N_i \ln N) \\
> &= -N \ln c - \sum N_i \ln f_i \\
> &= -N \ln c - N \sum_{i=1}^{c} f_i \ln f_i
> \end{aligned}
> \tag{5.15}
> $$
>
> Hence we have found that
>
> $$
> \ln P(N) = -N \ln c + N H(f_1, \dots, f_c)
> \tag{5.16}
> $$
>
> As $-N \ln c$ is constant, it is clear that the distribution f which maximizes the SMI H also maximizes the state probability $P(N)$.
> ∎

Equation (5.16), together with (5.13), implies that

$$
W(N) = \exp[N H(f_1, \dots, f_c)]
\tag{5.17}
$$

Thus, for any state-distribution $N = (N_1, \dots, N_c)$ we can define a probability distribution $f = (f_1, \dots, f_c)$, with which we can compute the SMI $H(f_1, \dots, f_c)$. The SMI is related to the multiplicity of the distribution $W(N)$ and to the probability of the state-distribution

$$
P(f) = P(N) = \frac{W(N)}{c^N} = \frac{\exp[N H(f_1, \dots, f_c)]}{c^N}
\tag{5.18}
$$

Because the denominator in (5.18) is a (normalization) constant, it follows immediately that *the distribution f^* which maximizes the SMI, also maximizes the probability of the state.*

Let's now consider an "experiment" with a system of N marbles distributed in c cells. For simplicity, we start with an initial distribution $N_{\text{initial}} = (N, 0, \ldots, 0)$, i.e. with all the marbles in the first cell. This distribution has the lowest multiplicity $W(N_{\text{initial}}) = 1$, hence also the lowest state probability $P(N) = c^{-N}$ and the lowest SMI: $H = 0$.

> **Exercise 5.2.** Check the values of multiplicity, probability and SMI for the initial distribution, and explain why they have minimal values for this distribution.

If we perform the experiment with real marbles and cells, we must physically shake the system in such a way that the marbles can freely move from one cell to another. Alternatively, we can use a computer simulation[4], in which we "shake" the system using the following procedure:

(1) Choose a specific marble at random, i.e. pick any number between 1 and N with equal probabilities.
(2) Choose a specific cell at random, i.e. pick any number between 1 and c with equal probabilities.
(3) Place the chosen marble in the chosen cell.
(4) Repeat the operations 1 to 4.

Each move of a marble will be referred to as a step. At each step we have a new distribution N, for which we can calculate the values of $W(N)$, $P(N)$ and SMI.

If we perform the physical experiment with some gas, we can start with any arbitrary distribution of particles distributed among different compartments (cells). Then we remove the partitions between the compartments. There is no need to "shake" the system: the random motion of the particles does the job for us.

A computer simulation is the best choice, if we have only few marbles and want to closely follow the evolution of the system. For example, we could simulate $N = 10$ marbles in $c = 10$ cells, and "shake" the system

---

[4]The reader may test different simulations on the web site of Ben-Naim (2010).

for a large number of steps. We know already that the maximal value of H is reached for the uniform distribution $f_i^* = \frac{1}{10}$. Using the base 2 for the logarithms we find $H_{max} = -\sum f_i^* \log_2 f_i^* = \log_2 10 = 3.32$ bits. The multiplicity is also maximized, $W_{max} = 10! = 3\,628\,800$, as well as the state probability $P_{max} = 10!/10^{10} = 3.63 \times 10^{-5}$.

Figure 5.5 shows the evolution of the values of the SMI H for 10 marbles and different number of available cells. The top plot shows the behavior of the SMI when $c = 10$. Despite from the fluctuations from step to step, the overall trend of the SMI (proportional to the logarithm of the multiplicity) is a monotonic increase until a plateau is reached. At the plateau, fluctuations are still visible. However, if we increase the number of available cells to 100 or 1000, the relative magnitude of these fluctuations decreases (middle and bottom plots in figure 5.5). As the number of steps is related to the time elapsed since the preparation of our system, we can also conclude that, independently from the initial state, the system will evolve toward a situation in which the SMI is very close to its maximum value, and will remain so for a long time.

If we consider more and more complex systems, the relative magnitude of the fluctuations away from the maximum SMI are smaller and smaller. For macroscopic systems they are completely invisible. We can now deduce that, *as the time elapses, the system approaches a macroscopic state at which the values of H and W remain practically constant.* We shall refer to this macroscopic state as an ***equilibrium state***. In this state, the distribution of marbles in the cells is uniform. In addition, the SMI of the equilibrium state is maximal, and so is also the value of the state-distribution probability. This implies that *the system will spend most of its time in the equilibrium state.*

Actually, the equilibrium state is not a single state: we always have some fluctuations in the distribution. However, when the system is very large the relative fluctuations are very small and we can ignore these fluctuations.[5]

---

[5]A more rigorous treatment would be to start from eq. (5.18) and compute the Taylor expansion of H about the maximum. As the first derivative is zero at this state, one must keep all terms up to the second one, obtaining a parabolic approximation to H. This implies that $P(N)$ is approximately Gaussian, and it can be shown that its variance decreases with increasing N. In practice, all the density is concentrated about the state-distribution with maximum probability: only the state-distributions very close to that have non-negligible probability but, from the

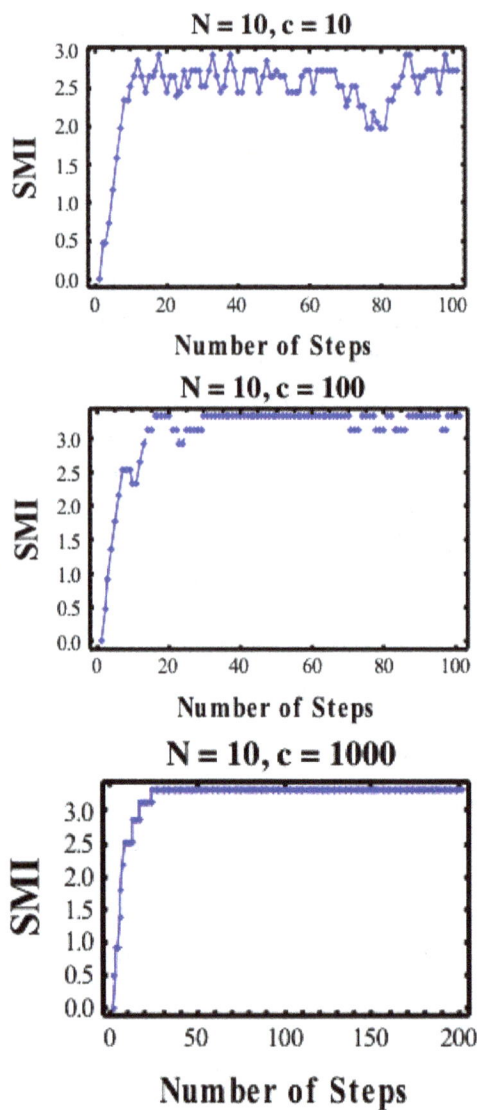

Figure 5.5: System of $N = 10$ marbles in $c = 10, 100, 1000$ cells (Ben-Naim, 2010).

The important conclusion is that, when we have *large* number of particles distributed in cells, *the system will always evolve toward an equi-*

---

macroscopic point of view, they are indistinguishable from the peak.

librium state, characterized by a uniform distribution of marbles in the cells. Furthermore, the equilibrium (i.e. the final) distribution is the distribution that maximizes *both* the SMI and the multiplicity (as well as the state probability).

In chapter 8 we will learn about the laws of thermodynamics. Here, we anticipate how we can formulate the second law of thermodynamics for a real system of N particles in c compartments. For simplicity, we assume that all cells have the same volume V/c, with V being the total volume of the system. Starting with any initial distribution of particles among the compartments, we remove the partitions between the cells. While the particles will randomly move everywhere in the accessible space (while keeping the total volume and the total energy constant), the system will evolve toward an equilibrium state characterized by a uniform distribution (i.e. N/c particles per cell). The equilibrium distribution is also the distribution that maximizes the SMI of the system.

> **The maximum (i.e. the equilibrium) value of the SMI will be identified (up to a multiplicative constant) with the entropy of the system.**

This is very important. Being identified as the SMI at equilibrium (up to a multiplicative constant, corresponding to the freedom of choosing the units), *the entropy does not change with time*. Of course, it is true that the SMI evolves in time, and for an isolated system the SMI tends to a maximum at equilibrium. But the entropy is defined only for equilibrium states, hence by definition it is constant (as well as all other thermodynamic quantities at equilibrium).

Furthermore, the probability of the equilibrium state is the maximal probability over all generic configurations of the N particles in the c cells. This explains *why* the system will evolve from *any* initial distribution to the final uniform distribution.

Even though fluctuations always happen, their relative size becomes quickly negligible for increasing N. In addition, the maximal state probability becomes nearer and nearer to one, such that the ratio with the probability of any other generic configuration becomes larger and larger. Hence, for N big enough, the measurable departures from the equilibrium state are so unlikely that, in practice, they are never observed: the system will stay in the equilibrium state "forever".

Finally, we note that the equilibrium state is not a single distribution, but a small set of states. Their probabilities sum up to almost exactly one, hence the system will be found in the equilibrium state for sure at any time. For more details, see Ben-Naim (2008).

5.4   The Second Theorem: The Exponential Distribution

The second theorem proved by Shannon is the following.[6] We start again with the functional H as defined in (4.9) and seek the maximal value of H over all possible functions $f(x)$ subjected to the conditions

$$\int_0^\infty f(x)\,dx = 1 \tag{5.19}$$

and

$$E[X] = \int_0^\infty x f(x)\,dx = a \tag{5.20}$$

with $a > 0$ being an arbitrary (but finite) positive constant.

This problem is of the same kind as the one discussed in section 5.1 but with some difference. First, now the support of the function is unlimited: $f(x)$ is defined in $\mathbb{R}^+$ and x can assume any positive value. Second, in addition to the normalizatin condition (5.19), we have now one additional condition: we require that the mean of the density function $f(x)$ is fixed. Thus, we are searching for the maximum value of H over all possible normalized functions whose mean (5.20) is given.

The solution is found again with the Lagrange method of undetermined multipliers. Again, we do not need to be familiar with the mathematical details. The result of this procedure is that the function $f^*(x)$ which maximizes the SMI is

$$f^*(x) = \frac{1}{a}\exp\left(-\frac{x}{a}\right) \tag{5.21}$$

known as the ***exponential distribution***. The mean of this distribution is $E[X] = a$, the ***median*** (i.e. the point dividing the domain into two partitions each covering 50% probability) is $a\ln 2$, and the variance is $V[X] = a^2$.

---

[6]The results of this and the next section are important in statistical mechanics but are not used in this book. Hence the impatient reader may skip these two sections and jump to section 5.6.

Exercise 5.3. Verify that this function fulfills the two conditions (5.19) and (5.20).

The value taken by H for the particular function $f^*(x)$ is[7]

$$
\begin{aligned}
H_{\max} &= -\int_0^\infty f^*(x)\ln f^*(x)\,dx \\
&= -\frac{1}{a}\int_0^\infty \left(-\frac{x}{a}-\ln a\right)\exp\left(-\frac{x}{a}\right)dx \\
&= 1+\ln a = \ln(ae)
\end{aligned}
\tag{5.22}
$$

The exponential density in (5.21) is a fundamental distribution in statistical mechanics, although we shall not need it in this book.

Quantum mechanics tells us that a confined system with a fixed *average* energy can be found in different **energy levels**. In other words, at the microscopic level not all values of the energy are allowed, but only a set of discrete values that depend on the potential. For example, an electron around a nucleus can only assume well defined energy values, and can "jump" from a level to a higher one by absorbing a photon whose energy exactly matches the difference between the two levels. Alternatively, an electron can move to a lower level causing the emission of a photon whose energy equals the difference between two levels. Atoms are stable because of the existence of a discrete set of energy levels, which possesses a minimum value. Electrons with this energy populate the so-called **fundamental state** and can only have transitions to higher levels (at the condition of absorbing one photon). Thus they cannot get arbitrarily close to the positively charged nucleus, which would happen in about a nanosecond if all real values were allowed for the energy.

When we study the thermodynamics of a system with discrete energy levels, we obtain the discrete analog of (5.21), which is referred to as the **Boltzmann distribution**:

$$
f_i^* = \frac{1}{Q}\exp\left(-\frac{E_i}{k_B T}\right)
\tag{5.23}
$$

where T is the absolute temperature, k_B is the Boltzmann constant, and

$$
Q = \sum_i \exp\left(-\frac{E_i}{k_B T}\right)
\tag{5.24}
$$

---

[7]Here again we use the natural base for the logarithms.

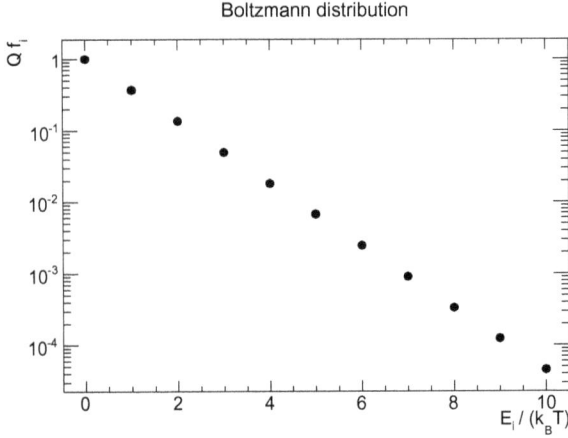

Figure 5.6: Boltzmann distribution.

is the normalization constant. $Q = k_B T$ but when written in the form (5.24) is called in statistical mechanics the **partition function**. The distribution (5.23) is shown in figure 5.6.

Note that f_i^* in (5.23) is obtained by the Lagrange method of undetermined multipliers. The distribution $\{f_1^*, f_2^*, \ldots\}$ is the distribution that maximizes the value of H subject to the two conditions

$$\sum_{i=1}^{\infty} f_i^* = 1 \tag{5.25}$$

and

$$\sum_{i=1}^{\infty} E_i f_i^* = \langle E \rangle \tag{5.26}$$

where $\langle E \rangle$ is the average energy of the system in an ensemble of system characterized by a fixed temperature. We omit the details here, but the interested reader can find them in Ben-Naim (2008).

In the next section we show that f_i^* can also be obtained as the limiting distribution, when the system evolves towards equilibrium.

5.5   Reinterpretation of f_i^* as an Equilibrium Distribution

As in section 5.3, the purpose of this section is to show that a distribution that maximizes the SMI is also the distribution that is obtained when the system evolves towards equilibrium. In other words, this is also the distribution with maximal probability.

In section 5.3 we saw that, if we start with any arbitrary distribution of N particles in c cells and let the system undergo random changes, then the system evolves towards equilibrium, and the equilibrium distribution of the particles among the cells will be uniform.

Here we have again N particles, but now they are distributed among an infinite number of cells. The "cells" actually represent energy levels, in the sense that we assume that each marble can assume any value in an infinite set $\{E_i, i = 1, 2, \ldots\}$, and say that a particle is in cell i when it has energy E_i.

We can start again with any arbitrary but known distribution of particles in all cells, such that the total number of particles is fixed

$$\sum_{i=1}^{\infty} N_i = N \tag{5.27}$$

and the total energy of the system is fixed

$$\sum_{i=1}^{\infty} E_i N_i = E_{\text{tot}} \tag{5.28}$$

Next we define the distribution $f_i \equiv N_i/N$ and rewrite (5.27) and (5.28) as the normalization condition

$$\sum_{i=1}^{\infty} f_i = 1 \tag{5.29}$$

and as the requirement of a known average energy

$$\sum_{i=1}^{\infty} E_i f_i = \langle E \rangle \equiv E_{\text{tot}}/N \tag{5.30}$$

We now "shake" the system so that the particles can change energy level, but in such a way that the conditions (5.27) and (5.28), or equivalently (5.29) and (5.30), are maintained.

We can perform a simulation with a computer program similar to the program we used in section 5.3, but with one difference. When we move a particle from one cell to another, the energy of the system will change, therefore a single move at each step will not conserve the energy of the system. Thus, the modified procedure is the following, where each cycle is considered as one step[8]:

---

[8]Simulations are available on the web site of Ben-Naim (2010).

1. Choose two marbles at random.
2. Choose two cells at random.
3. Place the two chosen marbles in the chosen cells.
4. Check the total energy of the new configuration: if it did not change in the move, then accept the new configuration, otherwise reject the new configuration.
5. Repeat the cycle from action 1.

We can now start with any initial distribution of marbles in cells, randomly generate a change, and calculate the total energy of the system with eq. (5.28). Whenever the change is accepted, we can calculate the value of the SMI and the multiplicity W of the distribution, as we have done in section 5.3. However, the probability of the distribution cannot be calculated as in (5.14), since now we do not accept all possible distributions. Instead, we need to calculate the conditional probability of the configuration N (or equivalently of $f_1, f_2, \ldots$) given $\langle E \rangle$:

$$P(N \,|\, \langle E \rangle) = \frac{W(N \,|\, \langle E \rangle)}{\sum_N W(N \,|\, \langle E \rangle)} \tag{5.31}$$

Here, $W(N \,|\, \langle E \rangle)$ is the multiplicity of the distribution N having average energy per particle $\langle E \rangle$. In the denominator we sum over all the distributions N, keeping $\langle E \rangle$ fixed.

If we denote by $W_{\text{tot}} = \sum_N W(N)$ the total number of specific distributions, that is the same quantity we used in (5.14), we can rewrite (5.31) as

$$P(N \,|\, \langle E \rangle) = \frac{W(N \,|\, \langle E \rangle)}{W_{\text{tot}}} \frac{W_{\text{tot}}}{\sum_N W(N \,|\, \langle E \rangle)} = \frac{P(N, \langle E \rangle)}{P(\langle E \rangle)} \tag{5.32}$$

Here $P(N, \langle E \rangle)$ is the probability of finding the distribution N having average energy $\langle E \rangle$, while $P(\langle E \rangle)$ is the probability of finding a system with any distribution having average energy $\langle E \rangle$. The same result can be obtained by considering N and $\langle E \rangle$ as two random variables and by applying eq. (3.11).

The evolution of SMI and $\log_2 W$ are similar to what we have seen in section 5.3. However the equilibrium values of SMI and W and the equilibrium distribution are different.

In section 5.4 we have seen that the distribution f that maximizes the SMI with the constraint of a fixed average energy per particle is the exponential distribution

$$f_i^* = \frac{\exp(-\beta E_i)}{Q} \tag{5.33}$$

where β is a positive parameter (the undermined multiplier in the Lagrange method[9]), and Q is the normalization constant defined by

$$Q = \sum_i \exp(-\beta E_i)$$

with the sum running over all possible states.

Now we show that the distribution that maximizes the SMI subject to the condition of fixed $\langle E \rangle$ is the same distribution that maximizes the probability $P(N | \langle E \rangle)$.

> **Proof** — The relationship between the probability $P(N | \langle E \rangle$ and the SMI is obtained by taking the natural logarithm of eq. (5.32), inserting $W(N)$ from eq. (5.13), and using the Stirling approximation as we have done in eq. (5.15) above:
>
> $$\ln P(N | \langle E \rangle) = \ln W(N | \langle E \rangle) - \ln \sum_N W(N | \langle E \rangle)$$
>
> $$= \ln \frac{N!}{\prod_{i=1}^{\infty} N_i!} - \ln \sum_N W(N | \langle E \rangle)$$
>
> $$= N \sum_{i=1}^{\infty} f_i \ln f_i - \ln \sum_N W(N | \langle E \rangle)$$
>
> $$= N H(f_1, f_2, \ldots) - \ln \sum_N W(N | \langle E \rangle)$$
>
> Hence
>
> $$\ln P(N | \langle E \rangle) = N H(f_1, f_2, \ldots) - \ln \sum_N W(N | \langle E \rangle) \tag{5.34}$$
>
> This has the same form as in equation (5.16) except that the first term on the r.h.s. of (5.16) is now replaced by the term $-\ln \sum_N W(N | \langle E \rangle)$. Since the summation runs over all possible distributions N, this term is independent of N. Therefore, the same distribution that maximizes the SMI also maximizes the probability $P(N | \langle E \rangle)$. ∎

We repeat once more that, because it is the distribution with maximum probability, f^* is also the state in which the system remains for

---

[9]It turns out that $\beta = 1/k_B T$, but in this section we are not interested in the meaning of β, only in the form of the equilibrium distribution.

the biggest fraction of time. The probability distribution defined over the N's is so sharply peaked that, in practice, only the generic configurations with f very close to f^* have some chance to be "visited" by the system during its evolution. For macroscopic systems they are so close to f^* that in practice they are indistinguishable from it: once the system reaches thermodynamic equilibrium, it will remain there practically forever.

Unlike in section 5.3, where $H_{\max}$ was simply $\ln c$, here the maximal value of H is obtained by using the distribution in (5.33)

$$
\begin{aligned}
H_{\max}(f_i^*, f_2^* \ldots) &= -\sum_i f_i^* \ln f_i^* \\
&= -\sum_i f_i^* (-\beta E_i - \ln Q) \\
&= \beta \sum_i f_i^* E_i + \ln Q \\
&= \beta \langle E \rangle + \ln Q
\end{aligned}
\tag{5.35}
$$

At equilibrium we can identify the entropy S with $k_B H_{\max}$ hence we can rewrite (5.35) as

$$
S = k_B H_{\max} = k_B \beta \langle E \rangle + k_B \ln Q
\tag{5.36}
$$

If we set $\beta = 1/k_B T$, the term $-k_B T \ln Q$ is the **Helmholtz energy** of the system (which we will treat in section 8.10):

$$
A = -k_B T \ln Q = \langle E \rangle - TS
\tag{5.37}
$$

Note that the maximum of H is translated into minimum of A over all possible distributions. As we emphasized above, in the approach followed in this book it is not S that is maximized, as the entropy is defined only at equilibrium. What is maximized is the SMI, and the maximum is taken with respect to all possible distributions, according to their probability.

Although we did not define the value of A for any distribution, but only for the equilibrium distribution, we shall see in chapter 8 that the Helmholtz energy A plays an important role for the determination of equilibrium states for systems at constant temperature and volume. When we remove a constraint in a system with constant T, V, N, the system will evolve toward a new equilibrium state in which the Helmholtz energy will be smaller. The reason is again the overwhelmingly larger probability of the final equilibrium state.

The thermodynamic quantities S and A (as well as the Gibbs energy defined in 8.10) are defined for equilibrium states. When removing a constraint, they reach an extremum with respect to all possible constrained equilibrium states. Their extremum corresponds to the distribution with the highest probability.

5.6   The Third Theorem: The Normal Distribution

As in section 5.1 and 5.4, we use the functional H (i.e. the SMI viewed as a function of the function $f(x)$) as defined in eq. (4.9). We seek a maximum of H over all possible functions $f(x)$ subjected to the two conditions

$$\int_{-\infty}^{\infty} f(x)\,dx = 1 \tag{5.38}$$

and

$$E[X^2] = \int_{-\infty}^{\infty} x^2 f(x)\,dx = \sigma^2 \tag{5.39}$$

Note that the range of integration is now $(-\infty, \infty)$. The second condition is that the variance of the distribution is constant. Since the variance must be a positive number, we shall use the notation σ^2 for this constant.

Using the Lagrange method of undetermined multipliers, Shannon found that the distribution that maximizes H, subjected to the two conditions (5.38) and (5.39) is the normal (or Gaussian) distribution (3.33) centered on zero:

$$f^*(x) = \frac{1}{\sqrt{2\pi\sigma^2}} \exp\left(-\frac{x^2}{2\sigma^2}\right) \tag{5.40}$$

Being a symmetric distribution, one immediately finds that $E[X] = 0$, which in turn implies that the variance equals the square of the **root mean square** value (RMS) of X:

$$V[x] = \langle X^2 \rangle - \langle X \rangle^2 = \sigma^2 \tag{5.41}$$

Again, the mathematical details leading to this distribution are of no concern to us. The interested reader can find some more details in Ben-Naim (2008).

Exercise 5.4. Verify that this distribution fulfills the two conditions (5.38) and (5.39).

The value that H attains for the particular function $f^*(x)$ is

$$
\begin{aligned}
H_{\max} &= -\int_{-\infty}^{\infty} f^*(x)\ln f^*(x)\,dx \\
&= \int_{-\infty}^{\infty} \left[\frac{1}{2}\ln(2\pi\sigma^2) + \frac{x^2}{2\sigma^2} \right] f^*(x)\,dx \qquad (5.42) \\
&= \frac{1}{2}\ln(2\pi e\sigma^2)
\end{aligned}
$$

The normal distribution (5.40) is one of the most important distributions in the theory of probability. When used for the distribution of velocities of real particles in a macroscopic system at a given temperature, this distribution is known as the Maxwell-Boltzmann distribution, which will be derived in section 6.4.1. We shall use this distribution along with the uniform distribution derived in section 5.1 to build-up the entropy function of an ideal gas in the next chapter.

Maxwell was the first to show that the velocity distribution of the atoms or molecules has the normal form. Later, Boltzmann proved that if we start from any distribution of velocities, a system with constant temperature will evolve toward an equilibrium state in which the velocity distribution is the normal distribution. This is why it is now referred to as the Maxwell-Boltzmann distribution.

As we have found in sections 5.3 and 5.5 the distribution that maximizes the SMI also maximizes the state probability. We could prove this also for the case of the normal distribution. We could also perform a computer simulation by adapting the previous algorithm: we would start with a system of N marbles moving with some distribution of velocities in one dimension, then simulate the evolution of such a system with the conditions (5.38) and (5.39), etc.

Fortunately, we do not need to perform this simulation to obtain the normal distribution. The reason is that the condition of constant variance in (5.39) when applied to velocity distribution is equivalent to requiring a constant average kinetic energy of the particles. Therefore, we can use the result of section 5.5 and change variable, moving from a distribution in energy levels to a distribution of velocities.

5.7 Summary of What We Have Learned in this Chapter

The SMI H is a functional, i.e. a function of $f(x)$, which itself is a function of x. We have derived three distributions by maximizing the SMI under different conditions. The method is always the same: one seeks a maximum of H over all possible *functions* $f(x)$ that satisfy certain conditions, unsing Lagrange's method of undetermined multipliers.

Even without getting into the details of the derivations, the reader is advised to remember few important results. First, the uniform distribution is the one that maximizes H if the normalization is the only condition given. Second, if there are two conditions of the form (5.38) and (5.39), which fix the normalization and the variance, then the resulting distribution is the normal distribution.

In all cases, the distribution that maximizes H also maximizes the state probability, hence it is also the equilibrium distribution.

We shall use two results to build-up the entropy function in the next chapter:

(1) If we have N non-interacting particles moving in one dimensional box of length L, the system will tend to an equilibrium state, where the density of particles at each point between 0 and L will be constant: $\rho = N/L$.

(2) Imposing a fixed total kinetic energy on the particles moving in a 1-dimensional system, the system will evolve towards equilibrium where the velocity distribution is normal.

Chapter 6

The Entropy Function of a Classical Ideal Gas

This chapter is a bridge between information theory and thermodynamics. We shall derive an explicit expression for the entropy of an ideal monoatomic classical gas, as a particular case of Shannon's measure of information (SMI).

The system under study contains N indistinguishable and non-interacting (ideal) particles. In the classical description the particles are described by their locations and their momenta. We assume that the particles have no internal structure, hence no rotation, vibration, or electron energy levels are considered. In this case the total energy consists only of the kinetic energy of the particles due to their translation in space. This picture of "structure-less" particles is approximately true for monoatomic gases.

The equation we shall obtain in section 6.6 was first derived independently by Otto Sackur (1880–1914) and Hugo Martin Tetrode (1895–1931) in 1912. It is usually discussed in the context of statistical mechanics, typically taught in more advanced courses. However, in this book we derive the same expression using the language of information theory. By doing so we shall obtain the *entropy as a particular case of SMI*. This approach was first proposed by Ben-Naim (2008) to interpret entropy. Here we use this method to define and derive the entropy function. This approach has the obvious advantage that at each stage of the derivation we know the meaning of the various terms that contribute to the entropy. The four contributions to the entropy of an ideal gas are the following:

- The SMI associated with the locations of the particle, i.e. how much uncertainty there is in the locations of the particles.

- The SMI associated with the velocities (or momenta) of the particles.
- The reduction in SMI due to the indistinguishability of the particles.
- A correction to the SMI due to the quantum mechanical uncertainty principle.

Once the entropy function has been computed, we can apply it to the study of the thermodynamics of an ideal gas. This will be done in chapter 7. In the remaining parts of the book, we shall treat general thermodynamic systems such as liquids, solids, mixtures, etc., for which we do not have an explicit expression for the entropy function. The reader should be aware of the fact that in thermodynamics one never deals with the explicit function for the entropy. One only assumes the *existence* of such a function, and then proceeds to derive relationships between different thermodynamic quantities. However, in our approach we start with an explicit function for the entropy of one system (an ideal gas) as it is done in statistical mechanics, although we do not use the statistical mechanical formalism. For all other systems we use the thermodynamic approach, where we do not have an explicit expression for the entropy function.

Whatever we do, we shall always keep in mind that the entropy of any system has the *meaning* of a Shannon measure of information.

6.1   Some Comment on the Mathematical Notation

Before we proceed further, it is good to pause for a moment and consider what happens when we study the probability distribution of a continuous physical quantity. In the previous chapter, we have made use of continuous density functions, but we considered them only from a mathematical point of view. This means that every quantity was dimensionless. On the other hand, a physical quantity typically has units attached to it, which is another way of saying that it carries physical dimensions (like length, time, mass, etc.). For example, a velocity has the dimensions of a length divided by a time. *The dimensions are part of the quantity*, not just some auxiliary or optional addition.

For example, let us consider once again the Gaussian distribution of the velocity component along any arbitrary direction. We take a co-ordinate system with the x-axis along that direction and write equation (5.40) as

$$f^*(v_x) = \frac{1}{\sqrt{2\pi\sigma^2}} \exp\left(-\frac{v_x^2}{2\sigma^2}\right)$$

The first thing to notice is that, in general, one should make sure that *the argument of each mathematical function*, like the exponential in the previous equation, *is dimensionless*. In this case, both v_x and σ are velocities, hence the argument of the exponential is a pure number. Good.

The second thing is subtler. The probability is a pure number, i.e. it carries no physical dimensions. We get a probability by integrating over $f^*(v_x)$, which means that we always deal with elements written as $f^*(v_x)\mathrm{d}v_x$. In turn, $\mathrm{d}v_x$ has physical dimensions (it is a velocity). Hence, $f^*(v_x)$ must have the physical dimensions of an inverse of a velocity (i.e. a time divided by a length). Hence, *a probability density function of a physical quantity has the dimensions of the inverse of that quantity*. The reason is that the probability, which is a pure number, is obtained by "multiplying" the density by the differential.

In the previous equation the normalization constant in front of the exponential has the dimensions of an inverse of velocity, which means that $f^*(v_x)\mathrm{d}v_x$ is a pure number. So far, so good.

The issue appears when one is interested into the SMI, as we have defined H in eq. (4.9) as the expectation of (minus) the logarithm of the probability density function. Taking the logarithm requires either the PDF to be dimensionless or that we divide it by a constant with the same dimensions. Note that the discrete version (4.1) of H does not suffer from this problem, because it is defined in terms of probabilities.

Problems may be avoided by redefining each random variable in order to make it dimensionless. This can be done by dividing it by its physical units, although this makes the notation quite heavy.

To keep things simpler, in what follows we will treat all symbols connected to physical quantities as dimensionless numbers.

In practice, this is the same as implicitly dividing by the units, leaving only the numerical values. Although this is some sort of abuse of notation, it is simpler to understand and gives more compact formulae. When useful, we will remind the reader about this abuse of notation.

In all important results, the physical quantities will appear inside mathematical functions with dimensionless expressions (i.e. as products or ratios canceling the units). Hence our simplified notation won't cause any issue. Furthermore, in practical problems one is always interested into differences in SMI and this automatically cancels out the physical dimensions, as one gets logarithms of ratios between homogeneous quantitites. In addition, some kind of discretization is possible (as we will see below), and this again solves any problem with physical units.

6.2   The Locational SMI of an Ideal Gas

Our goal is to find the SMI of an ideal gas of non-interacting indistinguishable particles confined in some volume V. We start from the contribution arising from the uncertainty about the location of each particle inside V.

In section 5.1 we found that, if a particle is confined to a one-dimensional "box" of length L, the distribution that maximizes H is the uniform distribution

$$f^*(x) = \frac{1}{L} \qquad (6.1)$$

and that, using natural logarithms, the corresponding SMI is

$$H_{\max}(\text{location of one particle along the } x\text{-axis}) = \ln L \qquad (6.2)$$

(here L is dimensionless, it is not a physical length).

It should be noted that the results (6.1) and (6.2) were obtained for the function H as defined in (4.9). In practice we always use a discrete division of the segment L into a finite number of cells. However, in this and in the next section we shall use the continuous language for both the locations and the velocities (or the momenta). In section 6.4 we shall apply the quantum mechanical uncertainty principle, which imposes a discretization on the joint space of locations and momenta.

The generalization of (6.1) and (6.2) to the three dimensional case is straightforward. Suppose the particle is confined to a cubic box of edge L and volume $V = L^3$. Clearly, the SMI associated with the y-axis and the z-axis will be the same as in (6.2). Furthermore, we assume that the 3 events "being at some location x", "being at some location y", and "being at some location z" are independent events. Therefore, the equilibrium density is

$$f^*(x,y,z) = \frac{1}{L}\frac{1}{L}\frac{1}{L} = \frac{1}{V}$$

and the SMI associated with the location (x,y,z) within the cube of volume V is the *sum* of the SMI associated with the three axes (as we have seen in section 4.7). Thus, if we use the notation $H_{max}(x)$ for the quantity in (6.2) we can write

$$H_{max}(x,y,z) = H_{max}(x) + H_{max}(y) + H_{max}(z) = 3\ln L = \ln V \qquad (6.3)$$

(in which L and V are dimensionless).

We next extend the result (6.3) to the case of N *independent* and *distinguishable* (D) particles. We also use the shorthand notation: $\boldsymbol{R}_i \equiv (x_i, y_i, z_i)$ for the locational vector of particle i, and $\boldsymbol{R}^N \equiv \{\boldsymbol{R}_1, \boldsymbol{R}_2, \dots, \boldsymbol{R}_N\}$ for the locational vector of all the N particles.

Since the particles are *independent*, the SMI of the N particles is simply the sum of the SMI of all the single particles, and since the SMI for a single particle (6.3) is the same for each particle we have

$$H_{max}^{D}(\boldsymbol{R}^N) = NH_{max}(x,y,z) = N\ln V \qquad (6.4)$$

Note that we added the superscript D for *distinguishable* particles. We shall see in the next section that the fact that the particles are *indistinguishable* (ID) introduces a correlation between the particles which causes a *reduction* in the SMI of N particles. Note also that we still retain the subscript "max" and, as we have seen in the previous chapter, this correspond to the limiting value of H at equilibrium.

6.3 The Mutual Information Due to the Indistinguishability of the Particles

Consider two particles 1, 2 distributed in M equal cells. The particles are independent and distinguishable (D). The number of possible arrangements of one particle is simply M and the corresponding SMI is:

$$H(1) = \ln M$$

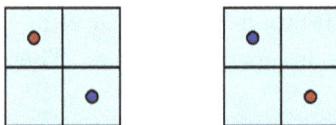

Figure 6.1: Two specific configurations of 2 marbles in 4 boxes obtained by swapping the marbles.

and the same is true for the other particle, $H(2) = \ln M$. The SMI of the two particles is the sum of the SMI associated to each particle:

$$H^{D}(1,2) = H(1) + H(2) = 2\ln M \tag{6.5}$$

If the particles are indistinguishable (ID), then the counting of the number of arrangements is different. The reason is quite simple: we can not distinguish between two configurations that are obtained by exchanging the particles.

Figure 6.1 shows two possible specific configurations of 2 particles in 4 cells. Clearly, when the two particles are distinguishable, these arrangements are *different*, and therefore should be counted as 2 specific configurations. However, if the two particles are indistinguishable, which amounts to erasing the colors, the two configurations coalesce into a single one. Thus, in general the total number of specific configurations is *reduced* when we "un-label" the particles. For example, when erasing the color of the marbles shown in figure 5.3 on page 110 we "loose" 6 configurations out of the 16 arrangements which are possible with distinguishable particles.

In the general case of N particles in M cells, the counting of the total number of configurations of N distinguishable particles is simply M^N. On the other hand, the number of configurations of N indistinguishable particles is more complicated to compute. However, for the case that $N \ll M$, i.e.n when the number of cells is so large that the occurrence of more than one particle in a cell is a rare event, the number of configurations is reduced from M^N to $M^N/N!$

Exercise 6.1. Calculate the total number of configurations of $N = 2$ in $M = 4$ cells, when the particles are distinguishable and when they are indistinguishable. Keep $N = 2$ fixed and change M to $10, 100, 1000$ and finally extrapolate very large M.

In the example of figure 5.3, when the number of cells M is very large, the number of configurations of the system of 2 indistinguishable particles is $M/2$, and the corresponding SMI is

$$H^{\text{ID}}(1,2) = \ln \frac{M^2}{2}$$

The difference between $H^D(1,2)$ and $H^{\text{ID}}(1,2)$ can be interpreted in terms of *mutual* information (see section 4.7):

$$H^{\text{ID}}(1,2) = H(1) + H(2) - I(1;2)$$

where

$$I(1;2) = \ln 2 > 0$$

In the more general case of N particles on M cells, with $N \ll M$, we have

$$H^D(1,2,\ldots,N) = \sum_{i=1}^{N} H(i) = \ln M^N \qquad (6.6)$$

and

$$H^{\text{ID}}(1,2,\ldots,N) = H^D(1,2,\ldots,N) - \ln N! = \ln \frac{M^N}{N!} \qquad (6.7)$$

The mutual information due to the indistinguishability of the particles is

$$I(1,2,\ldots,N) = \ln N! \qquad (6.8)$$

We conclude that the *indistinguishability of the particles introduces a correlation between the particles*, which causes a *reduction of the SMI*. We have calculated the mutual information of indistinguishable particles, by calculating the change in the number of configurations of the system of distinguishable particles caused by "erasing the labels" on the particles. Alternatively one could also give a probabilistic interpretation of the mutual information, in terms of correlation between events (for details see chapter 4 of Ben-Naim (2008)). We shall not need this interpretation here. The reader should be convinced by checking a few examples that, whenever we "un-label" the particles, the number of configurations or arrangements are reduced, and as a result the value of the SMI of the system is also reduced.

6.4   The Momentum SMI

As we have explained in section 5.6, the distribution that maximizes H subject to the two conditions (5.38) and (5.39)

$$\int_{-\infty}^{\infty} f(x)\,dx = 1 \qquad \text{and} \qquad E[X^2] = \int_{-\infty}^{\infty} x^2 f(x)\,dx = \sigma^2$$

is the normal distribution (5.40):

$$f^*(x) = \frac{1}{\sqrt{2\pi\sigma^2}} \exp\left(-\frac{x^2}{2\sigma^2}\right)$$

We now apply this distribution to the velocity of particles moving in one dimension, say v_x:

$$f^*(v_x) = \frac{1}{\sqrt{2\pi\sigma^2}} \exp\left(-\frac{v_x^2}{2\sigma^2}\right) \tag{6.9}$$

Before we proceed to calculate the velocity distribution in three dimension, we first express the variance σ^2 in terms of the absolute temperature T. The average kinetic energy of a particle moving in one dimension is

$$\frac{1}{2}m\left\langle v_x^2\right\rangle = \frac{m}{2}\sigma^2 \tag{6.10}$$

Thus, the average kinetic energy of a particle moving in three dimensions is given by

$$\begin{aligned}
\langle E_k\rangle &= \frac{m\left\langle v^2\right\rangle}{2} = \frac{m\left\langle v_x^2 + v_y^2 + v_z^2\right\rangle}{2} \\
&= \frac{m\left\langle v_x^2\right\rangle}{2} + \frac{m\left\langle v_y^2\right\rangle}{2} + \frac{m\left\langle v_z^2\right\rangle}{2} = \frac{3}{2}m\sigma^2
\end{aligned} \tag{6.11}$$

where v is the magnitude of the 3-dimensional velocity vector of the particles. By using the relation (2.10) between the absolute temperature and the average kinetic energy, $\langle E_k\rangle = \frac{3}{2}k_B T$, we find

$$\sigma^2 = \frac{k_B T}{m} \tag{6.12}$$

As the variance σ^2 is a measure of the dispersion of the distribution, we see that the larger the temperature, the larger is the *spread* of the distribution of the velocities. In other words, the absolute temperature T is also a measure of the dispersion of the velocity distribution.

Normal distribution $f^*(v_x)$

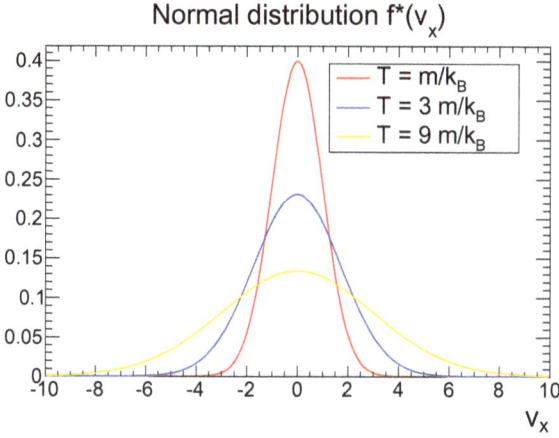

Figure 6.2: Normal distribution for one velocity component, with three different temperatures: $T = m/k_B$, $T = 3m/k_B$, and $T = 9m/k_B$.

Inserting eq. (6.12) into the distribution of velocities (6.9) we get

$$f^*(v_x) = \sqrt{\frac{m}{2\pi k_B T}} \exp\left(-\frac{mv_x^2}{2k_B T}\right) \tag{6.13}$$

Figure 6.2 shows the distribution $f^*(v_x)$ for three values of T.

In terms of the temperature, we can rewrite the SMI (5.42) associated with the distribution $f^*(v_x)$ as

$$H_{\max}(v_x) = \frac{1}{2}\ln(2\pi e k_B T/m) \tag{6.14}$$

As we have done in section 6.2, we now assume that the velocities v_x, v_y, v_z along the three axes are independent. Therefore, the SMI for a single particle moving with velocity vector (v_x, v_y, v_z) is

$$H_{\max}(v_x, v_y, v_z) = H_{\max}(v_x) + H_{\max}(v_y) + H_{\max}(v_z)$$
$$= \frac{3}{2}\ln(2\pi e k_B T/m) \tag{6.15}$$

For the purpose of constructing the entropy of ideal gas, we shall need the distribution of the momenta, which carries essentially the same information as the velocity distribution. As $p_x = mv_x$, we obtain the momentum distribution along the x-axis from (6.13), after the variable transformation $v_x \mapsto p_x/m$.

When transforming variables in continuous probability density functions, one must remember that also the differential undergoes a transformation. In our case $dv_x = dp_x/m$, which ensures that the probability is conserved: $\int_{-\infty}^{\infty} f(v_x) dv_x = \int_{-\infty}^{\infty} f(p_x) dp_x$. Hence we obtain:

$$f^*(p_x) = \frac{1}{\sqrt{2\pi m k_B T}} \exp\left(-\frac{p_x^2}{2m k_B T}\right) \tag{6.16}$$

The corresponding SMI in one dimension is

$$H_{\max}(p_x) = \frac{1}{2} \ln(2\pi e m k_B T) \tag{6.17}$$

and in three dimensions we have

$$H_{\max}(p_x, p_y, p_z) = \frac{3}{2} \ln(2\pi e m k_B T) \tag{6.18}$$

For a system of N independent particles the SMI is simply the sum of the SMI of each particles, hence we have

$$H_{\max}(\boldsymbol{p}_1, \ldots, \boldsymbol{p}_N) = \frac{3N}{2} \ln(2\pi e m k_B T) \tag{6.19}$$

This is the SMI associated with the momentum of the particles. Together with the locational SMI (6.4), this will be used in section 6.6 to construct the entropy of an ideal gas.

6.4.1 *The Maxwell-Boltzmann distribution*

We compute here the distribution of the magnitude $v = \sqrt{v_x^2 + v_y^2 + v_z^2}$ of the 3-dimensional velocity vector. Sometimes v is referred to as the **speed** of the particle. Since the motions along the three axes are independent we can write

$$\begin{aligned}
f^*(v_x, v_y, v_z) &= f^*(v_x) f^*(v_y) f^*(v_z) \\
&= \left(\frac{m}{2\pi k_B T}\right)^{3/2} \exp\left[-\frac{m(v_x^2 + v_y^2 + v_z^2)}{2k_B T}\right]
\end{aligned} \tag{6.20}$$

Again, the requirement that the probability is conserved implies that the volume element $dv_x\, dv_y\, dv_z$ must also be transformed. This is easier in spherical coordinates, because $dv_x\, dv_y\, dv_z = v^2 \sin\theta\, dv\, d\theta\, d\phi$ and by spherical symmetry the integration on the polar and azimuthal angles is trivial, giving just a multiplicative constant factor 4π.

The normalization condition reads then

$$
1 = \iiint_{-\infty}^{\infty} \left(\frac{m}{2\pi k_B T} \right)^{3/2} \exp\left[-\frac{m(v_x^2 + v_y^2 + v_z^2)}{2k_B T} \right] dv_x \, dv_y \, dv_z
$$

$$
= \int_0^{\infty} \left(\frac{m}{2\pi k_B T} \right)^{3/2} \exp\left(-\frac{mv^2}{2k_B T} \right) 4\pi v^2 \, dv
$$

from which we finally obtain the PDF of the velocity magnitude:

$$
f^*(v) = \sqrt{\frac{2}{\pi}} \left(\frac{m}{k_B T} \right)^{3/2} v^2 \exp\left(-\frac{mv^2}{2k_B T} \right) \tag{6.21}
$$

known as the ***Maxwell-Boltzmann distribution***. Mean $\langle v \rangle$, ***mode*** v_{mpv} (i.e. most probable value), and variance $V[v]$ are

$$
\langle v \rangle = \sqrt{\frac{8k_B T}{\pi m}}, \quad v_{mpv} = \sqrt{\frac{2k_B T}{m}}, \quad V[v] = \frac{(3\pi - 8)}{\pi} \frac{k_B T}{m} \tag{6.22}
$$

Note that the variance is different from the square of the RMS velocity, defined as

$$
v_{rms} \equiv \sqrt{\langle v^2 \rangle} = \sqrt{\frac{3k_B T}{m}} \tag{6.23}
$$

because the distribution (6.22) is asymmetric, with non-zero mean. Thus, we have (figure 6.3)

$$
v_{mpv} < \langle v \rangle < v_{rms} \tag{6.24}
$$

Recall that $f^*(v)dv$ is the probability of finding a particle with a velocity magnitude between v and $v + dv$. Note carefully the difference between the distribution (6.13) of the velocity *components*, which are defined on the whole real line, and the distribution (6.21) of the velocity *magnitude* (or speed), which is a non-negative quantity. A velocity component (e.g. v_x) can be either positive or negative, and is normally distributed with center at 0, which is both the average and the most probable value. It is a symmetric distribution, i.e. there is the same probability of moving with a velocity $v_x > v_0$ or $v_x < -v_0$. On the other hand, the speed distribution (6.21) is not symmetric, thus the average and the most probable values are not identical (figure 6.3).

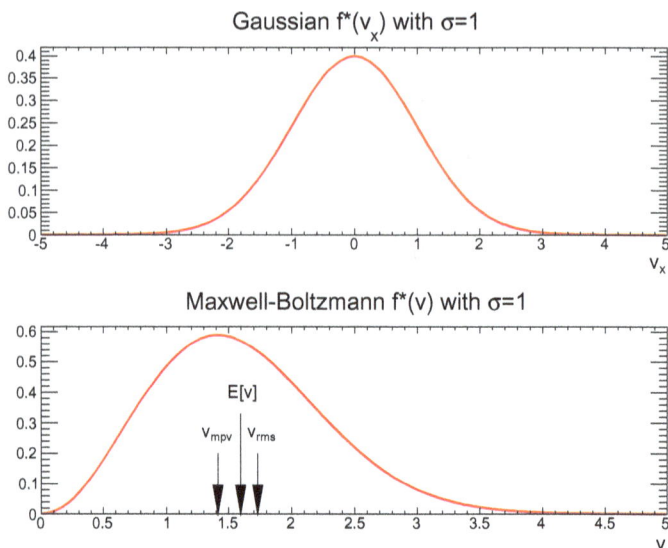

Figure 6.3: Normal distribution for one velocity component, and Maxwell-Boltzmann distribution for the velocity magnitude.

6.5 The Correction Due to the Uncertainty Principle

In section 5.1 we saw that, from the continuous SMI as defined in (4.9), the maximum SMI for the location of a single particle confined within the 1-dimensional "box" of length L is

$$H_{\max}(x) = \ln L \tag{6.25}$$

which is the same as eq. (5.5). We also noted that, if we are only interested to know the position with a precision h (which is equivalent to say that we only need to know which cell of length $h = L/n$ contains the particle), instead of eq. (6.25) we need to use its discrete analog, eq. (5.12), obtained from $H_{\max}(x)$ by subtracting $\ln h$:

$$H_{\max}(x) = \ln L - \ln h = \ln n \tag{6.26}$$

In practice, what we have done here is subtracting two quantities that are calculated from the continuous analog of the SMI, to obtain a discrete SMI associated with the finite number of cells. Since we always have a limit on the accuracy of determination of the location of the particle, eq. (6.26) is the only one meaningful quantity.

We will do the same thing when combining the locational and the momentum SMI, calculated in section 6.2 and 6.4, respectively.

Classical thinking would lead us to conclude that the SMI associated with *both* the location *and* momentum of a particle is the sum of the two SMIs. However, quantum mechanics dictates that the measurements of the location and of the momentum of the same particle are *not independent*, as the results depend on the order in which they are performed. Furthermore, the **Heisenberg uncertainty principle** sets a lower limit to the accuracy with which we can measure simultaneously position and momentum of the same particle. This limit is of order of the **Planck constant** $h = 6.626 \times 10^{-34}$ J s. The value of h is so small that from our macroscopic perspective this limitation seems completely irrelevant. However, it makes a big difference when looking at the microscopic scales.

This is the same situation we encountered when passing from the continuous segment $(0, L)$ to the discrete number of cells. Here, we also divide the entire space of locations and momenta (i.e. the so-called **phasespace**) into cells of size h, where now h is the Planck constant.

In terms of SMI, the uncertainty principle implies a correction $-\ln h$ to the classical picture, which one is tempted to interpret as the mutual information connected to the correlation between the accuracies of location and velocity (or momentum) for the same particle. However such correction actually comes from the discretization of the phase-space.[1]

For a particle in motion in a 1-dimensional space, the phase-space is discretized in cells of size $h_x h_p = h$ accordingly to the uncertainty principle. The locational SMI corrected for the discretization is given by eq. (6.26), which can be rewritten as $H_{\max}(x) = \ln L - \ln h_x$. A similar correction shall be also applied to the momentum SMI given by eq. (6.17), which becomes $H_{\max}(p_x) = \ln(2\pi e m k_B T)^{1/2} - \ln h_p$. Hence the sum of locational and momentum SMI is[2]

$$H_{\max}(x, p_x) = H_{\max}^{\text{cl}}(x) + H_{\max}^{\text{cl}}(p_x) - \ln h$$

$$= \ln \left[L \frac{(2\pi e m k_B T)^{1/2}}{h} \right] \qquad (6.27)$$

---

[1] Rigorously speaking, the correction $-\ln h$ is *not* equal to the average logarithm of the correlation function between the variables x and p_x. It derives from their measurement accuracy. In the original derivation by Ben-Naim (2008) the interpretation of this term is erroneous.

[2] Note that the argument of the logarithm function in (6.27) is a dimensionless number.

where $H_{\max}^{\text{cl}}(x)$ is the locational SMI in the classical description, given by eq. (6.25), and $H_{\max}^{\text{cl}}(p_x)$ is the classical momentum SMI is given by (6.17).

Equation (6.27) gives the SMI for one particle in one dimension. For the three dimensional case we repeat the same procedure, i.e. we subtract $\ln h$ for each degree of freedom, obtaining

$$H_{\max}(x,y,z,p_x,p_y,p_z) = H_{\max}^{\text{cl}}(x,y,z) + H_{\max}^{\text{cl}}(p_x,p_y,p_z) - 3\ln h$$

$$= \ln\left[L^3 \frac{(2\pi e m k_B T)^{3/2}}{h^3} \right]$$

Here $H_{\max}^{\text{cl}}(x,y,z) = 3\ln L$ and $H_{\max}^{\text{cl}}(p_x,p_y,p_z)$ is the classical momentum SMI in 3D space given by eq. (6.18).

Finally, putting together the contributions to the SMI from the location and momentum uncertainties, plus the mutual information related to the indistinguishability and the uncertainty princtiple, for N indistinguishable and non-interacting particles, we have

$$\boxed{H_{\max}^{\text{ID}}(N) = H_{\max}^{\text{D}}(\boldsymbol{R}^N) + H_{\max}^{\text{D}}(\boldsymbol{p}^N) - \ln N! - 3N\ln h} \qquad (6.28)$$

This is an important result. To obtain the SMI of N particles, described by their locations and momenta, we first treat the particles as being distinguishable and classical. In this case we can sum the SMI associated with all the location ($\boldsymbol{R}^N$) and all the momenta ($\boldsymbol{p}^N$) of the particles. Then we include two corrections due to the fact that the particles are not classical. One is the mutual information related to the indistinguishability of the particles, and the other comes from the discretization of the phase-space due to the Heisenberg uncertainty principle.

6.6   The Entropy of a Classical Ideal Gas

In the previous section, we have calculated the maximal value of the SMI of a system of N non-interacting and indistinguishable particles. Recall that the SMI may be defined for any distribution. It can be defined for any distribution of locations and any distribution of momenta. It can be defined for any number of particles and can be defined for distinguishable or indistinguishable particles.

In this section, as well as in the rest of the book, we shall be interested in very special distributions. These are the distributions of locations and momenta that maximize the corresponding SMI. We denote these special distributions by f^* or p^*, and the corresponding SMI by $H_{\max}$. However, we also know that, starting with any arbitrary distribution of locations and momenta, the system will tend to a limiting equilibrium distribution: the uniform distribution for locations, and the normal distribution for the momenta along each direction.

Up to this point you could rightfully regard the SMI as a quantity that measures the size of some 20Q game (where is the particle? what is its velocity?). In this section we make a huge conceptual leap, from the SMI of games to a fundamental concept of thermodynamics. As we shall soon see, this leap is rendered possible by recognizing that *the SMI associated with the equilibrium distribution of locations and momenta of a large number of indistinguishable particles is identical to the statistical mechanical entropy of an ideal gas.* Since the statistical mechanical entropy of an ideal gas has the same properties of the thermodynamic entropy as defined by Clausius, we can then declare that *this special SMI is identical to the entropy of an ideal gas.*

We recall that the SMI of a system of N particles at equilibrium (6.28) is the sum of two contributions due to location (6.4) and momentum (6.19), and two corrections due to the indistinguishability of the particles (6.8) and the uncertainty principle ($3N \ln h$). Thus, we have the following expression for the SMI of N non-interacting particles at equilibrium:

$$
\begin{aligned}
H_{\max}^{\mathrm{ID}}(N) &= N \ln V + \frac{3N}{2} \ln(2\pi e m k_{\mathrm{B}} T) - 3N \ln h - \ln N! \\
&= N \ln \left[\frac{V}{N} \left(\frac{2\pi m k_{\mathrm{B}} T}{h^2} \right)^{3/2} \right] + \frac{5}{2} N
\end{aligned}
\tag{6.29}
$$

In order to obtain the expression for the entropy of an *ideal gas* in the usual thermodynamic units, all we have to do is to multiply $H_{\max}^{\mathrm{ID}}(N)$ in (6.29) by the Boltzmann's constant k_{B}:

$$
\boxed{S \equiv k_{\mathrm{B}} H_{\max}^{\mathrm{ID}}(N) = k_{\mathrm{B}} N \left\{ \ln \left[\frac{V}{N} \left(\frac{2\pi m k_{\mathrm{B}} T}{h^2} \right)^{3/2} \right] + \frac{5}{2} \right\}}
\tag{6.30}
$$

This expression is identical with the equation obtained by Sackur and Tetrode in 1912, based on Boltzmann's definition of entropy. They first computed the number of states W, then applied Boltzmann's definition of entropy $S = k_{\mathrm{B}} \ln W$.

On the other hand, we did not compute W nor used Boltzmann's entropy. Instead, we have derived equation (6.30) from considerations based on the SMI corresponding to the locations and the momenta of the particles, with corrections due to the uncertainty principle and to the indistinguishability of the particles.

The multiplication by a constant k_B, as well as the usage of natural logarithms, determines the units in which we measure entropy. This does not affect the meaning of the entropy of an ideal monoatomic gas as *defined* in (6.30), i.e. as the SMI for a specific system of N non-interacting particles at equilibrium. In thermodynamics, both N and V are very large, and one could even take the limit in which both tend to infinity while keeping the density N/V constant. However, this does not change the meaning of (6.30) as a particular form of SMI.

We have assumed throughout that the particles do not interact with each other (ideal gas). Normally, the absence of interactions is considered to be equivalent to independence. However, as we have seen in this chapter, *dependence* can occur even between non-interacting particles. In particular, we have seen that un-labeling particles introduce some correlation. Another example is the correlation induced by Heisenberg's uncertainty principle. When the particles interact with each other, a new kind of dependence is introduced. This dependence brings in one additional correlation between the particles, which causes another correction in the SMI. This means that interactions can be included in our approach quite naturally, by adding the corresponding mutual information to (6.30). More information can be obtained in Ben-Naim (2008).

From now on *we shall define the entropy of an ideal gas by equation (6.30)*.

The *meaning* of the entropy is derived from the four terms listed in equation (6.29): we have the SMI associated with the locations of the particles, the SMI associated with the momenta of the particles, and the two corrections due to the uncertainty principle and the indistinguishability of the particles.

6.7   The *Disorder* and the *Spreading* Metaphors of the Entropy

The concept of entropy, along with the formulation of the second law of thermodynamics, was originally introduced by Clausius without any reference to atoms and molecules. As such, the concept of entropy did not reveal its meaning on a molecular level. The ever increasing entropy was accepted as an experimental fact: no molecular reason for this behavior was offered. It is not surprising that the entropy and the second law were enshrouded with a thick envelope of mystery.

In this chapter we have developed the concept of entropy of an ideal gas, based on the simple, precise and well defined concept of SMI. We have also seen that the mystery associated with the second law dispels once we recognize that the system proceeds from a low probability to a high probability state.

The probabilistic interpretation of the second law has its roots in Boltzmann's writing. Boltzmann also used the metaphor of "disorder" to describe entropy. A spontaneous process is governed by probability: the system proceeds from an *ordered*, hence less probable, state to a *disordered* and more probable state.

The metaphor of disorder for the entropy became very popular. Indeed, in some cases the description of entropy as a measure of disorder is quite appealing. We perceive the uniform distribution over locations as a state of disorder, compared with the states where all the particles are concentrated inside a small region of the system. It is however not easy to quantify the disorder, and there are many examples where the changes in entropy do *not* conform with our intuitive perception of changes in the extent of order or disorder, as ironically shown in figures 6.4 and 6.5, taken from Ben-Naim (2008).

We have also seen that the Boltzmann distribution and the normal distribution are a result of maximizing the SMI under certain constraints. It is difficult to describe these distributions in terms of the maximum disorder. Another example in which the disorder metaphor can be misleading is discussed in section 7.4.

Figure 6.4: The metaphor of disorder is not well suited to describe the entropy.

Another analogy that was proposed more recently is the "spreading" metaphor. This metaphor may sound reasonable when interpreting the attainment of the uniform distribution as a maximal spreading of particles in space. The spreading metaphor can also be applied to the variation of the speed distribution with temperature: the higher the temperature, the flatter the speed distribution. This can be perceived as a spreading of kinetic energy over a larger range of speeds. However, it is less obvious how to interpret a process of heat transfer from a hot to a cold body in terms of either spreading or disorder.

There are other limitations to these metaphors when used to describe entropy. We shall not discuss these here. For the specific system of an ideal gas considered in this chapter, we have clarified that *the entropy is a particular case of an SMI*.

Figure 6.5: The metaphor of disorder is not well suited to describe the entropy.

In this book we make a step further in this direction: *we shall assume that the entropy of any system*, not necessarily an ideal gas, *is also a SMI applied to that system*. Although we cannot prove this for any general system, we accept this as a postulate. We draw further confidence in this belief from the fact that any kind of intermolecular interactions can also be cast in the form of mutual information (Ben-Naim, 2008).

Throughout the rest of the book we shall use standard thermodynamic relationships between thermodynamic quantities. We shall not need to interpret the entropy whenever it appears in our discussion. Thermodynamics is not concerned with the value of the entropy, nor with its interpretation: it only deals with entropy changes. However, we believe that it is easier and more satisfying to have in the back of our minds a meaning for entropy, a meaning that we have derived only for ideal gases, although we accept to extend it also to any other thermodynamic system.

6.8   Fundamental Properties of the Entropy Function $S(E,V,N)$

This section could have been entitled "The properties of the *fundamental* function $S(E,V,N)$" instead. Indeed, both the entropy function $S(E,V,N)$ and its properties are fundamental in the sense that is clarified in the following.

In section 6.6 we derived the function $H_{\max}^{\mathrm{ID}}(N)$ which, when applied to a system of N non-interaction particles in a volume V and at temperature T, was identified with the *entropy* of the system. We can rewrite here the Sackur-Tetrode equation (6.30) as

$$S(T,V,N) = N k_{\mathrm{B}} \ln \left[\frac{V}{N} \left(\frac{2\pi m k_{\mathrm{B}} T}{h^2} \right)^{3/2} \right] + \frac{5}{2} N k_{\mathrm{B}} \tag{6.31}$$

For reasons discussed below, the fundamental function for the entropy is not $S(T,V,N)$, but the function $S(E,V,N)$, where E is the total energy of the particles.

For an ideal monoatomic gas, the total energy E of the system is simply the total kinetic energy E_{k} of the particles, obtained by inverting the relation (2.10) between the absolute temperature and the average kinetic energy of a single atom, and multiplying by the number N of atoms:

$$E_{\mathrm{k}} = \frac{3}{2} N k_{\mathrm{B}} T \tag{6.32}$$

One gets the same result by inserting the RMS velocity v_{rms} from eq. (6.23) into $E_{\mathrm{k}} = N m v_{\mathrm{rms}}^2 / 2$.

Writing $T = (2E)/(3N k_{\mathrm{B}})$ from (6.32) and substituting in (6.31) we obtain the **fundamental entropy function**

$$\boxed{S(E,V,N) = N k_{\mathrm{B}} \ln \left[\frac{V}{N} \left(\frac{E}{N} \right)^{3/2} \left(\frac{4\pi m}{3h^2} \right)^{3/2} \right] + \frac{5}{2} N k_{\mathrm{B}}} \tag{6.33}$$

We can rewrite (6.33) in a form which highlights the dependence on thermodynamic quantities and on the atomic properties:

$$S(E,V,N) = N k_{\mathrm{B}} \ln \left[\frac{V}{N} \left(\frac{E}{N} \right)^{3/2} \right] + \frac{3}{2} N k_{\mathrm{B}} \left[\ln \left(\frac{4\pi m}{3h^2} \right) + \frac{5}{3} \right] \tag{6.34}$$

Note that the argument of the logarithm in (6.33) is a pure number. However in (6.34) we have assumed that all symbols are dimensionless, as explained in section 6.1: the entropy is written in the form of a sum of a function of the number density N/V and of the energy per particle E/N, with a constant term which only depends on the mass m of the particles. This form better illustrates the dependence of the entropy on the thermodynamic variables, at the price of some abuse of notation.

The function (6.33) is fundamental for two reasons. First, one can derive all thermodynamic quantities of an ideal gas from this function. Second, the second law of thermodynamics formulated in terms of the entropy is valid *only* for this function, i.e. when S is expressed as a function of the variables E, V, N as in (6.33), and not in terms of any other set of variables, e.g. T, V, N as in (6.31).

Furthermore we can say that, *for any thermodynamic system*, the following two statements are valid and have fundamental importance:

(1) *For any thermodynamic system of N particles enclosed into a volume V and having total energy E, the entropy function $S(E, V, N)$ provides all the thermodynamic quantities of that system.*

(2) *For any thermodynamic system of N particles in a volume V with total energy E, the entropy function $S(E, V, N)$ has a maximum over all possible constrained equilibrium states.*

We will spend the rest of the section in clarifying these statements.

Definition of the system.   The volume is defined by the boundaries of the specific thermodynamic system under study. We always assume that the system is macroscopic, i.e. that its dimensions are very large compared to the molecular dimensions. We also assume that surface effects are negligible (actually, surface effects can be taken into consideration separately, but they will not be of any concern to us in this book). In addition, we assume that there are no external fields operating on the system, such that the force acting on any particle is the resulting vector sum of all mutual interactions (also called ***internal forces***). Strictly speaking, this is impossible to have in reality. For example, some gravitational field is always present. So, what we really assume is that gravitational effects can be neglected, compared to the strength of internal interactions.

Nature of the particles. If there are c components (in other words, c diffrent types of particles), we have simply to replace N with the vector $(N_1, \ldots, N_c)$ where N_i is the number of molecules (or moles) of species i (section 7.1). The particles in the system can actually be atoms or molecules, and they can have different contributions to the total energy, like the translational, vibrational, and rotational energies, etc. In the classical idealization of point-like masses, the only contribution to the total energy comes from the kinetic (i.e. translational) energy. However, for molecules one has also vibrational contributions, due to the oscillations of each atom about its average position, and rotational energy corresponding to the molecule rotations with respect to the axes of the non-rotating reference frame with origin in the center of momentum[3] of the molecule. When quantum effects are also taken into account, other types of internal energy contributions can also arise.

Energy of the system. The total energy E of the system includes all the internal energies of the atoms and molecules, as well as the potential energies of the (internal) interactions. We shall see in section 8.1 that E, referred to as the *internal energy* of the system, is introduced by the first law of thermodynamics. Actually, the internal energy of the system is always defined with respect to some arbitrary chosen zero: the first law is essentially a statement on the *changes* in the internal energy caused by the exchange of either heat or work with it surroundings.

Which function reaches a maximum? In some formulations of the second law of thermodynamics it is stated that the entropy of a system tends to a maximum at equilibrium. Such a statement is faulty in two respects: first, it does not specify which *function* has a maximum; second, it does not specify with respect to which *variable* the entropy has a maximum. Without such specifications, the generic statement that "entropy always reaches a maximum" is not valid.

In thermodynamics we assume that we have the liberty of choosing at will the independent variables. For instance, for a one-component system

---

[3]We recall that the center of momentum is defined as the position with respect to which the vector sum of all momenta is zero.

we may choose to describe the independent variables E, V, N or T, V, N or T, P, N, etc. For each of these independent variables, we can write the entropy function: $S(E, V, N)$, $S(T, V, N)$, $S(T, P, N)$, etc.

Clearly, there are many possible choices for the set of independent variables, and as many corresponding entropy functions, but *the second law of thermodynamics applies only to one entropy function: $S(E, V, N)$*. That is the reason why this particular function is referred to as the fundamental entropy function. *It is the fundamental entropy function $S(E, V, N)$ that reaches a maximum at equilibrium.*

Note that we are only considering isolated systems here. For other systems the second law is formulated in terms of the Helmoltz energy A, reaching a minimum for systems in equilibrium at constant T, V, N (heat flow is allowed), and the Gibbs energy G, reaching a minimum for systems in equilibrium at constant T, P, N (volume changes and chemical reactions are also allowed). These state functions will be defined in chapter 8.

Definition of a maximum.   In calculus, when we say that a function $y = f(x)$ has a single maximum, we mean that there is a value of x such that the value of y is maximal compared to all the values of y obtained for any other value of x. A function can have more than one maximum, in which case one needs to define each maximum locally. The mathematical requirements for a maximum of the function $y = f(x)$ at the point x^* are that the first derivative is null and the second derivative is negative:

$$\frac{df(x)}{dx}\bigg|_{x=x^*} = 0 \quad \text{and} \quad \frac{d^2 f(x)}{dx^2}\bigg|_{x=x^*} < 0 \tag{6.35}$$

These equations mean that the *slope* of the function at $x = x^*$ is zero, and that the *curvature* is negative (i.e. that the concavity is downward).

In the n-dimensional case, we say that there exists a maximum of the function $y = f(x_1, \ldots, x_n)$ at some point $(x_1^*, \ldots, x_n^*)$, if $f(x_1^*, \ldots, x_n^*)$ is the largest value of y when $x_1, \ldots, x_n$ are varied in the neighborhood of $(x_1^*, \ldots, x_n^*)$. The analog of the conditions (6.35) are that all first partial derivatives are zero, i.e. that the **gradient** ∇f is the null vector

$$\nabla f \equiv \left(\frac{\partial f}{\partial x_1}, \ldots, \frac{\partial f}{\partial x_n} \right) = (0, \ldots, 0) \tag{6.36}$$

and that the determinant of the **Hessian matrix**

$$\mathcal{H}(f) \equiv \begin{bmatrix} \dfrac{\partial^2 f}{\partial x_1^2} & \dfrac{\partial^2 f}{\partial x_1 \partial x_2} & \cdots & \dfrac{\partial^2 f}{\partial x_1 \partial x_n} \\[2ex] \dfrac{\partial^2 f}{\partial x_2 \partial x_1} & \dfrac{\partial^2 f}{\partial x_2^2} & \cdots & \dfrac{\partial^2 f}{\partial x_2 \partial x_n} \\[2ex] \vdots & \vdots & \ddots & \vdots \\[2ex] \dfrac{\partial^2 f}{\partial x_n \partial x_1} & \dfrac{\partial^2 f}{\partial x_n \partial x_2} & \cdots & \dfrac{\partial^2 f}{\partial x_n^2} \end{bmatrix} \tag{6.37}$$

is negative, which means that the concavity of f is downward.

Maximum with respect to which? In the mathematical examples given above, when we have written a function $f(x)$, we knew that $f(x)$ had a maximum with respect to the variable x. In the case of $f(x,y,z)$ the maximum is with respect to the variation of the independent variables x, y and z. In general, it is understood implicitly that the function has a maximum with respect to the *arguments* of the function, i.e. $x_1, \ldots, x_n$ in $f(x_1, \ldots, x_n)$. In thermodynamics we must first choose the independent variables to describe the system, say E, V, N or T, V, N. Second, unlike in the mathematical examples, *the entropy function does not have a maximum with respect to the independent variables E, V, N but, on the contrary, these variables must be kept constant.* What we vary to obtain the maximum of $S(E, V, N)$ is some *internal distribution* keeping E, V, N constant.

In chapter 4 we have seen that the SMI is maximal with respect to all possible *distributions* of the locations and momenta of the particles. The same is true for the function $S(E, V, N; \text{distribution})$. Here, we added the new argument "distribution" to the function $S(E, V, N)$, in order to make it explicit that there is some additional degree of freedom beyond the three thermodynamic variables.

We can formulate the second law of thermodynamics as follows:

For a system at equilibrium with fixed values of E, V and N, the entropy function $S(E, V, N)$ attains its maximal value over all possible distributions.

These are the distributions that can be achieved as constrained equilibrium states. A **constrained equilibrium state** defines a distribution of locations and momenta compatible with the constraint. Not all possible distributions correspond to a constrained equilibrium state.

Thus, another possible formulation is to say that *entropy is maximal over all possible constrained equilibrium states.* This will be clarified with several examples in chapter 8.

Note that this law is valid for any system, for an ideal gas as well as for a non-ideal gas. It is valid whether or not we know the explicit function $S(E,V,N)$. The important thing is that the variables E,V,N are *kept constant.* Such a system is referred to as an **isolated system**, i.e. a system that does not interact with its surroundings: it does not exchange energy, volume of matter with its surroundings.

We stress again that the law of maximal entropy is valid only when the entropy is viewed as a function of E,V,N, and it has a maximum with respect to all possible distributions when E,V,N are kept constant. This is not true if we express the entropy in the form $S(T,V,N;$distribution$)$ or $S(T,P,N;$distribution$)$ etc. In chapter 8, we shall discuss the Helmoltz energy $A(T,V,N;$distribution$)$ and the Gibbs energy $G(T,P,N;$distribution$)$, which reach a minimum at equilibrium when T,V,N and T,P,N are kept constant, respectively. However, before describing such systems we shall explore the fundamental properties of the function $S(E,V,N)$. We shall also point out which properties are specific to ideal gases and which have a general validity and pertains to any thermodynamic system.

6.9 Summary of What We have Learned so far

This chapter introduced the concept of entropy and the second law for an ideal gas, with a rather non-conventional approach. We have seen that the SMI (6.28) of N indistinguishable and non-interacting particles can be calculated by summing the SMI related to their locational distribution and to their momentum distribution, and accounting for the mutual information arising from the correlation due to their indistinguishability, plus a correction arising from the Heisenberg uncertainty principle. Apart from a multiplicative constant, the result is identical to the entropy function $S(T,V,N)$ for an ideal gas (6.30) obtained by Sackur and Tetrode in 1912 based on Boltzmann's definition of entropy.

This apparently surprising result is the bridge between information theory and thermodynamics, showing that we can identify the SMI of an ideal gas at equilibrium with its entropy. The next step is to assume that the entropy of any system at equilibrium, not necessarily an ideal gas, is also a SMI. This way, the second law of thermodynamics becomes a consequence of the properties of SMI, which evolves toward a maximum, achieved for the macroscopic configuration with the largest probability. For a system composed of a very large number of particles, any sizable departures from this equilibrium SMI are so rare that the life of the universe is too short for observing them. This makes the second law of thermodynamics "absolute" in a very precise sense, and dissolves any sense of mystery related to the entropy and the second law.

This approach is very recent: the first derivation of the entropy function for an ideal gas based on Shannon's measure of information was obtained by Ben-Naim (2006a), and the extension to interacting particles was provided by Ben-Naim (2008). Hence it is important to review the conceptual steps that led us to define entropy, from the very beginning.

First, we started with the general concept of information. We do not know how to define information, but we intuitively know what information is. Shannon sought out a measure of information, and he found one; not for any information, but for certain kinds of information, those contained in well defined probability distributions. We have referred to this measure as Shannon's Measure of Information (SMI).

Clearly, the kinds of information on which we can define the SMI form only a subset of a vast, perhaps infinite variety of information. However, it is by no means clear whether or not a "general information" can be measured, and certainly not clear if the SMI can be applied to such general information. Thus, at this stage we have "distilled" a concept of information on which we could apply the SMI.

We show that schematically in figure 6.6. The outer region denoted by "General information" is not a well defined region at all. It encloses a smaller set, called "SMI", which contains the types of information on which the SMI can be defined. All we want to convey in this figure that the SMI is defined on a (very small) subset of information. It should be noted that the subset denoted by "SMI" is not the SMI itself, but all the possible *distributions* on which the SMI can be defined and applied.

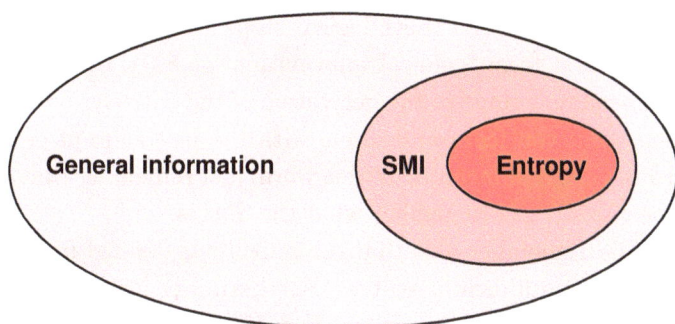

Figure 6.6: Information, SMI and entropy.

Although this subset is very small compared with all possible forms of information, it is still very large: it contains infinite possible distributions. For example, the distributions pertaining to tossing a coin, throwing a die, measuring the volume of an expanding gas, or the frequency of occurrence of a letter in the alphabet of a specific language.

Within this large subset on which we can define the SMI, we identify an even smaller subset denoted by "Entropy" in figure 6.6. This subset contains *all the distributions relevant to a thermodynamic system at equilibrium*. Clearly, these are only a tiny set of distributions, compared to all possible distributions on which the SMI is definable.

The SMI defined on the distributions pertaining to a system at equilibrium is, up to an additive constant and with a choice of special units, identical to the thermodynamic entropy introduced by Clausius.

Conceptually, the SMI and the entropy are completely different things. However, both are measures of the information contained in a distribution. In this sense, we can safely say that entropy is a particular case of SMI. On the other hand, one cannot refer to any SMI as entropy, although this terminology is unfortunately very common since it was also adopted by Shannon himself.

It is impressing and perhaps even astonishing that the entropy, a concept that was spawned out of considerations of heat engines, turned out to be nothing but a certain measure of information. This link between two apparently very different concepts is not less amazing than the connection we found between the temperature of a gas and the kinetic energy connected with the motion of its particles.

Once you have grasped these logical steps and the sequence of concepts leading first from "general information" to SMI, then to entropy, you do not need to memorize the derivation of the entropy.

We hope that you feel comfortable with the new concept of entropy. Disregard the original meaning of this word, just remember that entropy is one type of SMI, and remember what the SMI is.

You should remember also that the entropy is defined in thermodynamics only for equilibrium states. Furthermore, for thermodynamic systems at fixed values of the energy E, volume V and number of particles N (i.e. for isolated systems), the entropy attains a maximum value over all possible *distributions*, which can be realized as constrained equilibrium states.

Chapter 7

Thermodynamics of Ideal Gas

In this chapter we explore the thermodynamics of ideal gases or, more specifically, ideal *monoatomic* gas. We shall assume that we have N non-interacting particles whose internal energy is the sum of the kinetic energies of the particles, which have no internal degree of freedom.

An ideal gas is the only system for which we have an explicit expression for the entropy. Strictly speaking, the results obtained in this chapter pertains to ideal gases only. However, some of them have general validity, as we shall point out when this is the case.

After studying the shape of the entropy function, we shall examine the entropy change for some fundamental spontaneous processes which are at the heart of the second law of thermodynamics.

We shall also discuss processes of mixing and demixing. The mixing process of ideal gases has been wrongly considered a typical irreversible process when the entropy was viewed as a measure of disorder. However, our informational interpretation of entropy allows us to view it in a new light, providing at once a way of understanding the phenomenon and a solution of the puzzles related with its misinterpretation.

Here we take the point of view of statistical mechanics, where one counts particles. On the other hand, in thermodynamics it is more common to count the moles. If $N = n N_{Av}$ is the number of particles and n is the number of moles, with N_{Av} being Avogadro's number, switching from one view to the other implies replacing the Boltzmann constant k_B with the gas constant $R = N_{Av} k_B$ in all formulae.

7.1   The Additivity of the Function $S(E,V,N)$

Let us rewrite the equation (6.33) for the entropy in the form

$$S(E,V,N) = Nk_B \ln \left[\frac{V}{N} \left(\frac{E}{N} \right)^{3/2} \right] + 3Nk_B \alpha \qquad (7.1)$$

where $\alpha = \frac{1}{2} \left[\ln \left(\frac{4\pi m}{3h^2} \right) + \frac{5}{3} \right]$ is a numerical constant independent of E,V,N, which assumes a different value for each gas, as it depends on the mass m of a gas particle. Note that all quantities in (7.1) are treated as pure numbers as explained in section 6.1.

First, note that $S(E,V,N)$ is a homogeneous function of degree 1: if we multiply each of the variables E,V,N by a positive constant $\lambda > 0$, we have

$$S(\lambda E, \lambda V, \lambda N) = \lambda Nk_B \ln \left[\frac{\lambda V}{\lambda N} \left(\frac{\lambda E}{\lambda N} \right)^{3/2} \right] + 3\lambda Nk_B \alpha$$

$$= \lambda S(E,V,N) \qquad (7.2)$$

When λ is integer, this is equivalent to saying that the entropy of a set of λ replicas of the same isolated system is λ times the entropy of a single system. *The entropy is additive for independent systems.* In other words, the entropy is an **extensive property** of the system. The total energy is another extensive property, for example.

On the other hand, temperature and pressure are **intensive properties**: they do not add up. Instead, when two isolated systems are put in contact, temperature and pressure of the final system assume some intermediate value, that can be expressed as a weighted average of the initial values.

As we noted in section 6.8, the expression for the entropy of an ideal gas can be extended to a system of c components. Let $N_1, N_2, \ldots, N_c$ be the composition of the system, where N_i is the number of molecules of the species i.[1] Then the entropy of the mixture of the c components in a

---

[1] If one has the number n of moles instead of the number N of particles, all multiplicative factors Nk_B in the entropy formulae become nR.

volume V and with total energy E is

$$S(E,V,N_1,\ldots,N_c) = \sum_{i=1}^{c} S(E,V,N_i)$$

$$= \sum_{i=1}^{c} N_i k_B \left\{ \ln \left[\frac{V}{N_i} \left(\frac{E}{N_i} \right)^{3/2} \right] + 3\alpha_i \right\} \tag{7.3}$$

This expression can be easily obtained by taking into account that

- the locational and the momentum contributions are clearly additive;
- the correction due to the uncertainty principle is also additive;
- the correction due to indistinguishability is $\ln N_i!$ for each species.

Recall that the numerical constant α_i depends on the species i only through the mass m_i.

Equation (7.3) means that the entropy of a *single* system characterized by $(E,V,N_1,\ldots,N_c)$ is equal to the sum of the entropies of c systems characterized by $(E,V,N_1),(E,V,N_2),\ldots,(E,V,N_c)$.

We shall use equation (7.3) in section 7.2.3, whereas in the next two sections we shall deal with one-component systems only.

7.2 The Shape of the Function $S(E,V,N)$

Within the limits of its validity, the function $S(E,V,N)$ is concave downwards with respect to each of the variables E,V,N. We shall now examine the variation of $S(E,V,N)$ with respect to each variable separately.

7.2.1 *Dependence on E, keeping V and N constant*

From (7.1) we get the partial derivative with respect to the energy E, keeping V and N constant:

$$\left(\frac{\partial S}{\partial E} \right)_{V,N} = \frac{3Nk_B}{2E} \tag{7.4}$$

For an ideal monoatomic gas equation (6.32) gives $E = \frac{3}{2}Nk_B T$ which, inserted into (7.4), gives

$$\boxed{\left(\frac{\partial S}{\partial E} \right)_{V,N} = \frac{1}{T} > 0} \tag{7.5}$$

This is an important result. Although we obtained it from the fundamental expression for the entropy of an ideal gas, the relationship (7.5) is actually valid for any thermodynamic system at equilibrium.

The slope of the function $S(E,V,N)$ with respect to E, keeping V and N constant, is the inverse of the absolute temperature. This also means that the rate of change of entropy with the internal energy (at fixed V,N) is always positive.

Because of its general validity, we can also conclude from (7.5) that at very low temperature (formally when $T \to 0$) the slope must tend to infinity, whereas at very high temperatures (when $T \to \infty$) the slope will tend to zero. This conclusion is valid for *any* system. However, we should carefully avoid to apply the specific function $S(E,V,N)$ in (7.1), which is valid for an ideal classical gas, at very low temperatures: no real gas has an ideal behavior at very low temperatures.

The second derivative of the function $S(E,V,N)$ with respect to the internal energy is

$$\left(\frac{\partial^2 S}{\partial E^2}\right)_{V,N} = \frac{-3Nk_{\mathrm{B}}}{2E^2} < 0 \tag{7.6}$$

Thus, the curvature of the function $S(E,V,N)$ with respect to E is negative.

For the curvature of S with respect to E, we can get a more general expression, valid for any system, by taking the derivative of (7.5) with respect to E:

$$\left(\frac{\partial^2 S}{\partial E^2}\right)_{V,N} = \frac{\partial}{\partial E}\left(\frac{1}{T}\right) = -\frac{1}{T^2}\left(\frac{\partial T}{\partial E}\right)_{V,N} \tag{7.7}$$

Defining the **heat capacity at constant volume** as

$$C_V \equiv \left(\frac{\partial E}{\partial T}\right)_{V,N} \tag{7.8}$$

we can rewrite the second derivative (7.7) as

$$\boxed{\left(\frac{\partial^2 S}{\partial E^2}\right)_{V,N} = -\frac{1}{T^2 C_V} < 0} \tag{7.9}$$

The heat capacity of the system at constant volume, C_V, is always positive.[2] It relates the change of temperature and total energy, and a

---

[2] In statistical mechanics, C_V is expressed as a function of the fluctuations in the energy of the system: $C_V = \dfrac{\left\langle (E - \langle E \rangle)^2 \right\rangle}{k_{\mathrm{B}} T^2} > 0$.

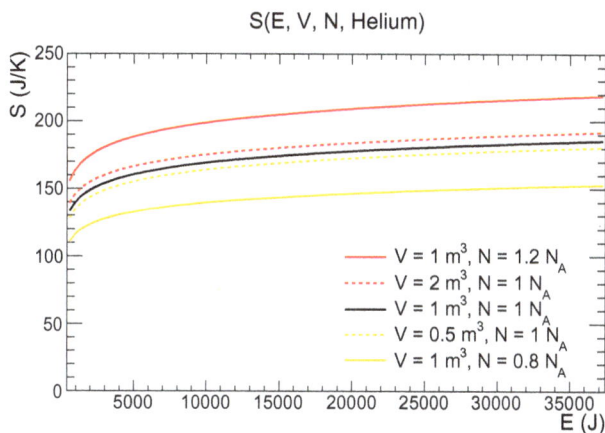

Figure 7.1: Entropy of Helium gas as a function of the energy, at constant volume and number of atoms. The energy corresponds to a temperature range of 30–3000 K.

positive change in T always correspond to an increase of E. Therefore, the curvature of the function $S(E,V,N)$ with respect to E is negative (for any system, not only for an ideal gas).

From equations (7.6) and (7.9) we get the expression for the heat capacity for an ideal gas at constant volume

$$C_V = \frac{3}{2}k_{\mathrm{B}}N \qquad \text{(ideal monoatomic gas)} \qquad (7.10)$$

Figure 7.1 shows the shape of the entropy as a function of the energy for different values of V and N. These curves were drawn using equation (7.1) for helium (He), a noble gas with mass 4.002602 amu, where the **atomic mass unit** is defined as the twelfth of the mass of one atom of carbon-12 (^{12}C):

$$1\,\mathrm{amu} \equiv \frac{m(^{12}\mathrm{C})}{12} = 1.660\,538\,921(73) \times 10^{-27}\,\mathrm{kg} \qquad (7.11)$$

(the digits in parentheses represent the uncertainty, which is less than 0.04 parts-per-million). The curves are all concave downwards. As we noted above, the concavity of the function is a property of the function $S(E,V,N)$ for any thermodynamic system. The entropy monotonically increases for larger values of the energy. It is also larger for bigger volumes and increasing number of particles, as we will see below.

7.2.2 Dependence on V, keeping E and N constant

The slope of the function $S(E,V,N)$ with respect to the volume V, at fixed values of E and N is

$$\left(\frac{\partial S}{\partial V}\right)_{E,N} = \frac{Nk_B}{V} > 0 \tag{7.12}$$

For an ideal gas, we can use the equation of state (2.8), $PV = Nk_BT$, to rewrite (7.12) as

$$\boxed{\left(\frac{\partial S}{\partial V}\right)_{E,N} = \frac{P}{T} > 0} \tag{7.13}$$

Again, we note that although (7.13) was obtained for an ideal gas, it is valid for any thermodynamic system at equilibrium. The slope of the entropy with respect to the volume at constant E and N is always positive.

The second derivative of $S(E,V,N)$ with respect to the volume V of an ideal gas is obtained from (7.12):

$$\left(\frac{\partial^2 S}{\partial V^2}\right)_{E,N} = -\frac{Nk_B}{V^2} < 0 \tag{7.14}$$

Thus, the curvature of the function $S(E,V,N)$ with respect to the volume is negative. The result (7.14) is valid only for an ideal gas. However,

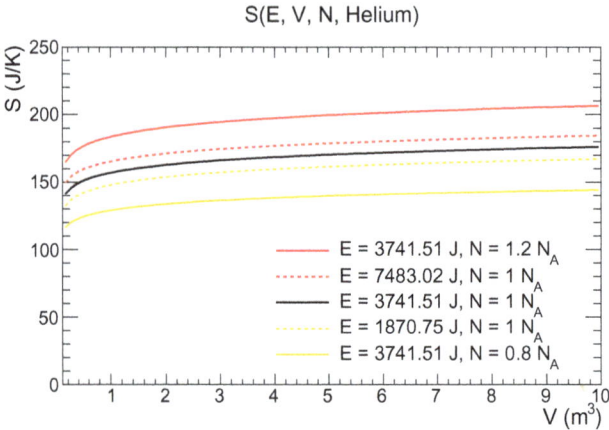

Figure 7.2: Entropy of Helium gas as a function of the volume, at constant energy and number of atoms. The energy values correspond to temperatures of 150, 300 and 600 K.

the curvature of the entropy of *any* system with respect to the volume with fixed E, N is always negative:

$$\boxed{\left(\frac{\partial^2 S}{\partial V^2}\right)_{E,N} < 0} \tag{7.15}$$

Figure 7.2 shows the shape of the entropy as a function of the volume, for different values of E and N. These curves were drawn using equation (7.1) for He gas. The entropy monotonically increases with increasing volume.

7.2.3 *Dependence on N, keeping E and V constant*

The slope of the function $S(E, V, N)$ with respect to the number of particles N at fixed values of E and V is

$$\left(\frac{\partial S}{\partial N}\right)_{E,V} = k_B \ln\left[\sqrt{\frac{64}{27}\frac{V}{N}\left(\frac{\pi m E}{N h^2}\right)^{3/2}}\right] \tag{7.16}$$

Here we formally treat N as a continuous variable, although it is a discrete quantity in reality. This provides a good approximation for any thermodynamic system, because N is typically very very large (of the order of the Avogadro's number), such that even tiny relative changes of N involve a huge number of particles.

Substituting $E = \frac{3}{2}N k_B T$ we obtain the following useful relationship

$$\left(\frac{\partial S}{\partial N}\right)_{E,V} = -k_B \ln(\rho \Lambda^3) \tag{7.17}$$

where $\rho = N/V$ is the number density of the gas, and

$$\Lambda \equiv \frac{h}{\sqrt{2\pi m k_B T}} \tag{7.18}$$

is known in Quantum Mechanics as the **thermal de Broglie wavelength**.

An important criterion for the validity of classical statistical mechanics is that $\rho \Lambda^3 \ll 1$, which basically means that the density has to be so small that the typical distance between particles is much bigger than the scale at which quantum effects become important. Accepting that $\rho \Lambda^3$ is smaller than unity, we see that the slope of $S(E, V, N)$ with respect to N for an ideal gas is positive.

We now define the **chemical potential** of an ideal monoatomic gas by

$$\mu^{\text{ig}} \equiv k_B T \ln(\rho \Lambda^3) \tag{7.19}$$

For an ideal gas $\rho \Lambda^3 < 1$, hence the chemical potential is always negative: $\mu^{\text{ig}} < 0$.

Please note that the phyiscal dimensions of the chemical potential are the same as energy: μ is a form of potential energy. In addition, in thermodynamics the Boltzmann constant k_B in (7.19) is often replaced by the gas constant R, when the quantity of matter is given in terms of moles.

In thermodynamics we usually express the chemical potential of an ideal gas in terms of temperature and pressure, the quantities which are typically measured while the gas is enclosed by a container with fixed volume. In order to transform to these variables, we use the equation of state (2.8) to get $\rho = P/(k_B T)$ which, inserted into eq. (7.19), gives[3]

$$\mu^{\text{ig}} = k_B T \ln \frac{P \Lambda^3}{k_B T} \tag{7.20}$$

$$= \mu^{\text{ig},0} + k_B T \ln P$$

where

$$\mu^{\text{ig},0} \equiv k_B T \ln \frac{\Lambda^3}{k_B T} \tag{7.21}$$

is referred to as the **standard chemical potential**.

Coming back to eq. (7.17), for an ideal gas we have

$$\left(\frac{\partial S}{\partial N}\right)_{E,V} = -\frac{\mu^{\text{ig}}}{T} > 0 \tag{7.22}$$

In general, the slope of S with respect to N for *any* system is always positive and is given by

$$\left(\frac{\partial S}{\partial N}\right)_{E,V} = -\frac{\mu}{T} > 0 \tag{7.23}$$

If the system is a multi-component system, the entropy function can be easily generalized (see equation 7.3). The derivative of S with respect

---

[3]The first expression is dimensionally correct, whereas in the second expression and in (7.21) the quantities inside the logarithm are taken as dimensionless numbers.

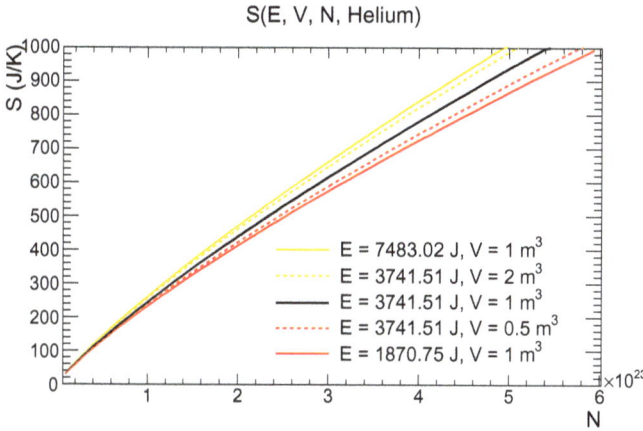

Figure 7.3: Entropy of Helium gas as a function of the number of particles, at constant energy and volume. The energy values correspond to temperatures of 150, 300 and 600 K.

to N_i is

$$\left(\frac{\partial S}{\partial N_i}\right)_{E,V} = -\frac{\mu_i}{T} > 0 \qquad (7.24)$$

For a mixture of ideal gases one writes $\mu_i = \mu_i^{ig} = k_B T \ln(\rho_i \Lambda_i^3)$ where $\rho_i = N_i/V$ and Λ_i is defined as in (7.18) for each species of mass m_i.

The second derivative of $S(E,V,N)$ with respect to N may be obtained from (7.16):

$$\left(\frac{\partial^2 S}{\partial N^2}\right)_{E,V} = -\frac{5k_B}{2N} < 0 \qquad (7.25)$$

This result is valid only for an ideal gas, but the curvature of S with respect to N is always negative for *any* system.

Figure 7.3 shows the shape of the entropy as a function of the number of particles, for different values of E and V. These curves were drawn using equation (7.1) for He gas. The entropy monotonically increases with increasing N, and it does so more quickly than with increasing E and V, because on the latter it has only a logarithmic dependence.

7.2.4 *The total change in entropy*

We started with an explicit function $S(E,V,N)$ and calculated the partial derivatives with respect to all variables. Now can we write the total differential of S as

$$dS = \frac{\partial S}{\partial E} dE + \frac{\partial S}{\partial V} dV + \sum_{i=1}^{c} \frac{\partial S}{\partial N_i} dN_i$$

which lead to the general expression

$$\boxed{dS = \frac{1}{T} dE + \frac{P}{T} dV - \sum_{i=1}^{c} \frac{\mu_i}{T} dN_i} \qquad (7.26)$$

In chapter 8 we shall use this expression for the variation of S for any system (not necessarily an ideal gas) and we shall show in section 8.8 that T, P and μ_i behave as "potentials" that govern the flow of heat, volume and particles of species i.

7.3 Two Spontaneous Processes Involving Ideal Gases

In this section we take advantage of the availability of the explicit expression for the entropy function $S(E,V,N)$ of an ideal gas to examine the change in entropy for two spontaneous processes: the expansion of a gas, and the heat transfer from a hot to a cold body. The result we derive here are valid for ideal gases, but the conclusions hold for any thermodynamic system.

7.3.1 *Expansion of an ideal gas*

Consider the simplest process of expansion depicted in figure 7.4. Initially, we have an ideal gas consisting of N particles in a compartment of volume V and having a total (kinetic) energy E. We remove the partition between the two compartments. As the system is isolated, both E and N remain constant. What we shall observe is that the gas will always expand to occupy the two compartments, i.e. a volume $2V$.

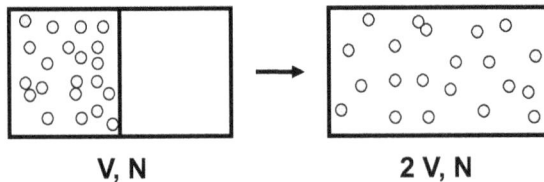

V, N   **2 V, N**

Figure 7.4: Expansion of an isolated ideal gas from volume V to volume $2V$.

What is the change of entropy in this process? We can use either the expression for the entropy in (7.1) or integrate the differential of the entropy in (7.26) at constant E and N_i. For this particular process we find

$$\Delta S = N k_B \ln(2V/V) = N k_B \ln 2 \qquad (7.27)$$

This result is well known in thermodynamics, which however does not offer any interpretation of it, nor does it provide any reason why this process always occurs in this particular direction, and never in the reverse direction.

The informational interpretation of the result (7.27) is very simple. To interpret ΔS in (7.27) in terms of the SMI, we only need to change the basis for the logarithms using the identity $\ln x = \log_2 x / \log_2 e$, and remember to divide by k_B:

$$\Delta H = \frac{\log_2 e}{k_B} \Delta S = N \log_2 2 = N \qquad (7.28)$$

Thus, ΔH is simply equal to N. The interpretation of this result is straightforward. Initially, each particle was confined to a compartment of volume V, but in the final state each particle can be found either in one compartment or in the other. Hence our uncertainty on the location increased. We are interested only in the question: in which compartment each particle is? In terms of the 20-questions game, we have two possible boxes in which each particle may be found. Therefore, we need *one* question to determine in which of the two compartments any given particle is. Since the particles are independent, we need in total N questions to determine in which compartment each of the particles is.

Thus, the change in the SMI for this process is simply N. Multiplying by k_B and taking the natural logarithm do not affect this interpretation: this only change the units in which we measure the changes in the SMI, which in this case we call the entropy change.

As we have noted in section 4.8, one could also give a plausible interpretation of the entropy increase in this process in terms of a tendency towards more disorder, or in terms of more spreading of the particles (or the energy they carry with them). However, we have already emphasized that this is not possible in the general case, hence will refrain from doing so.

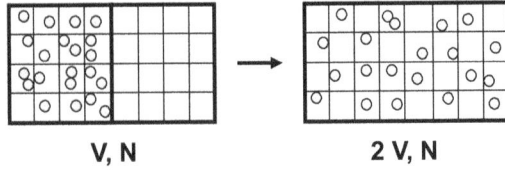

V, N   **2 V, N**

Figure 7.5: Expansion of an isolated ideal gas from volume V to volume $2V$. In this case, the question is about the location of each particle within the small cells.

Note that we have chosen to specify the location of each particle only in terms of compartments. The reader might justifiably ask what if we were interested in the location of the particles within smaller cells. The following exercise should answer this question.

> **Exercise 7.1.** Suppose we divide the volume V into M cells each of volume V/M (figure 7.5). We are interested in which cell each particles is. We do the same division into cells for the two compartments. Calculate H for the initial and the final states. Calculate the change in H for the process of expansion from M to $2M$ cells.

> **Exercise 7.2.** Suppose that the volume of the left compartment is V, but that of the right compartment is $5V$. Calculate the change in entropy in this expansion process, and interpret this result in terms of SMI.

We have seen that in a spontaneous expansion process the entropy always increases. The next question is: *why* the process has occurred in this particular direction?

The answer to this question is beyond the scope of thermodynamics. However, now we know that the answer to this question is probabilistic: the system will always proceed from a state of lower, to a state of higher probability. As we have seen in chapter 5, at equilibrium the system spends all the time in the family of specific configurations with the maximum probability, and the SMI is also maximum in this case.

> **Exercise 7.3.** Suppose that we start with N_A molecules of type A in the left compartment and N_B molecules of type B in the right compartment (figure 7.6). Calculate the change in entropy for the process that occurs when we remove the partition between the two

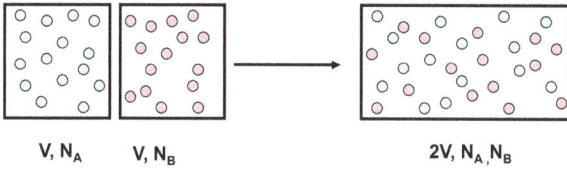

Figure 7.6: Expansion of a mixture of two isolated ideal gases from volume V to volume $2V$.

compartments. Interpret the entropy change in this process.

Note that the entropy is defined for the initial (E, V, N) and the final $(E, 2V, N)$ equilibrium states. Once we remove the partition, the locational SMI will tend to a maximum at the new equilibrium state. It is only once such equilibrium state has been reached that we have a new entropy (during the transition the latter is not defined). The new entropy is a maximum with respect to all constrained equilibrium states, i.e. all possible locations of the particles compatible with the new accessible volume.

7.3.2 *Heat transfer from a hot to a cold body*

The second process we discuss in this section is one of the processes for which the second law was originally formulated: heat always flows spontaneously from a hotter to a colder body. However, the interpretation of the change in entropy is not as simple as in the process of expansion.

Consider again two systems each consisting of N molecules in a container of volume V (figure 7.7). The initial temperatures are T_1 and T_2, with $T_2 > T_1$. The system as a whole is assumed to be isolated, i.e. the energy, the volume and the number of particles of the combined systems are constant.

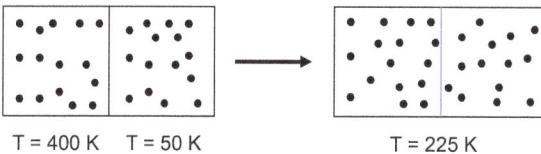

Figure 7.7: The heat transfer between two systems continues until they reach the thermal equilibrium at the same temperature.

We now replace the partition between the two containers by a diathermal, i.e. a heat conducting partition (note that the combined system is still isolated). What we shall observe is that the temperature of the hotter body will decrease and the temperature of the colder body will increase. At equilibrium the two systems will have the same temperature T.

The entropy change for this process can be calculated from the expression (6.31) for the entropy $S(T,V,N)$. We write the initial entropy of the combined system as

$$
\begin{aligned}
S_{\text{in}} &= S(T_1,V,N) + S(T_2,V,N) \\
&= Nk_{\text{B}} \ln\left[\left(\frac{V}{N}\right)^2 \left(\frac{2\pi m k_{\text{B}}}{h^2}\right)^3 (T_1 T_2)^{3/2}\right] + 5Nk_{\text{B}}
\end{aligned}
\tag{7.29}
$$

and the final entropy as

$$
\begin{aligned}
S_{\text{fi}} &= S(T,V,N) + S(T,V,N) \\
&= Nk_{\text{B}} \ln\left[\left(\frac{V}{N}\right)^2 \left(\frac{2\pi m k_{\text{B}}}{h^2}\right)^3 T^3\right] + 5Nk_{\text{B}}
\end{aligned}
\tag{7.30}
$$

The change of entropy $\Delta S \equiv S_{\text{fi}} - S_{\text{in}}$ is then

$$
\Delta S = Nk_{\text{B}} \ln\frac{T^3}{(T_1 T_2)^{3/2}} = Nk_{\text{B}} \ln\left(\frac{T}{T_1}\right)^{3/2} + Nk_{\text{B}} \ln\left(\frac{T}{T_2}\right)^{3/2}
\tag{7.31}
$$

In this example we can calculate the final temperature T simply from the principle of the conservation of the total kinetic energy of the system. Before the transfer of heat we had

$$
E_{\text{in}} = \frac{3}{2}Nk_{\text{B}}T_1 + \frac{3}{2}Nk_{\text{B}}T_2
\tag{7.32}
$$

from equation (6.32). The final kinetic energy is

$$
E_{\text{fi}} = \frac{3}{2}2Nk_{\text{B}}T
\tag{7.33}
$$

When imposing $E_{\text{fi}} = E_{\text{in}}$ we find

$$
T = \frac{T_1 + T_2}{2}
\tag{7.34}
$$

Inserting this into equation (7.31) we find

$$
\Delta S = 3Nk_{\text{B}} \ln\frac{(T_1 + T_2)/2}{\sqrt{T_1 T_2}} \geq 0
\tag{7.35}
$$

where the equality holds when $T_1 = T_2$ and the inequality follows from the theorem saying that the arithmetic average $(T_1 + T_2)/2$ is always larger than the geometric average $\sqrt{T_1 T_2}$.

Thus, we have found that the entropy never decreases in the spontaneous process of heat flow from a hotter to a colder body. This is true even though the entropy of the hotter system decreases, because this change is more than compensated by the increase of entropy of the colder body.

Exercise 7.4. Repeat the calculation of the entropy change for the process of heat transfer as above but when the first system has N particles and the second has $2N$ particles.

As we have noted for the process of expansion, thermodynamics does not offer any interpretation for the change of the entropy in this process. Certainly it does not provide an answer to the question of why the process will occur only in this direction, and never in the reverse direction.

In the expansion process, the change of volume is simply related to the redistribution of the locations of the particles: from uniformly distributed in a volume V to a uniformly distributed in a larger volume $2V$. In the present example of heat transfer, the change in temperatures of the two systems from T_1 to T and from T_2 to T conceal changes in the distribution of speeds of the molecules. Let us see how.

Because the overall system is isolated, the total energy (which in this example is just the kinetic energy) is constant. This implies, by virtue of eq. (6.11), that the variance of the velocity distribution is also constant.

From Shannon's third theorem (section 5.6), we know that if we start with any initial 1-dimensional distribution, say $f(v_x)$ for the velocity component along one axis, and let the system evolve towards equilibrium while keeping the variance σ^2 fixed, we shall obtain a Gaussian distribution (5.40). In addition, we have shown that a Gaussian distribution on each of the three components of the velocity vector gives the Maxwell-Boltzmann distribution (6.21) for the magnitude v of the velocity vector.

In the particular example shown in figure 7.7, we can then write the overall initial velocity distribution for the two systems as the average $f_{\text{in}}(v) = \frac{1}{2}[f_1(v)+f_2(v)]$ between the Maxwell-Boltzmann distributions

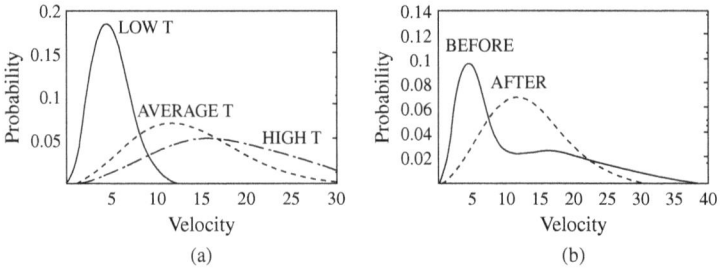

Figure 7.8: The Maxwell-Boltzmann distributions of the initial and final states.

(6.21) corresponding to the initial temperatures

$$f_1(v) = \sqrt{\frac{2}{\pi}} \left(\frac{m}{k_B T_1} \right)^{3/2} v^2 \exp\left(-\frac{mv^2}{2k_B T_1} \right)$$
$$f_2(v) = \sqrt{\frac{2}{\pi}} \left(\frac{m}{k_B T_2} \right)^{3/2} v^2 \exp\left(-\frac{mv^2}{2k_B T_2} \right) \tag{7.36}$$

(figure 7.8). The reason is that, before replacing the athermal partition, we have waited long enough to have each individual system at constant temperature (i.e. at thermal equilibrium). Hence, each of them is characterized by a Maxwell-Boltzmann distribution, which maximizes the SMI.

In the final state, when the two systems are in thermal equilibrium, the velocity distribution is a *single* Maxwell-Boltzmann

$$f_{fi}(v) = \sqrt{\frac{2}{\pi}} \left(\frac{m}{k_B T} \right)^{3/2} v^2 \exp\left(-\frac{mv^2}{2k_B T} \right) \tag{7.37}$$

where the temperature T at this stage is still unknown. However, we know that the final variance must be equal to the initial variance (because of conservation of energy). Hence

$$\sigma_{fi}^2 = \sigma_{in}^2 = \int_0^\infty v^2 \frac{f_1(v) + f_2(v)}{2} \, dv$$
$$= \frac{1}{2} \left[\int_0^\infty v^2 f_1(v) \, dv + \int_0^\infty v^2 f_2(v) \, dv \right] \tag{7.38}$$
$$= \frac{\sigma_1^2 + \sigma_2^2}{2}$$

We have found that the variance is proportional to the temperature, eq. (6.12). Hence (7.38) implies that the final temperature is $T = \frac{1}{2}(T_1 + T_2)$, as we have already found in eq. (7.34) above.

Immediately after replacing the partition, when the overall system has a velocity distribution that is the average between the 2 individual Maxwell-Boltzmann distributions, the SMI is not equal to the maximum value. The latter is achieved later on, at thermal equilibrium, when the velocity distribution is a single Maxwell-Boltzmann, with the result that the final temperature is the average between the two initial temperatures.

Thus, it follows from Shannon's third theorem that the passage from the initial to the final state must be such that the final velocity distribution, corresponding to the final temperature T, has the largest value of SMI (achieved by the Maxwell-Boltzmann distribution). This provides an interpretation of the positive change in entropy in the spontaneous process of heat transfer from the hot to the cold body.

Can we find a satisfactory interpretation also in terms of disorder or spreading?

Looking at the distributions drawn in figure 7.8a, one can give a plausible interpretation of the temperature dependence of the shape of the distribution in terms of disorder or in terms of spreading. The higher the temperature, the larger is the *spread* of energy over larger range of velocities. This can also be conceived as a higher degree of disorder. On the other hand, comparing the distributions of speeds in the initial and the final states shown in figure 7.8b, it is difficult to claim that one is more disordered than the other, or that the spread on one is larger than the spread in the other (indeed, the spread as it is quantified by the variance is the same in both cases).

As we have seen, Shannon's third theorem provides not only an explicit form for the velocity distribution, but also an interpretation of the change in entropy from the initial to the final state in terms of SMI, which is hard to obtain using the metaphors of spread or disorder.

As to the question of why the heat will always flow from the hot to the cold body, again the answer comes from probability. The final state has a velocity distribution which is (much) more probable than the initial distribution. In section 5.6 we have seen that, under the constraint of a constant variance of the magnitude of the velocity along any direction, the PDF with maximum probability is the Gaussian distribution. In section 6.4.1 we found that the corresponding PDF for the magnitude of the velocity vector is the Maxwell-Boltzmann distribution. The corresponding temperature is the average between the initial temperatures, which means that the colder system has to heat up while the hotter system has to cool down: the heat flow is from the hotter to the colder system.

7.3.3 *Clausius' definition of entropy*

We end this section with the derivation of Clausius' definition of entropy for an ideal gas. Suppose we have an ideal gas with a given energy E, confined in a volume V. The energy consists of the total kinetic energy of the N particles

$$E = E_{\mathrm k} = \frac{Nm\langle v^2\rangle}{2} = \frac{3N}{2}m\,\langle v_x^2\rangle \tag{7.39}$$

where

$$\langle v_x^2\rangle = \int_{-\infty}^{\infty} v_x^2 f(v_x)\,\mathrm dv_x \tag{7.40}$$

For simplicity, we now treat the velocity distribution as a discrete distribution $\{f_i | i = 1,2,\ldots,M\}$ with M a very very large number, satisfying the normalization condition $\sum_i f_i = 1$:

$$\langle v_x^2\rangle = \sum_i v_{xi}^2\,\mathrm df_i \tag{7.41}$$

This is a basic technique in numerical integration: a uniform sampling with better and better precision for increasing M.

Suppose that a small amount of heat is transferred to the system, such that the temperature increases by a negligible amount. This corresponds to an infinitesimal increase of the kinetic energy of the gas particles:

$$\mathrm dE_{\mathrm k} = \frac{3Nm}{2}\sum_i v_{xi}^2\,\mathrm df_i \tag{7.42}$$

Since f_i has the form

$$f_i = C\exp\!\left(-\frac{mv_{xi}^2}{2k_{\mathrm B}T}\right)$$

where C is a normalization constant, we can write

$$v_{xi}^2 = \frac{-2k_{\mathrm B}T}{m}\ln\frac{f_i}{C} = \frac{-2k_{\mathrm B}T}{m}(\ln f_i - \ln C)$$

from which we obtain

$$
\begin{aligned}
dE_k &= \frac{3Nm}{2} \sum_i \frac{-2k_B T}{m} (\ln f_i - \ln C)\, df_i \\
&= -3Nk_B T \sum_i \ln f_i\, df_i + 3Nk_B T(\ln C) \sum_i df_i
\end{aligned}
$$

Because the changes in shape of the distribution f_i must preserve the total area ($\sum_i f_i = 1$), their sum must be zero. Hence the second term on the r.h.s. cancels out and we find

$$
dE_k = -3Nk_B T \sum_i \ln f_i\, df_i \tag{7.43}
$$

Now, we consider the differential

$$
\begin{aligned}
d\left(\sum_i f_i \ln f_i\right) &= \sum_i \frac{f_i}{f_i} df_i + \sum_i \ln f_i\, df_i \\
&= \sum_i df_i + \sum_i \ln f_i\, df_i
\end{aligned}
$$

The condition $\sum_i df_i = 0$ implies that first term on the r.h.s. cancels out and we find that the infinitesimal change of the entropy per particle per dimension is

$$
dS_1 = d\left(-k_B \sum_i f_i \ln f_i\right) = -k_B \sum_i \ln f_i\, df_i \tag{7.44}
$$

Therefore, we can rewrite (7.43) as

$$
dE_k = 3NT\, dS_1 \tag{7.45}
$$

When we consider the motion in a 3-dimensional space, the entropy per particle changes by $3\,dS_1$, hence the total change in entropy for N particles is $dS = 3N\,dS_1$. This means that eq. (7.45) can be written in the form

$$
dS = \frac{dE_k}{T} = \frac{\delta Q}{T} \tag{7.46}
$$

where we have identified the small change in kinetic energy dE_k with the small flow of heat δQ into the system. Equation (7.46) is Clausius' definition of entropy associated with heat transfer into the system at constant temperature.

7.4   Spontaneous Mixing and Demixing

We have already seen two spontaneous processes: the expansion of a
gas in section 7.3.1, and the heat transfer from a hot to a cold body in
section 7.3.2. The first involves the redistribution of the locations of the
particles, while the second involves redistribution in the velocities of the
particle.

Both of these processes are "driven" by the higher probability of the
final states, although the distributions that are changed are different.
Both of these processes are said to be inherently irreversible, meaning
that they occur spontaneously only in one direction: a gas *always* expands from a smaller to a larger volume; heat *always* flows from a system
at higher temperature to a system with lower temperature (assuming
that the entire system is isolated).

In this section we examine another process, the mixing of two ideal
gases, which had also been viewed as "inherently irreversible", although
we will show that this is not always true. The example originally studied
by Gibbs (and since then described in many thermodynamics textbooks)
is shown in figure 7.9, where an isolated system initially has two compartments of equal volume V containing different gases. When the partition is removed, the two gases mix by expanding in the entire available
volume.

This is essentially the same process of exercise 7.3 on page 168, and
is indeed associated with a positive change in entropy: $\Delta S = (N_A +
N_B)k_B \ln 2$. Hence it is a spontaneous irreversible process. However, the
"driving force" of this process is not the gas mixing. Instead, it is the
expansion of both gases. The entropy increases because the uncertainty
on the location of each particle is bigger in the final state.

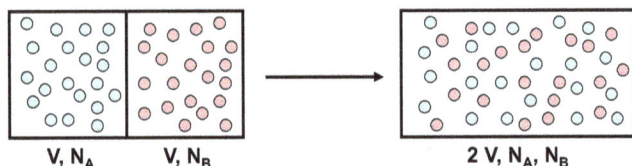

Figure 7.9: Irreversible mixing coupled with expansion.

The fact that mixing can happen spontaneously in both directions can be shown with one simple example. Figure 7.10 shows a system in which both gases are initially contained in the small volume in the middle. Then two walls of this small compartment are removed, leaving two semipermeable walls that can only be crossed by particles of type A (dashed line) but not B, or can be traversed by particles of type B (dotted line) but not A. Each gas is then free to expand in a different part of the container, with A particles occupying the upper volume and B particles occupying the lower volume, in addition to the central compartment.

If the central compartment is much smaller than the other two rooms, then the final result is an almost complete demixing. The process is spontateous and irreversible, because the entropy change is positive. Once again, the "driving force" is the expansion: the uncertainty on the location of each gas increases, because both types of particles can explore a much larger volume in the final configuration, compared to the initial state.

Gas mixing may also happen without change in entropy. Two examples are shown in figure 7.11, where there is no expansion. The second example can be implemented with semipermeable walls, and can be reversed.

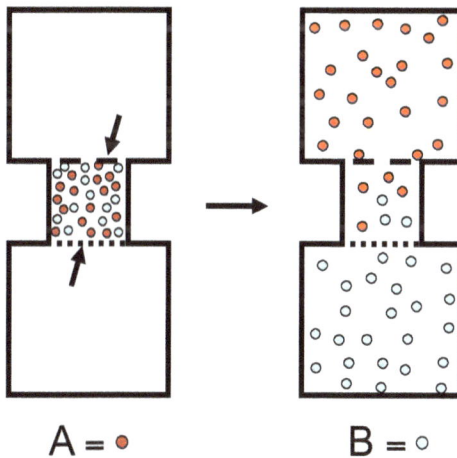

$$A = \bullet \qquad B = \circ$$

Figure 7.10: Irreversible demixing coupled with expansion.

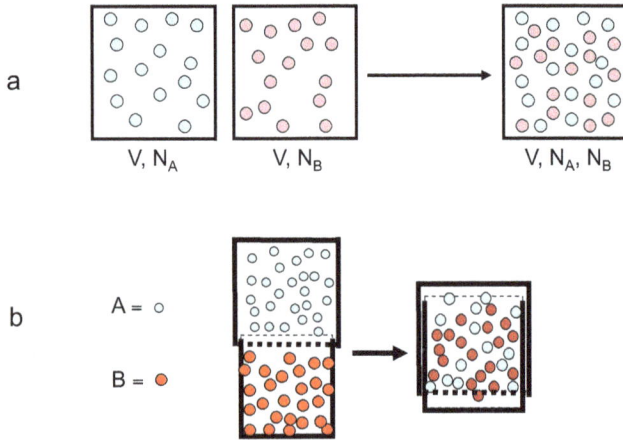

Figure 7.11: Pure mixing of two components A and B. The second version is reversible.

Coming back to the process shown in figure 7.9, it is easier to compute the entropy change if we consider it as the sum of two independent expansion processes, as represented by figure 7.12. The entropy change of each expansion is given by eq. (7.27): $\Delta S_{(a)} = N_A k_B \ln 2$ and $\Delta S_{(b)} = N_B k_B \ln 2$. Hence the total entropy change is $\Delta S_{tot} = (N_A + N_B) k_B \ln 2$.

The sort of false syllogism that induced Gibbs (and others) to conclude that mixing, rather than expansion, is the irreversible process, can be summarized as follows:

(i) mixing of two or more gases is perceived as a process that *increase* disorder;

(ii) increase in disorder is associated with increase in entropy.

Therefore from (i) ad (ii) it follows that

(iii) mixing of different gases is associated with an increase in entropy.

The above reasoning sounds quite intuitive, although it is wrong. Indeed, many textbooks following Gibbs reach the conclusion that mixing of ideal gases is an inherently irreversible process; two gases will always *mix* spontaneously, we never see the reverse of this process occur spontaneously. The fallacy of these conclusions has been discussed in great details few decades ago (Ben-Naim, 1987, 2006a): it results from the as-

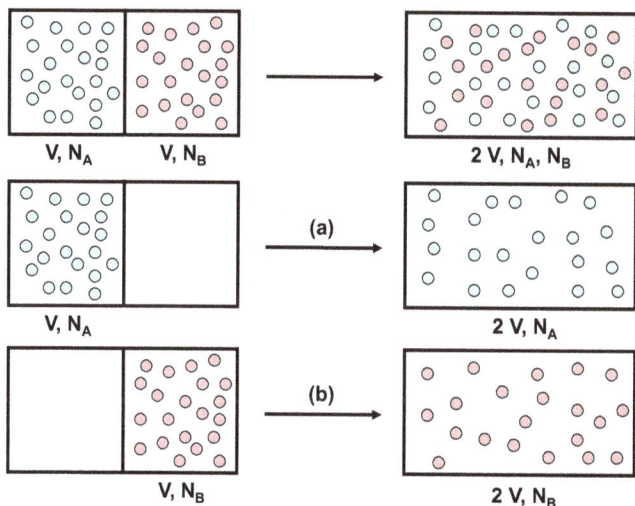

Figure 7.12: The mixing process is equivalent to two expansion processes.

sociation of entropy with disorder.

Although there exist no precise definition of order or disorder, we all feel intuitively that the mixing of two things, say two different gases or two different balls or mixing apples and oranges, is a disordering process. Therefore, the first statement (i) is accepted qualitatively and intuitively as correct.

The second statement was also commonly accepted as part of the interpretation of entropy as a measure of disorder, something which dates back to Gibbs himself (Gibbs, 1906). Clearly, it is not possible to check the validity of the second statement, either qualitatively or quantitatively, simply because we do not have a precise measure of order or disorder. Instead, we can check the validity of the concluding statement (iii), and indirectly infer on the validity of statement (ii).

As we have already shown examples in which the mixing of two gases is associated to positive, negative and even null entropy changes, it is clear that the conclusion (iii) is wrong. This implies that the assumption (ii) that entropy is associated with disorder is incorrect. Thus, we conclude that an increase in disorder can not, in general, be associated with an increase in entropy.

Why so many textbooks on thermodynamics reach the conclusion that mixing is an irreversible process and that mixing involves an increase in entropy?

The probable answer is that in daily life we do see many mixing processes that are spontaneous. A drop of ink dropped into a glass of water will mix with the water, and we never see the reversal of these process.

Admittedly, demixing processes occurring spontaneously are rare, but they do occur. If one vigorously mixes water with oil and let it settle, the two liquids will demix spontaneously.

Gibbs, who probably was the first to analyze the thermodynamics of mixing, was apparently more puzzled by the fact that the entropy of mixing of the two different gases is *independent* of the type of gases, than by the fact that when "mixing" two gases of the same kind, there is no change in entropy. Yet it seems that Gibbs failed to see that in the process that he referred to as a "mixing of gases of different kinds" is simply an expansion process and that the mixing in this process is only incidental.

7.5   Summary of What We Have Learned in this Chapter

The reader should keep in mind that the entropy function $S(E,V,N)$ is a monotonic increasing function with respect to each of the independent variables E,V,N. In addition, the curvature is negative with respect to these three variables, which implies that the function is concave downwards.

We have shown the properties of the function $S(E,V,N)$ for an ideal gas only. However, we assume that an entropy function with the same properties exists for any thermodynamic system.

We have also examined few fundamental spontaneous processes for which the entropy increases. The change in entropy in these processes may be interpreted in terms of SMI. In addition to clarifying the meaning of the second law of thermodynamics, this interpretation makes it also easier to understand mixing and demixing of ideal gases. These processes may exist with positive, null, or negative change in entropy. When they happen spontaneously in one direction, we have seen that their "driving force" is the expansion of a gas.

Chapter 8

The Fundamental Principles of Thermodynamics

In this chapter we present the first and second laws of thermodynamics, its most important building blocks. The first law is basically an extension of the principle of conservation of energy to include also the form of energy that we call *heat*. This law introduces the concept of the *internal energy E* of a system as a *state function*, i.e. as a function that depends only on the *thermodynamic state* of the system, and not on the manner the system was brought to that state.

The second law introduces another state function S, referred to as the *entropy*. Since we have already shown that the entropy of the system is a particular case of Shannon's measure of information (SMI), we can regard S as a shorthand notation for SMI, provided that we remember that the latter is a more general entity whose specific form for a thermodynamic system at equilibrium is S. The second law is a statement on the *non-conservative* nature of the entropy: in any spontaneous process carried out in an isolated system, the entropy can only increase. This principle is essentially the same as what we have discussed in Chapter 6: the value of the SMI reaches a maximum with respect to all possible *distributions*.

The second law also applies to more general systems, in terms of the Helmoltz energy A, reaching a minimum for systems in equilibrium at constant T, V, N (heat flow is allowed), and the Gibbs energy G, reaching a minimum for systems in equilibrium at constant T, P, N (volume changes and chemical reactions are allowed).

The combination of the first and the second laws has been very useful for describing, interpreting and sometimes also in understanding a wide range of phenomena. In particular, the existence of an extremum (maximum entropy, or minimum Gibbs energy) leads to criteria for a host of equilibrium conditions (such as those discussed in section 8.7) and chemical equilibrium (discussed in chapter 11).

In this book we shall not discuss the other two laws of thermodynamics in detail, beyond what we have already written in section 2.5. The zeroth law is essentially a statement of the transitivity property of thermal equilibrium, and is an empirical law.[1] It states that if A is in thermal equilibrium with B, and B is in thermal equilibrium with C, then A is also in thermal equilibrium with C. This principle is used to establish the *existence* of temperature, which provides the basis for thermometry.

The third law is a statement on the limiting value of the entropy as the temperature goes to the absolute zero. The principle of the unattainability of the absolute zero is derived from the third law. We can lower the temperature of a system indefinitely, but it will only asymptotically approach the absolute zero, which can not be reached in a finite time.

Traditionally these four laws are considered to be the four cornerstones of thermodynamics. However, they are not equal in their significance, generality and in their range of applications. Therefore, we shall only focus on the first and second laws, the two most far-reaching ones, and we shall discuss their significance and their applications rather than their historical development.

8.1   Work and Heat: The First Law of Thermodynamics

The notions of force, pressure, energy and work are defined in classical mechanics. We accept the force as a primitive concept: qualitatively, it is the factor that causes a change in the motion of a body. Newton's second law connects the force (F) acting on a body to the acceleration a it imparts on the body along the direction of the force[2]:

$$F = \frac{\mathrm{d}}{\mathrm{d}t}(mv) = ma \qquad (8.1)$$

where m is the (constant) mass of the body.

---

[1] In fact, the transitivity property also applies for mechanical and material equilibrium.

[2] More generally, the second law of dynamics states that the force equals the change in time of the momentum. It is only when the mass of the body is constant, that the change in momentum becomes proportional to the acceleration.

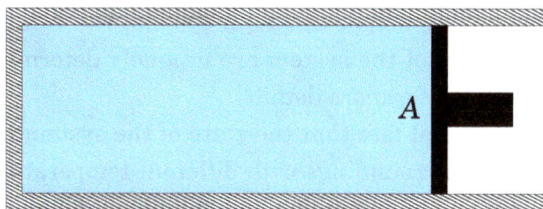

Figure 8.1: Cylinder sealed by a piston and enclosing a gas or a liquid at equilibrium.

The pressure P is defined operationally as the force acting per unit area. In section 2.2 we have seen that the microscopic interpretation of P is connected to the myriad of collisions of the molecules on the container's walls in any small time interval.

We now apply the concept of work as known in mechanics to a thermodynamic system at equilibrium. Consider a system inside a cylinder sealed with a movable piston, as in figure 8.1. The mechanical work done *on* the system is the product of the force exerted by the piston and the displacement of the piston: $F\,dx$. If the area of the piston is A, then the work done *on* the system when the piston moves by a small amount is

$$\delta W = F\,dx = PA\,dx = P\,dV \qquad (8.2)$$

Recall that P is the pressure exerted on the piston by the molecules and dV is the change in volume of the system. We take only a small change in volume dV rather than a finite change, because the pressure of the system might change when the system is compressed. We write δW, rather than dW, for a *small* amount of work, in order to remind ourselves that W is not a function of state (for which a differential dW could be defined).

Suppose that the system, say a gas or liquid enclosed by the container, has a *fixed* amount of matter. Such a system is referred to as a **closed system**, and its state can be characterized by a few parameters, say P and T. We shall always assume that the system is at equilibrium and that its thermodynamic state can be described by a small number of parameters.

The latter assumption is also known as the **state postulate**, saying that a thermodynamic state of a system is fully identified by the values of a suitable (and small) set of parameters known as **state variables** or **thermodynamic variables**. Once such a set of values of thermodynamic variables has been specified for a system, the values of all thermodynamic properties of the system are uniquely determined (see also sections 8.10.1 and 9.2 for more details).

It is an experimental fact that the state of the system might change when it is placed in surroundings with different temperatures. The rate of change depends on the material of the container. There are some materials, known as ***thermal insulators***, that preclude the change of the state of the system, or at least slow it down dramatically. Rigorously speaking, there is no perfect thermal insulator, although there are materials which are very good insulators, such that the exchange of heat between the system and its surroundings can be neglected while performing the experiment. We call a wall that is practically impermeable to the heat flow an ***adiabatic*** or an ***athermal*** wall.

It is also an experimental fact that one can change the state of the system enclosed by adiabatic walls by performing different kinds of work, like mechanical stirring, electrical heating, etc. It is found that the amount of work required to change the state of the system enclosed by adiabatic walls, from an initial state (P_1, T_1) to a final state (P_2, T_2), is independent of the type of the work done on the system and on the path along which the state of the system is changed from the initial to the final state.

This fact leads to the definition of the ***internal energy*** E of system, which depends only on the *state* of the system, and not on the way in which that state is reached. We write this experimental finding as

$$W_{\mathrm{ad}} = E_2 - E_1 = \Delta E \tag{8.3}$$

where E_2 and E_1 are the values of the internal energy of the system in the final and initial states, respectively, and W_{ad} is the adiabatic work done on the system.

A function that depends only on the state of the system and not on the way the system was brought to that state is called a ***state function***. Thus, the internal energy as defined above is a state function. The internal energy is defined up to an arbitrarily chosen additive constant. However, in thermodynamics we are always interested in differences in the internal energies between two equilibrium states. Hence, the additive constant is insignificant and therefore is not specified.

When the walls of the system are not athermal, we find that the work required to change the state of the system from the initial to the final state might depend on the environment in which the system is placed. The intuitive reason is that, when the system is adiabatic, the work done on the system is the only energy that is transferred to the system, and therefore the change in the internal energy is exactly equal to W_{ad}. When the walls of the system allow the flow of some kind of energy different from mechanical work W, we say that *heat* or *thermal energy* flows in (or from) to the system. In other words, we *define* the heat exchanged between the system and its environment by

$$Q \equiv \Delta E - W \qquad (8.4)$$

Note carefully the order in which the three quantities in eq. (8.4) are introduced. First, we start with the notion of mechanical work, with which we are familiar from classical mechanics. Second, we recognize the existence of adiabatic walls, which we intuitively understand as being impermeable to some kind of energy that flows through the walls.

We find experimentally that, if we have a system characterized by some initial state and we transfer a given amount of work under adiabatic conditions, the final state of the system is independent of the type of work we have done, or on the path leading from the initial to the final state: it depends only on the *amount* of work (which we have denoted W_{ad} to remember us that this is true only with adiabatic walls). This led us to introduce a *state function E*, that is the internal energy.

Finally we recognize that, when the system is not adiabatic, the change ΔE of internal energy is *not* equal to the mechanical work W. We suspect intuitively that another form of energy has been transferred through the walls. We call *heat* this form of energy and define the amount of heat transferred in this process as the difference between the total change in internal energy and the mechanical work, i.e. eq. (8.4).

It should be clear by now that whereas E is a *state function* and is defined for any state of the system, the quantities Q and W are *not* state functions. Both represent forms of energy that are exchanged between the system and its environment, and this in general depends on the details of the system and its environment, and not just on the initial and final states of the system.

Once we transferred either work or heat into the system, they are "stored" in the system as changes ΔE of internal energy. There is no such thing as the "work of the system" or the "heat of the system".

The following example should clarify the distinction between a state function and a quantity that is *exchanged* with the system. Suppose we fill a tub with water from two different faucets, call these A and B. At any given time, the level of the water in the tub is measured and the amount of water can be determined.

When we open the two faucets we can measure how much water flows into the tub by faucet A, call it Q, and how much water was flows into the tub by faucet B, call it W. However, once the water is in the tub only the *total* change in the amount of water can be determined. We cannot *recognize* the water in the tub as being of "type Q" or "type W": only the sum of these is recognized as state function, the total amount of water.

One can raise the level of the water in the tub by a given quantity by different amounts of water from both faucets, provided that the sum of the two amounts is the same. The same it true for the internal energy of the system: a change in the internal energy ΔE can be caused by different amounts of work and heat flow, such that the sum $Q + W$ is equal to ΔE.

To conclude, the first law of thermodynamics is simply an extension of the principle of conservation of energy that we already know from classical mechanics, to include another form of energy that we call heat. It is expressed by the equation

$$\Delta E = Q + W \tag{8.5}$$

which is equivalent to eq. (8.4). It is important to note that our convention is that $Q > 0$ when heat flows *into* the system and $W > 0$ when work is performed *on* the system.

8.2   Work in an Expansion Process

From classical mechanics we know that the work is the integral (over the path) of the force acting along the displacement. By virtue of eq. (8.2), the work performed by the system when expanding from volume V_1 to

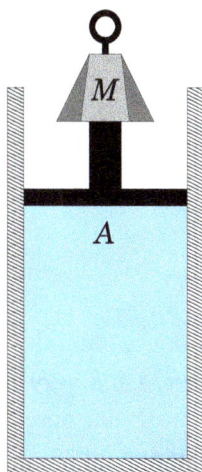

Figure 8.2: Cylinder sealed by a piston and enclosing a gas or a liquid at equilibrium. A weight exerts a force Mg on the piston.

$V_2 > V_1$ (figure 8.1) is then

$$W = - \int_{V_1}^{V_2} P \, dV \tag{8.6}$$

The minus sign means that we *define* as positive the work done *on* the system, and as *negative* the work done *by* the system. Please note that in chemistry the work done on the system is usually defined as a negative quantity, and work done by the system is positive. We adopt here the opposite convention.

In general, the pressure of the system may not be defined during the entire expansion process, therefore in this case we cannot perform the integration in (8.6). For example, suppose that some process like a spontaneous chemical reaction occurs within the system depicted in figure 8.2, with the result that the piston is pushed upwards. During this process the pressure within the system will change, and it might do so without mantaining the system in equilibrium, which means that the pressure of the system can not be defined. On the other hand, if the *external* pressure P_{ex} on the piston is constant (being the atmospheric pressure plus Mg/A for the example shown in figure 8.2), then we can write

$$W = - \int_{V_1}^{V_2} P_{ex} \, dV = -P_{ex}(V_2 - V_1) \tag{8.7}$$

Since P_{ex} is well defined and is a positive number, if $V_2 > V_1$ then W is negative, i.e. the work is performed *by* the system on its surroundings.

In general, the pressure P within the system is not constant during the process. We wish to relate the work W to the pressure P *within* the system. In order to do, so we need to know how the pressure depends on the volume. In the following two sections we discuss expansion processes for which we know the function $P(V)$, and therefore we can calculate the work done by the system.

8.3 Isothermal Quasi-static Expansion Process

Consider a system of ideal gas enclosed in a cylinder as depicted in figure 8.3. The system is immersed in a ***thermostat***, i.e. a very large bath maintained at constant temperature T. As the walls are not athermal, this keeps the temperature of the system constant, at the same value T of the surrounding thermostat.

Suppose that initially the system is characterized by the pressure P_1 and the volume V_1, and that a weight of mass M is placed on the piston, exactly balancing the pressure of the gas. In the initial state we have thermodynamic equilibrium, hence the external pressure is also equal to P_1:

$$P_1^{ex} = P(M) = 1\,\mathrm{atm} + Mg/A = P_1$$

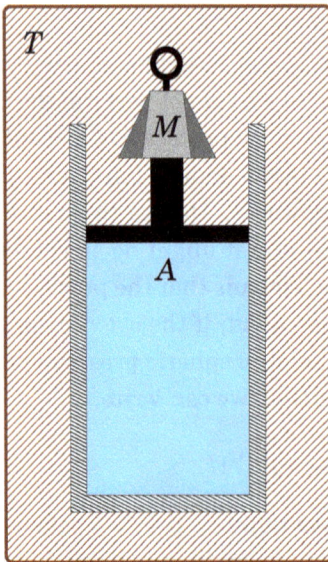

Figure 8.3: Cylinder sealed by a piston and enclosing a gas or a liquid at equilibrium. The container is inside a thermostat with temperature T. A weight exerts a force Mg on the piston.

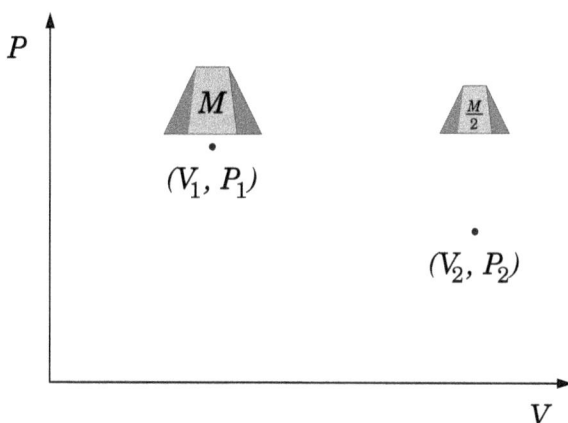

Figure 8.4: Initial and final states of the system shown in figure 8.3.

If we replace the weight of mass M by a lighter weight of mass $M/2$ in a very short time[3], the new external pressure becomes

$$P_2^{\text{ex}} = P(M/2) = 1\,\text{atm} + \frac{1}{2}Mg/A$$

Clearly, the gas will rapidly expand, the pressure of the gas will be reduced to, say, P_2, and the new volume will be $V_2 > V_1$. After a while, we have a new equilibrium state with $P(M/2) = P_2$. As the temperature of the gas is constantly equal to T for the entire process (which is then called an **isothermal** process), with the help of the equation of state of the ideal gas (2.6), we can write

$$P_1 V_1 = nRT = P_2 V_2$$

or equivalently

$$\frac{P_2}{P_1} = \frac{V_1}{V_2} \tag{8.8}$$

In this process, we know the initial and the final states of the gas. These two states can be described by the two points (V_1, P_1) and (V_2, P_2) represented in figure 8.4, showing a so-called "PV diagram".

By replacing the initial weight of mass M by a new weight of mass $M/2$, the external pressure on the piston has changed from $P(M)$ to

---

[3]It is easier to prepare the initial state with two equal weights with masses $M/2$, one of which is quickly removed.

$P(M/2)$, and the new pressure is maintained during the entire process. From eq. (8.7), the work done by the system in lifting the weight is then

$$W = -\int_{V_1}^{V_2} P(M/2)\,\mathrm{d}V = -P(M/2)(V_2 - V_1) \qquad (8.9)$$

If we wish to express the work done by the system in terms of the parameters P and V of the system, we encounter a problem. As we abruptly change the weight on the piston, the system expands rapidly. During this process the pressure in the system is not uniform. If we could measure the local pressure at different distances from the piston, we would find that initially, as the piston moves upward, the local pressure just below the piston gets reduced, while the pressure at the bottom of the system remains almost unchanged at P_1. It takes some time to reach a new equilibrium state where the pressure will be the same at every point within the gas.

Looking at figure 8.4, we can say that the system has moved from the point (V_1, P_1) to the new point (V_2, P_2), but we do not know the *path* along which the system has changed in this diagram. In fact, there exists no path that describes the process in the PV diagram. The volume of the system is defined at each moment of the process, but *the pressure of the system is undefined while the system is not at equilibrium*. Furthermore, we can reverse the process by placing a heavier weight on the piston. However, it makes no sense to talk about reversing the *path* of the process, as we do not know the pressure at each time.

Now, suppose that we start with a weight of mass M and replace it with a weight of mass $2M/3$, and then wait until the system reaches an equilibrium state at (V_3, P_3). Next, we replace it with with a weight of mass $M/2$ to reach a new equilibrium state at (V_2, P_2), as shown in figure 8.5. Initial and final states are the same as before. Again, we cannot describe the states of the system at any intermediate point in the process, because the pressure is defined only at thermodynamic equilibrium. Therefore, as before we can not describe the *path* followed by the system. All we know is three points.

We can make many intermediate steps, by changing the weight only by small amounts at each stage. If we wait long enough to allow the system to reach the equilibrium at each step, we end up with a sequence of points like figure 8.6, which looks almost like a "path" in the PV diagram.

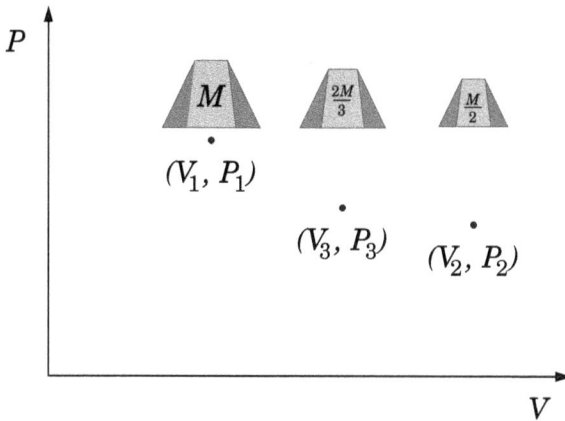

Figure 8.5: The three states of the system shown in figure 8.3.

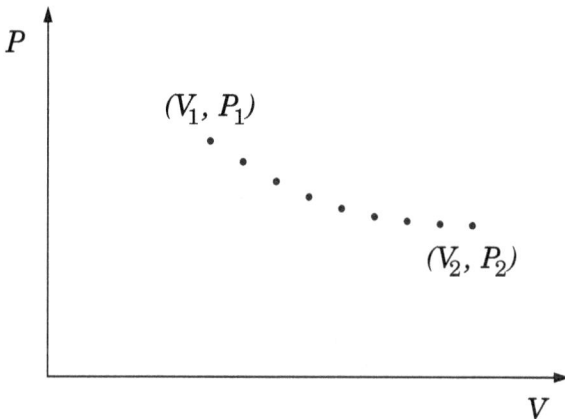

Figure 8.6: Several intermediate equilibrium states.

We can imagine even a slower process where at each step we reduce the pressure on the piston from the previous value P to $P - dP$. The work done by the system in this very small step is

$$\delta W = -(P - dP)dV \simeq -P\,dV \qquad (8.10)$$

If we perform the expansion process in many small steps, such that at each step we reduce the external pressure by a very small amount dP, we shall obtain a dense sequence of equilibrium states. In the limit of infinitesimal steps, such that the equilibrium pressure of the system changes almost continuously, we can describe the *path* of the process in the PV diagram as a continuous line leading from the initial to the final state. Along this path we know that the pressure is related to the volume (at constant temperature T) by $P = nRT/V$. Therefore, the work done by the system is

$$W(\text{isothermal expansion}) = -\int_{V_1}^{V_2} \frac{nRT}{V}\, dV = -nRT \ln \frac{V_2}{V_1} \qquad (8.11)$$

We shall refer to the limiting process described above as a **quasi-static** process. A quasi-static process is a sequence of very small steps such that at every moment the equilibrium of the system changes almost continuously. Therefore a continuous path may be drawn in the PV diagram leading from the initial to the final state.

Clearly, such a process is only an idealization of a real process. In any real process the change in the external pressure will cause a change in the internal pressure in the system with some delay: this change takes time.

Having a path leading from the initial to the final state, it becomes meaningful to speak about reversing the process along the *same path*. Simply add very small weights on the piston and wait enough at each step.

Because of the possibility of reversing the process along a well-defined path, a quasi-static process is also referred to as a reversible process. We shall refrain from using this terminology, and reserve the adjective "reversible" to the processes for which the entropy does not change.

It should be noted that in this isothermal process of expansion the internal energy of the system does not change: $\Delta E = 0$. This follows from the fact that a constant temperature means a constant average kinetic energy of the particles, and in this system the internal energy of the system is simply the total kinetic energy of the particles. Thus, by setting $\Delta E = 0$ in eq. (8.4)

$$\boxed{Q = -W = nRT \ln \frac{V_2}{V_1}} \qquad \text{(isothermal expansion, ideal gas)} \qquad (8.12)$$

Thus, the work done by the system (in lifting the weights) is equal to the heat flown from the bath into the system. Note that the heat entering the system is positive whereas work done by the system is negative.

8.4   Work in a Quasi-static Adiabatic Process

In an *adiabatic process* the walls of the system are athermal: no heat can be exchanged between the system and its surroundings, $Q = 0$. Hence

$$\Delta E = W \tag{8.13}$$

In order to calculate the work for an ideal gas in an adiabatic process, we need to know the function $P(V)$ for such process. For an infinitesimal step, the work is $-P\,dV$ hence the corresponding change in the internal energy is

$$dE = -P\,dV \tag{8.14}$$

We view the internal energy E as a function of the two variables T, V and write the total differential[4]

$$dE = \left(\frac{\partial E}{\partial T}\right)_V dT + \left(\frac{\partial E}{\partial V}\right)_T dV \tag{8.15}$$

We recognize that the first term features the *heat capacity at constant volume* (to be defined in the next section)

$$C_V = \left(\frac{\partial E}{\partial T}\right)_V$$

while the second term is zero, because for an ideal gas at constant temperature E does not change:

$$\left(\frac{\partial E}{\partial V}\right)_T = 0 \quad \text{(ideal gas)}$$

hence

$$dE = C_V\,dT \tag{8.16}$$

Therefore, from (8.14) and (8.16) we get

$$C_V\,dT = -P\,dV \tag{8.17}$$

If we denote by C_V the heat capacity for one mole of gas (or *molar heat capacity at constant volume*), i.e. $n = 1$, and use the equation of state (2.6) for one mole of ideal gas to write $P = RT/V$, we can rewrite (8.17) as

$$C_V\frac{dT}{T} + R\frac{dV}{V} = 0 \tag{8.18}$$

---

[4]The variable at the bottom of each parenthesis is kept constant while a variation in the other variable is considered.

Integrating this equation we get

$$C_V \ln T + R \ln V = \text{constant} \tag{8.19}$$

or equivalently

$$\boxed{TV^{R/C_V} = \text{const.}} \quad \text{(adiabatic process)} \tag{8.20}$$

For an ideal monoatomic gas, it is found that $C_V = \frac{3}{2}R$, therefore $R/C_V = \frac{3}{2}$. In general, for a gas C_V is a constant, different from $3R/2$.

Thus, when an ideal gas system expands from an initial to the final state we have

$$T_i V_i^{R/C_V} = T_f V_f^{R/C_V} \tag{8.21}$$

or equivalently

$$\frac{T_i}{T_f} = \left(\frac{V_f}{V_i}\right)^{R/C_V} \tag{8.22}$$

As C_V is always positive, if $V_f > V_i$ (adiabatic expansion) then eq. (8.22) implies that $T_f < T_i$: *this adiabatic expansion causes a cooling of the system.* As we will shortly see, the reason is that the work is performed at the expense of the internal energy.

With the help of relation (8.17), the work done by the system in a quasi-static adiabatic process is found to be

$$W(\text{adiabatic}) = -P \int_{V_i}^{V_f} dV = C_V \int_{T_i}^{T_f} dT$$

which gives

$$\boxed{W = C_V(T_f - T_i)} \quad \text{(adiabatic process)} \tag{8.23}$$

and is negative for an expansion. From eq. (8.13), the energy change is also negative in an adiabatic expansion. Indeed, the internal energy is proportional to the temperature and the latter decreases. The reduction of internal energy is the price the system pays for performing the work.

We can also express the work in terms of the change of the volume. From (8.22) and (8.23) we have

$$W(\text{adiabatic}) = C_V T_i \left(\frac{T_f}{T_i} - 1\right) = C_V T_i \left[\left(\frac{V_i}{V_f}\right)^{R/C_V} - 1\right] \tag{8.24}$$

Since $V_f > V_i$, the work W is negative for an adiabatic expansion, which means that it is performed *by* the system at the expense of its internal energy.

8.5 Heat Capacity at Constant Volume and at Constant Pressure

In most of this book we shall deal with only one kind of work: expansion work (sometimes referred to as PV-work).

When the volume is constant there is no expansion and the mechanical work $W = -\int P\,dV = 0$, hence the first law (8.5) becomes

$$\Delta E_V = Q_V \qquad (8.25)$$

where we have added the suffix V to remember that the volume is constant. If we limit ourselves to systems which can perform work only by moving a piston, when keeping the volume constant the entire change in the internal energy is due to the heat flow.

The ***molar heat capacity at constant volume*** is defined as

$$C_V \equiv \left(\frac{\partial E}{\partial T}\right)_V \approx \frac{\Delta E_V}{\Delta T} = \frac{Q_V}{\Delta T} \qquad (8.26)$$

and is operationally defined as the amount of heat required to raise the temperature of 1 mole of a substance by 1 degree, provided that no work is done on the system or by the system.

As the heat capacity depends on the temperature, when reporting its experimental value it is usual to annotate the temperature at which it was measured. Actually, for liquids and gases it also depends on the pressure at which the measurement was performed, although this is typically 1 atm = 101 325 kPa. For example, the heat capacity of liquid water at constant volume and ambient pressure is 74.539 J mol^{-1}K^{-1} at 25°C.

One may also define in a similar way the ***molar heat capacity at constant pressure*** C_P, as the amount of heat required for a unit increase of the temperature for one mole of substance.

When heating the substance at constant pressure, it will change its volume. Hence part of the heat is converted into mechanical work, which means that a larger heat flow is required to rise the temperature by 1 degree, compared to the measurement at constant volume. Therefore $C_P > C_V$ (usually by 30–60%). For example, for liquid water at 15°C and 1 atm, $C_P = 4.1855$ J g^{-1}K^{-1} or 1 calorie per gram per degree. One mole of water weighs 18.0152 grams, which means that its molar heat capacity at the same temperature and pressure is 75.403 J mol^{-1}K^{-1}.

It can be shown that

$$C_P - C_V = T \left(\frac{\partial P}{\partial T} \right)_{V,n} \left(\frac{\partial V}{\partial T} \right)_{P,n} \tag{8.27}$$

where the partial derivatives are taken at constant volume and constant number of moles, and constant pressure and constant number of particles, respectively. This can also be rewritten as

$$C_P - C_V = VT \frac{\alpha^2}{\kappa} \tag{8.28}$$

where

$$\alpha \equiv \frac{1}{V} \left(\frac{\partial V}{\partial T} \right)_P \tag{8.29}$$

is the **coefficient of thermal expansion**, and

$$\kappa = -\frac{1}{V} \left(\frac{\partial V}{\partial P} \right)_T \tag{8.30}$$

is the **isothermal compressibility**. For an ideal gas, one obtains **Mayer's relation**:

$$C_P - C_V = R \tag{8.31}$$

At constant pressure, the first law (8.5) becomes

$$\Delta E_P = Q_P - P\Delta V = Q_P - \Delta(PV) \tag{8.32}$$

It is now convenient to define an auxiliary function called the **enthalpy** and usually denoted by H (not to be confused with the symbol used by Shannon for the SMI):

$$H \equiv E + PV \tag{8.33}$$

From now on, H will denote the *enthalpy*, not the SMI.

At *constant pressure* we have from (8.32) and (8.33)

$$\Delta H_P = \Delta E_P + P\Delta V = Q_P \tag{8.34}$$

Thus, the heat exchange with a system at constant pressure is equal to the enthalpy change of the system. Note that since E, P and V are functions of state, also H is a function of state. The **molar heat capacity at constant pressure** can then be defined as

$$C_P \equiv \left(\frac{\partial H}{\partial T} \right)_P \approx \frac{\Delta H_P}{\Delta T} = \frac{Q_P}{\Delta T} \tag{8.35}$$

8.6 Carnot Cycle and Efficiency of a Heat Engine

The Carnot cycle is an example of an idealized heat engine. It is named after Sadi Carnot (1796–1832) who studied the efficiency of heat engines, a work that led to the formulation of the second law of thermodynamics.

The Carnot cycle consists of four steps and is shown in figure 8.7. We start with one mole of ideal gas at thermodynamic equilibrium contained in a cylinder of volume V_1, temperature T_1 and pressure P_1, and perform the following steps:

a) quasi-static isothermal expansion from state 1 to state 2;
b) quasi-static adiabatic expansion from state 2 to state 3;
c) quasi-static isothermal compression from state 3 to state 4;
d) quasi-static adiabatic compression from state 4 to state 1 (the initial state).

The work performed at the first step, while the temperature is constant ($T_2 = T_1$), is found by applying eq. (8.11) with $n = 1$:

$$W_a = -RT_1 \ln \frac{V_2}{V_1} < 0 \qquad (8.36)$$

Since the process is isothermal, there is no change in the internal energy in this step, $\Delta E_a = 0$, hence eq. (8.4) gives

$$Q_a = -W_a = RT_1 \ln \frac{V_2}{V_1} > 0 \qquad (8.37)$$

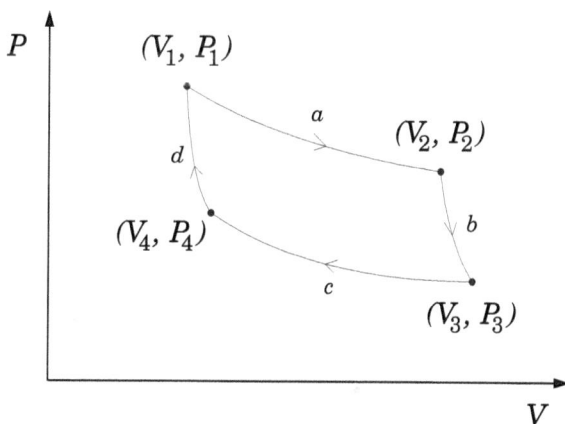

Figure 8.7: Carnot's cycle.

Since $V_2 > V_1$, the work is done by the system (negative by convention) and the heat flows *into* the system (hence it is positive).

The second step is an adiabatic expansion, hence the work is obtained from eq. (8.23):

$$W_b = C_V(T_3 - T_2) = C_V(T_3 - T_1) < 0 \qquad (8.38)$$

where C_V is the molar heat capacity of the gas. Here, again, the work is negative because it is done *by* the system, and the temperature drops from T_1 to $T_3 < T_1$ because without heat flow ($Q_b = 0$) the work is performed at the expense of a reduction in internal energy:

$$\Delta E_b = W_b = C_V(T_3 - T_1) < 0 \qquad (8.39)$$

The third transformation is a compression ($V_4 < V_3$) at a constant temperature ($T_4 = T_3$) for which the work is positive

$$W_c = -RT_3 \ln \frac{V_4}{V_3} > 0 \qquad (8.40)$$

As the work is done *on* the system and the internal energy is constant, the heat flows *out* of the system:

$$Q_c = -W_c = RT_3 \ln \frac{V_4}{V_3} < 0 \qquad (8.41)$$

In the final adiabatic ($Q_d = 0$) compression ($V_1 < V_4$) we have

$$W_d = C_V(T_1 - T_4) = C_V(T_1 - T_3) > 0 \qquad (8.42)$$

In a compression the work is done *on* the system, hence it is positive. Because there is no heat exchange, the work increases the internal energy, which implies that $T_1 > T_3$:

$$\Delta E_d = W_d = C_V(T_1 - T_3) > 0 \qquad (8.43)$$

Note that since the internal energy is a state function, it depends only on the state of the system, and not on the way that state is reached. Therefore, in a closed cycle the total change of internal energy must be zero. It is easy to verify that this is indeed the case for the Carnot cycle: there is no energy change during isothermal transformations ($\Delta E_a = \Delta E_c = 0$), and from eq. (8.39) and eq. (8.43) we see that $\Delta E_b = -\Delta E_d$.

We can now calculate the total work and the total heat exchanged between the system and the surroundings in the entire cycle. The work done during the two adiabatic transformations is opposite and cancels out, as eq. (8.38) and eq. (8.42) show. Therefore the total work is obtained by summing the work performed during the isothermal transformations, eq. (8.36) and eq. (8.40):

$$W_{\text{cycle}} = W_a + W_c = -RT_1 \ln \frac{V_2}{V_1} - RT_3 \ln \frac{V_4}{V_3} \tag{8.44}$$

Now consider the adiabatic transformation from state 2 to state 3: from eq. (8.21) and the equalities $T_1 = T_2$ and $T_3 = T_4$ we have

$$T_2 V_2^{R/C_V} = T_3 V_3^{R/C_V} \quad \Longrightarrow \quad T_1 V_2^{R/C_V} = T_4 V_3^{R/C_V} \tag{8.45}$$

Similarly, for the fourth transformation we have

$$T_1 V_1^{R/C_V} = T_4 V_4^{R/C_V} \tag{8.46}$$

By dividing these equations term by term we get

$$\frac{T_1 V_2^{R/C_V}}{T_1 V_1^{R/C_V}} = \frac{T_4 V_3^{R/C_V}}{T_4 V_4^{R/C_V}} \quad \Longrightarrow \quad \frac{V_2}{V_1} = \frac{V_3}{V_4} \tag{8.47}$$

Substituting this equality we into (8.44) we find

$$\boxed{W_{\text{cycle}} = R(T_3 - T_1)\ln \frac{V_2}{V_1} < 0} \tag{8.48}$$

Since $T_3 < T_1$ and $V_2 > V_1$ the work in the entire Carnot cycle is *negative*: a net amount of work is done *by* the system *on* the environment. This work depends linearly on the temperature difference and logarithmically on the ratio of the volumes. If one reverses the cycle, by changing across the four states in the reverse order, one obtains that the same amount of work has to be performed on the system (i.e. it is positive).

A system operating in repeated cycles such as the one described above is referred to as **heat engine**. In this particular cycle, the net work done by the system is W_{cycle}. The **efficiency** η of the heat energy is defined as the ratio of the absolute value of the work done by the system, and the heat absorbed by the system in contact with the hot reservoir at the initial temperature T_1:

$$\eta \equiv \frac{|W_{\text{cycle}}|}{Q_a} \tag{8.49}$$

From (8.37) and (8.48) we get

$$\eta = \frac{T_1 - T_3}{T_1} = 1 - \frac{T_3}{T_1} < 1 \tag{8.50}$$

since $T_1 > T_3$.

Now, by taking the ratio of the heat flows in the isothermal processes [equations (8.37) and (8.41), with the help of eq. (8.47)] we obtain another result of great importance in the history of the second law of thermodynamics:

$$\frac{Q_a}{Q_c} = -\frac{T_1}{T_3} \tag{8.51}$$

If we denote as $T_a = T_1$ the temperature of the first isothermal transformation, and as $T_c = T_3$ the temperature of the other isothermal process, then we can rewrite the previous equation in the form

$$\frac{Q_a}{T_a} + \frac{Q_c}{T_c} = 0 \tag{8.52}$$

In this book, we define the entropy as the SMI value at equilibrium (with natural logarithms and multiplied by Boltzmann's constant; see chapter 6). However, historically the entropy *difference* was defined for a small quantity δQ of heat transferred at constant temperature T as the ratio $dS \equiv \delta Q/T$ between the heat exchanged with the system and the temperature at which this happens. Any finite change in entropy could be then computed with the integral over any[5] quasi-static transformation of all infinitesimal heat transfers δQ divided by the instantaneous temperature. Hence eq. (8.52) says that *the entropy does not change in a Carnot cycle*.

We will call a process for which the entropy is constant a **reversible process**. Carnot's cycle is an example of a reversible cyclic process. Its efficiency (8.50) achieves the maximum possible value for a cycle in which heat is exchanged between two thermostats, and is higher for a bigger ratio between their temperatures. However, as no real system can reach 0 K, this efficiency is always strictly smaller than 100%. Whenever the entropy changes, the efficiency is smaller than (8.50). In addition, for real systems (8.52) should be replaced by an inequality, stating that the entropy change is always positive.

---

[5] S is a state function, hence its change only depends on the initial and final states, and not on the details of the transformation.

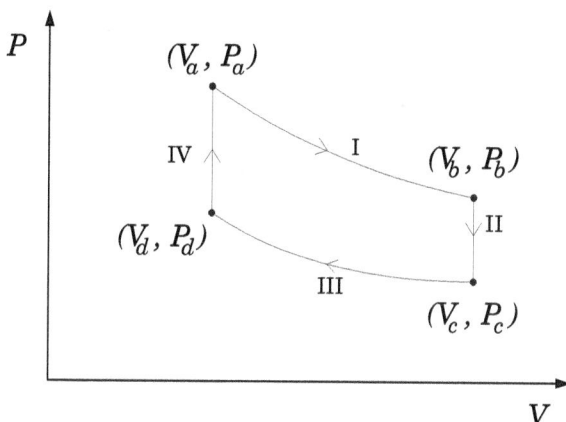

Figure 8.8: Stirling's cycle.

A process occurring at constant volume is called an ***isochoric*** or an ***isovolumetric*** process. In the following exercise, a cycle made of isothermal and isochoric processes, known as ***Stirling cycle***, is taken into account.

Exercise 8.1. Calculate the efficiency of a heat engine, whose cycle consists of the following four steps (figure 8.8):

I) quasi-static isothermal expansion from a to b;
II) quasi-static isochoric process (i.e., a constant volume) from b to c;
III) quasi-static isothermal compression from c to d;
IV) quasi-static isochoric process from d to a.

8.7 Entropy and The Second Law of Thermodynamics

The traditional derivation of the entropy and the second law is lengthy and tedious, formal and abstract. It also leaves the student baffled as to the meaning of entropy, and frustrated from any attempt to understand the second law.

In this section we shall introduce the entropy of any system at equilibrium as a generalization of the concept of entropy we have constructed for an ideal gas in chapter 6. Unlike what we did in chapter 6, we shall not be able to construct an explicit expression for the function $S(E,V,N)$ in the general case. Fortunately, such an explicit function is not needed in thermodynamics.[6] We only need to postulate is the *existence* of such a function, and that this function has the same *properties* that we have already seen in the special case of ideal gases. Also, we retain the *meaning* of entropy as the Shannon measure of information (SMI) at equilibrium as for an ideal gas.

The most fundamental postulate of thermodynamics is the existence of an equilibrium state.

A thermodynamic system consists of a huge number of particles, something of the order of 10^{23}. A microscopic description of such a system requires a huge number of parameters. Nevertheless, it is an experimental fact that a macroscopic system, characterized by the internal energy E, volume V, and composition $N_1, \cdots, N_c$ where N_i can be either the number of particles or the number of moles of species i, when left undisturbed for a sufficiently long period of time, reaches an equilibrium state. Once at equilibrium, such a system can be described by a small number of parameters.

How long it takes for a system to reach an equilibrium state depends on the system. Most systems dealt with by thermodynamics and in this book, reach an equilibrium state within a short time, seconds or minutes. There are systems, such as glasses, for which the evolution towards the equilibrium state might take a very long time. Such systems are not in equilibrium states, and will not be of any concern to us.

There is no formal definition of the state of equilibrium of a real system, nor a criterion to determine whether or not the system has reached an equilibrium state. In practice, an equilibrium state can be characterized as a state where no changes can be observed. For instance, if we measure the temperature, the pressure or the density at any point in the system, we find that these do not change with time. This is not *sufficient* for the complete characterization of the equilibrium state, as we might measure all these parameters in a piece of glass or in a solid at very low temperature, which are not systems at equilibrium, without detecting

---

[6]The explicit function $S(E,V,N)$ is the concern of statistical mechanics. However, even within statistical mechanics there are only a very few systems for which the entropy function can be written explicitly.

any changes in these parameters.

Operationally, one can "define" an equilibrium state of a system whenever measurements made on the system are consistent with those predicted by thermodynamics. However, since thermodynamic relationship applies only for systems at equilibrium, the above "definition" is circular.

We next *postulate* the existence of an entropy function defined for any system at equilibrium, which possesses the properties of the entropy function of an ideal gas: we *assume* that the function $S(E,V,N)$ is continuous and differentiable, and that it is a concave (downwards) function of the variables E,V,N. We further *assume* that this function is additive: the entropy of a composite system at equilibrium is the sum of the entropies of the constituent subsystems

$$S = \sum_k S^{(k)} \qquad (8.53)$$

where $S^{(k)}$ is the entropy of the macroscopic subsystem k at equilibrium.

In chapter 6 we derived the entropy of an ideal gas, based on the concept of SMI, and we have found that the SMI attains a maximum over all *distributions* subjected to some constraints. The distribution that maximizes the SMI was also found to be the distribution that is attained at equilibrium, and the equilibrium state was identified as the state having maximal probability.

We have identified the entropy of an ideal gas as the value of the SMI of the thermodynamic system at equilibrium. Next we wish to extend the principle of maximum entropy to any macroscopic system. Instead of the molecular distributions of locations and momenta, we shall deal with internal distributions of volumes, energies and number of moles of the different species.

Suppose that we have a macroscopic system of fixed energy E, volume V and $N = \sum_k N_k$ moles. Next, suppose that we divide this system into r macroscopic compartments. Each compartment is characterized by an energy $E^{(i)}$, volume $V^{(i)}$ and composition $N_1^{(i)}, \cdots, N_c^{(i)}$ (figure 8.9). The compartments are separated by walls that are rigid, athermal and impenetrable to material. Thus, each of the compartments is a macroscopic system that will reach an equilibrium state, for which its entropy $S^{(i)}$ is determined by the corresponding variables $E^{(i)}, V^{(i)}, N_1^{(i)}, \cdots, N_c^{(i)}$.

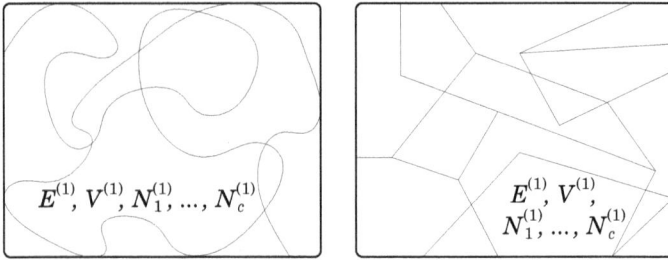

Figure 8.9: Two possible partitions of a system with energy E, volume V and $N = \sum_k N_k$ moles.

We also *require* that

$$E = \sum_{i=1}^{r} E^{(i)} \tag{8.54}$$

$$V = \sum_{i=1}^{r} V^{(i)} \tag{8.55}$$

$$N_k = \sum_{i=1}^{r} N_k^{(i)} \quad \forall k = 1, 2, \ldots, c \tag{8.56}$$

together with the additivity requirement (8.53) which we have already formulated as

$$S = \sum_{i=1}^{r} S^{(i)}$$

We shall refer to a specific subdivision of the energy E, the volume V and the composition $N_1, \cdots, N_c$ among all the compartments as a **constrained equilibrium distribution**. It is an equilibrium state because each subsystem, as well as the entire combined system, is at equilibrium. The *constraint* comes from the fact that, once a wall is removed, the system will reach a new equilibrium state.

Clearly, we have infinite ways of dividing the total energy, volume and material into r compartments. Each of these compartments is a macroscopic subsystem and is assumed to be in a state of equilibrium. Figure 8.9 shows a couple of possible partitions.

We now replace the principle of maximum entropy over all molecular distributions (of locations or momenta) by a principle of maximum over all possible *constrained equilibrium distributions*. This means that if we remove some or all of the constraints, the system will evolve toward a

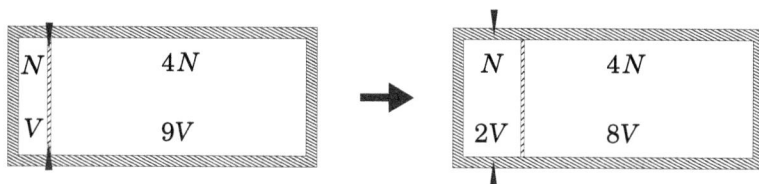

Figure 8.10: After releasing the volume constraint, the moving wall finds a new position which balances the pressure of both compartments.

new equilibrium state for which the entropy is larger (more precisely, cannot be lower) than in the initial constrained equilibrium state.

In the discussion of Shannon's theorems we have considered *two* kinds of molecular distributions, locational and momentum. In the macroscopic formulation of the principle of maximum entropy we use three kinds of constraints, on the energy, on the volume, and on the number of particles in each subsytem. The partition of the energies among the subsystems is the analog of the distribution of momenta in ideal gas. On the other hand, removing either the constraint on volumes or on the number of particles is equivalent to the locational redistribution of the molecules. The following exercises should clarify this point (further details in sections 8.8.2 and 8.8.3).

> **Exercise 8.2.** The system depicted in figure 8.10 consists of two subsystems (1) and (2) at thermal equilibrium. One subsystem contains N molecules in volume $V^{(1)} = V$, and the other contains $4N$ molecules in volume $V^{(2)} = 9V$. The partition is now allowed to move freely, until a new equilibrium position is reached. Calculate the final volumes of the two subsystems.

> **Exercise 8.3.** Consider the same initial system as in the previous exercise. Now we keep the volumes of the subsystems fixed, and we let particles flow between the two subsystems. Calculate the final distribution of particles.

We have seen from the last two exercises that allowing either a change of volume (movable partition) or an exchange of particles leads to a redistribution in the locations of the particles.

In the next section we shall examine a few simple examples of constrained equilibrium state. Before we move on, we add to our list of postulates one further *requirement*, that the total differential of the entropy of any thermodynamic system has the same form as the one we have already derived in section 7.2.4 for an ideal gas:

$$dS = \left(\frac{\partial S}{\partial E}\right)_{V,N} dE + \left(\frac{\partial S}{\partial V}\right)_{E,N} dV + \sum_{i=1}^{c} \left(\frac{\partial S}{\partial N_i}\right)_{E,V,N_i'} dN_i$$

$$= \frac{1}{T} dE + \frac{P}{T} dV - \sum_{i=1}^{c} \frac{\mu_i}{T} dN_i$$

(8.57)

Note that here we indicate with the subscript N that the vector $(N_1,\ldots,N_c)$ is constant, and with N_i' that we keep constant the vector with $c-1$ components $(N_1,\ldots,N_{i-1},N_{i+1},\ldots,N_c)$ which does not include the i-th component.

We shall also see in the next section that T, P and μ_i have the same meaning as in chapter 6. In the next section they will also appear as the "potentials" that cause flow of heat, volume and particles between different subsystems.

8.8 Examples of Internal Parameters and Conditions of Equilibrium

We present here four examples of constrained equilibrium distributions described by a single internal parameter in a system, with respect to which the entropy has a maximum.

8.8.1 *Spontaneous transfer of heat between two subsystems: thermal equilibrium*

Consider a system characterized by E, V, N, where N can be either the number of particles (or the number of moles) in the case of one component system, or represent the vector $(N_1,\ldots,N_c)$ when the system contains c components.

Suppose that the system consists of two subsystems, separated by a rigid, athermal and impermeable partition, as shown in figure 8.11. The two subsystems are in an equilibrium state, characterized by the variables $E^{(1)}, V^{(1)}, N^{(1)}$ and $E^{(2)}, V^{(2)}, N^{(2)}$. The entropy of the combined system is given by

$$S(E,V,N) = S^{(1)}(E^{(1)},V^{(1)},N^{(1)}) + S^{(2)}(E^{(2)},V^{(2)},N^{(2)})$$

(8.58)

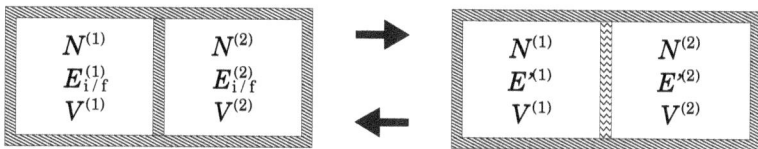

Figure 8.11: Constrained equilibrium distribution. Heat flow is allowed for a short time (right panel), such that the initial and final energies of each subsystem change by a small amount ($E_f^{(1)} = E_i^{(1)} + dE^{(1)}$ and $E_f^{(2)} = E_i^{(2)} + dE^{(2)}$).

We now replace the athermal partition between the two subsystems by a diathermal wall for a very short time. The only process that can happen is flow of heat from one subsystem to the other. We assume that the amount of heat flow is small enough, such that the spontaneous process that occurs is quasi-static. The change in entropy of the combined system as a result of this spontaneous change is[7]

$$dS = dS^{(1)} + dS^{(2)} = \frac{\partial S^{(1)}}{\partial E^{(1)}} dE^{(1)} + \frac{\partial S^{(2)}}{\partial E^{(2)}} dE^{(2)}$$

$$= \frac{1}{T^{(1)}} dE^{(1)} + \frac{1}{T^{(2)}} dE^{(2)}$$

(8.59)

We now define the parameter ξ by:

$$\xi \equiv \frac{E^{(1)}}{E}$$

(8.60)

where $E = E^{(1)} + E^{(2)}$ is the total energy, which is constant as the overall system is isolated. Clearly, the parameter ξ defines a specific partition of the total energy E into two contributions $E^{(1)} = \xi E$ and $E^{(2)} = (1 - \xi)E$. Next, we rewrite (8.59) as

$$dS = \frac{1}{T^{(1)}} E \, d\xi - \frac{1}{T^{(2)}} E \, d\xi = \left[\frac{1}{T^{(1)}} - \frac{1}{T^{(2)}} \right] E \, d\xi$$

(8.61)

The second law states that the entropy has a maximum over all possible constrained equilibrium distributions. This principle can be translated for this particular process into the statement that, for any spontaneous change in the internal parameter ξ, the entropy change can not be negative. In other words,

$$\left[\frac{1}{T^{(1)}} - \frac{1}{T^{(2)}} \right] E \, d\xi \geq 0$$

(8.62)

---

[7]To keep the notation as simple as possible we omit to denote the variables that are kept constant in these partial derivatives.

Since each of the subsystems is closed and maintained at constant volume, $dE^{(1)} = E \, d\xi$ is the amount of heat $dQ^{(1)}$ exchanged between the two systems (which in this specific case happens to be an exact differential). Thus, from (8.62) we have

$$dS = \left[\frac{1}{T^{(1)}} - \frac{1}{T^{(2)}} \right] dQ^{(1)} \geq 0 \tag{8.63}$$

If $dQ^{(1)} > 0$, i.e. if heat flows into subsystem (1), then the square bracket can not be negative, which implies

$$T^{(1)} \leq T^{(2)} \tag{8.64}$$

The reverse is true if $dQ^{(1)} < 0$, i.e. if heat flows into subsystem (2).

Thus, we see that the second law for this particular process is equivalent to the statement that heat flows spontaneously from the subsystem with higher temperature to that at lower temperature. We recognize this statement as Clausius' formulation of the second law.

Note that the parameter T introduced in the last passage of eq. (8.59) is still unidentified at that stage. It is only when we find that it governs the heat flow by means of (8.64) that we can recognize it as the absolute temperature. Recall that in chapter 7 we have identified the partial derivative of S with respect to E as the inverse of the absolute temperature of an ideal gas. Here we consider general systems, not only ideal gases, and we find that such derivative can be always interpreted in terms of the absolute temperature.

We can repeat the same process of figure 8.11 again and again, allowing at each step only a small exchange of heat between the two systems. At each step, when a spontaneous change occurs, dS must positive, until we reach a point of thermal equilibrium *between* the two subsystems connected by a diathermal wall. At this point S reaches a maximum with respect to ξ, i.e.

$$\left(\frac{\partial S}{\partial \xi} \right)_{E,V,N} = 0 \tag{8.65}$$

and the second derivative is negative[8]

---

[8]The second derivative is of importance for the stability of the system with respect to thermal fluctuations in the system. We shall not be concerned with this topic.

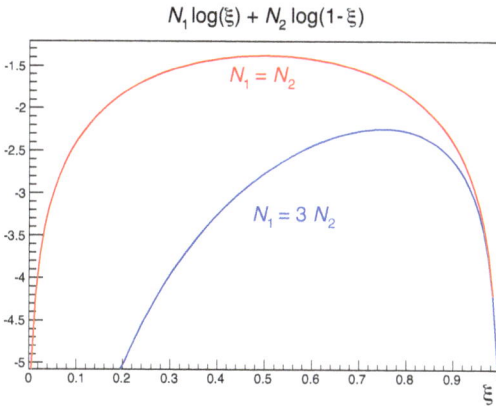

$$N_1 \log(\xi) + N_2 \log(1 - \xi)$$

Figure 8.12: Shape of the entropy function for the two systems of figure 8.11.

In order to plot the shape of the entropy as a function of ξ, one can insert the form (7.1) of the entropy of each system into eq. (8.58), and replace $E^{(1)} \mapsto \xi E$ and $E^{(2)} \mapsto (1 - \xi)E$, where E is the total energy, which is constant. By grouping together all invariant quantities into one constant term, one finds

$$S(\xi) = \text{const.} + \frac{3}{2} k_\text{B}[N_1 \ln \xi + N_2 \ln(1 - \xi)] \qquad (8.66)$$

The only variable piece is proportional to the simple function of ξ shown in figure 8.12 for $N_1 = N_2$ and $N_1 = 3N_2$. The result (8.66) holds for ideal gases, but the qualitative shape of figure 8.12 and the existence of a maximum are general properties, valid for any pair of systems.

Suppose that we start with two subsystems (1) and (2) such that initially $T^{(2)} > T^{(1)}$. We know that heat will flow from (2) to (1). Since $S^{(1)}(E^{(1)})$ is a monotonic increasing function of $E^{(1)}$, then $S^{(1)}(\xi E)$ will be a monotonically increasing function of ξ. Similarly, $S^{(2)}((1 - \xi)E)$ will be a monotonically decreasing function of ξ. In figure 8.13 we sketch the three functions $S^{(1)}(\xi E)$, $S^{(2)}((1 - \xi)E)$ and $S(\xi) = S^{(1)}(\xi E) + S^{(2)}((1 - \xi)E)$, by showing only the variable pieces computed for an ideal gas (compare with eq. 8.66). Again, the qualitative behavior is a general property and is the same for any pair of systems.

We have seen that the function $S(E, V, N; \xi)$ for constant E, V, N has a maximum at some value of ξ_eq. At the maximum the entropy change is locally null (i.e. $dS = 0$), hence from equation (8.63) it follows that at

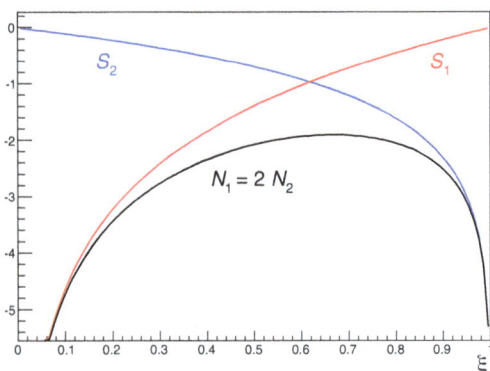

Figure 8.13: Shape of the entropy functions for the two systems of figure 8.11, with $N_1 = 2N_2$.

thermal equilibrium $T^{(1)} = T^{(2)}$, i.e. the two systems reach the same temperature.

It should be noted that we do not view the entropy as a function of time. The process we carry out in changing ξ is quasi-static, i.e. it follows a sequence of equilibrium points. In this sense it is meaningful to think about the *reverse* process by mentally tracing back the system along the same sequence of points. However, the process occurs spontaneously in one direction only. It will never proceed in the opposite direction *spontaneously*. For this reason we call such a process ***irreversible***. It is irreversible only in the sense that it does not occur in the opposite direction spontaneously (while maintaining E, V, N constant).

It is often said that a spontaneous process occurs only in one direction *because* the entropy has to increase. A more precise statement is that a spontaneous process occurs in one direction *because the final state is far more probable than the initial state*. The probability of the equilibrium state is almost one. Therefore the system will stay in the equilibrium state "forever", not because the entropy is maximum, but *because the probability of deviating from the equilibrium state is almost zero*. It turns out, however, that the entropy and the probability of the equilibrium state are related by a monotonic function (this is a general property of the SMI: see eq. (5.17) in section 5.3). Therefore, it is legitimate to say that the system will always proceed towards a state of maximum entropy.

But if we want to emphasize the *cause* that drives the process towards the new equilibrium state, we have to use the language of probability.

The ultimate cause of the process is the kinetic energy of the particles: without their motion, no process would occur. On the other hand, the *direction* of change is determined by the overwhelming higher probability of the new equilibrium state. The entropy change is the *result* of the evolution of the system toward equilibrium, not the cause.

8.8.2 Spontaneous "transfer" of volume between two subsystems: mechanical equilibrium

We present here another example of a constrained equilibrium distribution. Consider again a system characterized by E, V, N consisting of two subsystems characterized by $E^{(1)}, V^{(1)}, N^{(1)}$ and $E^{(2)}, V^{(2)}, N^{(2)}$. Initially, as in the section (8.58), the two subsystems are separated by athermal, rigid and impermeable partition. We shall assume that the temperatures of the two subsystems are equal.

We next release the constraint on the rigid partition. We let the partition move spontaneously leftwards or rightwards as in figure 8.10 on page 205.

We introduce an *internal* parameter η defined by

$$\eta \equiv \frac{V^{(1)}}{V} \tag{8.67}$$

such that $V^{(2)} = (1-\eta)V$. Next, we let the partition move by infinitesimal steps and follow the function $S(\eta)$ defined by [9]

$$S(\eta) = S^{(1)}(\eta V) + S^{(2)}((1-\eta)V) \tag{8.68}$$

whose differential is

$$
\begin{aligned}
dS &= \left(\frac{\partial S^{(1)}}{\partial V^{(1)}} \frac{\partial V^{(1)}}{\partial \eta} + \frac{\partial S^{(2)}}{\partial V^{(2)}} \frac{\partial V^{(2)}}{\partial \eta} \right) d\eta \\
&= \left(\frac{P^{(1)}}{T^{(1)}} - \frac{P^{(2)}}{T^{(2)}} \right) V \, d\eta
\end{aligned}
\tag{8.69}
$$

where we have made use of (7.13).

---

[9] As in the previous section we omit from the notation the independent variables which are not changed in the process.

The second law states that in a spontaneous process at constant E, V, N, the change in entropy can not be negative. If $d\eta > 0$, that is volume "flows" into subsystem (1), from (8.69) we must have

$$\frac{P^{(1)}}{T^{(1)}} - \frac{P^{(2)}}{T^{(2)}} \geq 0 \qquad (8.70)$$

Since we have assumed that $T^{(1)} = T^{(2)}$ it follows that $P^{(1)} > P^{(2)}$: the expansion of the first subsystem is due to its larger pressure. Similarly, if $d\eta < 0$ the volume "flows" from subsystem (1) to subsystem (2) because the latter exerts larger pressure.

Thus, volume "flows" from the subsystem at lower pressure to the one at higher pressure. More precisely, as the partition is pushed with different pressures, it moves toward the region at lower pressure, until the two opposite forces balance. When this happen, the two pressures are equal: $dS = 0$ implies $P^{(1)} = P^{(2)}$. This is the condition of *mechanical* equilibrium between the two systems.

As we have done above, we could plot the qualitative behavior of $S(\eta)$ by taking the entropy of ideal gases. After grouping together the constant pieces, the only function of η is a linear combination of $\ln \eta$ and $\ln(1 - \eta)$, from the first and second subystem, respectively. The qualitative shape of all terms appearing in eq. (8.68) is the same as in figure 8.13.

8.8.3 *Spontaneous transfer of matter between two subsystems: matter equilibrium*

Consider again a system characterized by E, V, N consisting of two subsystems, initially separated by an athermal, rigid and impermeable partition. As before, we assume that initially we have thermal equilibrium between the two subsystems. Thus, initially $T^{(1)} = T^{(2)} = T$.

In this experiment we *open* the subsystems, so that matter can flow between the two subsystems, while keeping the combined system as a whole closed (figure 8.14).

We introduce a new internal parameter x defined by

$$x \equiv \frac{N^{(1)}}{N} \qquad (8.71)$$

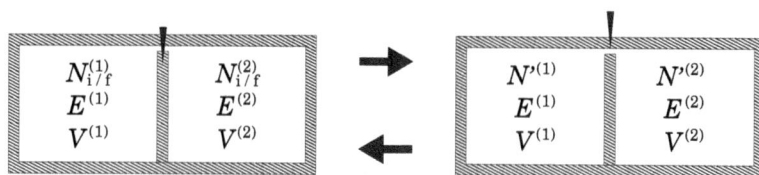

Figure 8.14: Constrained equilibrium distribution, when matter flow is allowed for a short time (right panel).

with $N = N^{(1)} + N^{(2)}$, such that $N^{(2)} = (1-x)N$. We let molecules flow from one subsystem to the other in a quasi-static process. For instance we open a very small hole in the partition separating the two subsystems. The idea is again to proceed very slowly towards the final equilibrium state by a sequence of equilibrium states.

The overall entropy function can be written in the form

$$S(x) = S^{(1)}(xN) + S^{(2)}((1-x)N) \tag{8.72}$$

The differential of S with respect to x is

$$\begin{aligned}
dS &= \frac{\partial S^{(1)}}{\partial N^{(1)}} \frac{\partial N^{(1)}}{\partial x} + \frac{\partial S^{(2)}}{\partial N^{(2)}} \frac{\partial N^{(2)}}{\partial x} \\
&= \left(\frac{-\mu^{(1)}}{T} - \frac{-\mu^{(2)}}{T} \right) N \, dx
\end{aligned} \tag{8.73}$$

with the help of (7.24).

Removing the constraint will cause a spontaneous process in the system, at constant total values of E, V, N. Therefore, the entropy must increase: $dS \geq 0$. Thus if $dx > 0$, which means that $N^{(1)}$ has increased, then the chemical potentials must satisfy

$$\mu^{(1)} < \mu^{(2)} \tag{8.74}$$

i.e. matter will flow from the subsystem at higher chemical potential into the one at lower chemical potential. If $dx < 0$ then $\mu^{(1)} > \mu^{(2)}$: matter *always* flows from high to low chemical potential, until they balance. Indeed, at equilibrium we have $dS = 0$ or equivalently

$$\mu^{(1)} = \mu^{(2)} \tag{8.75}$$

In order to plot the shape of the entropy (8.72) as a function of x, one can insert the form (7.1) of the entropy of each system into eq. (8.72). The

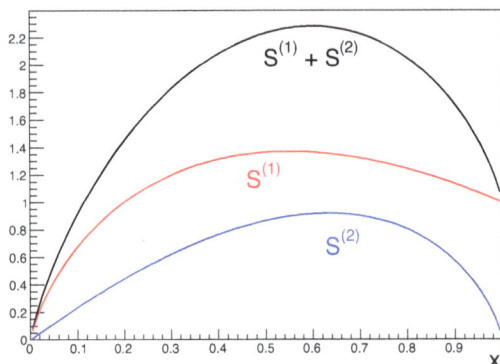

Figure 8.15: Shape of the entropy functions for the two systems of figure 8.11, when material flow is allowed (values $C_1 = 1.0$, $C_2 = 1.2$ have been chosen).

result, valid for ideal gases, is

$$S(x) = Nk_B[f_1(x) + f_2(1-x)]$$
$$= Nk_B\left\{x\left(C_1 - \frac{5}{2}\ln x\right) + (1-x)\left[C_2 - \frac{5}{2}\ln(1-x)\right]\right\} \quad (8.76)$$

where we omitted to indicate the invariant variables, and the parameters C_1 and C_2 group together constant terms depending only on volume and energy (invariant in the process considered here). Figure 8.15 shows the two functions $f_1(x)$ and $f_2(1-x)$ and their sum, for a particular choice of constant values. The qualitative shape of these functions is the same, independently from the composition of the subsystems (which do not need to be ideal gases): it is a general property of the entropy function.

We have discussed above the case of a one-component system. In general, if we have a system with c species and composition $N_1, \ldots, N_c$, the total variation dS is the sum of expressions like (8.73) for each component (by the rule of the total derivative). We can follow each species individually and repeat the previous treatment for each component by imagining that we open the subsystem only with respect to one component at each time. We will then obtain the result (8.75) for each component: matter of species i will flow from the subsystem where *that* species has higher chemical potential to the one at lower chemical potential. Each species follows its own chemical potential, eliminating any difference at equilibrium (once the constraint is removed).

8.8.4 Spontaneous transformation of molecules from one component to another: chemical equilibrium

In the previous subsection we discussed the spatial flow of matter from one location to another. Matter (of each specific species) flows from a location with high chemical potential to a location with low chemical potential. In this section we derive a similar result for another type of "flow" of matter: the transformation from one species into another. We shall discuss chemical reactions and chemical equilibrium in chapter 11. Here we only address a very simple chemical reaction, the conversion between two isomers of the same molecule.

Suppose we have a molecule that has two isomers, denoted A and B. For instance, dichloroethylene may be found in two isomers, called "cis" and "trans" (figure 8.16).

We also assume that we have some technique which can inhibit the conversion between these two isomers. The new constraint in the system could be an inhibitor for the conversion reaction or the absence of a catalyst, in case this is required for the conversion to happen. In either case, the presence of an inhibitor or the absence of a catalyst defines a constrained equilibrium distribution. Unlike the constraint imposed in section 8.8.3 (preventing molecules to change container), here we preclude the transformation of matter from one state to another. In section 8.8.3, the "states" were defined by the locations of the molecules in different compartments. Here, all molecules are in the same compartment, but they occupy different states, A or B, corresponding to different geometries.

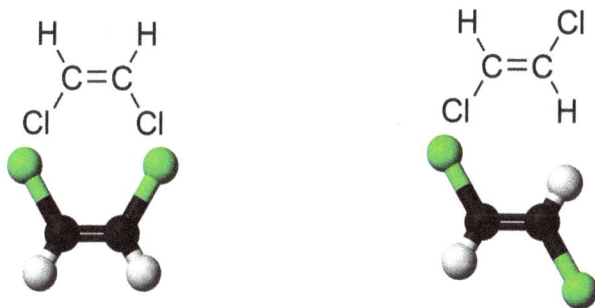

Figure 8.16: Dichloroethylene has two isomers, called cis-1,2-dichloroethene (left) and trans-1,2-dichloroethene (right).

Initially, we prepare the system with N_A molecules (or moles) of A and N_B molecules (or moles) of B, in the presence of an inhibitor (or in absence of a catalyst). The system is a two-component system described by $S(E,V,N_A,N_B)$, initially at equilibrium: the temperature, the pressure and the two chemical potentials μ_A and μ_B are fixed at any point in the system and do not change with time (similarly to the previous sections).

Next, we release the constraint imposed on the fixed values of N_A and N_B. We can do that either by adding a catalyst that enables the conversion between A and B or by removing an inhibitor for this conversion. We also allow the system to evolve spontaneously towards equilibrium in very small steps, i.e. in a quasi-static process.

We define the **mole fraction** of A as

$$x_A \equiv \frac{N_A}{N_A + N_B} \tag{8.77}$$

x_A is the internal parameter, analogous to ξ, η and x of the previous sections, which describes the entropy behavior. We can now write the function $S(E,V,N_A,N_B)$ as $S(E,V,x_A N,(1-x_A)N)$ or simply $S(x_A)$. Note that, when we allow the conversion between A and B, N_A and N_B are not fixed: only the sum $N = N_A + N_B$ is constant. The differential of S with respect to x_A at constant values of E,V,N is

$$\begin{aligned} dS &= \frac{\partial S}{\partial N_A} dN_A + \frac{\partial S}{\partial N_B} dN_B \\ &= -\frac{(\mu_A - \mu_B)}{T} N \, dx_A \end{aligned} \tag{8.78}$$

As dS can not be negative in a spontaneous process, if $dx_A > 0$ (which means that the number of A molecules increases) then $\mu_A < \mu_B$, whereas $dx_A < 0$ implies $\mu_A > \mu_B$. Thus, there is a "flow" of molecules from the state at higher chemical potential to the state at lower chemical potential. At equilibrium ($dS = 0$) we have $\mu_A = \mu_B$. This result and its generalization for any chemical reaction will be discussed in more details in chapter 11.

8.9   Combining the First and the Second Laws

In section 8.1 we have written the first law (8.5) as

$$\Delta E = Q + W$$

For a quasi-static process it reads

$$dE = \delta Q + \delta W \tag{8.79}$$

where we use the symbol δ for a very small variation of some quantity to remind ourselves that such quantity is not a function of state (hence the variation is not an exact differential).

If the only kind of work which the system is allowed to perform is a mechanical expansion work, then the infinitesimal work is an exact differential:

$$dW = -P\,dV \tag{8.80}$$

Although other kinds of works may be exchanged with the system, such as electrical and magnetic work, for the present discussion we shall assume that our system can perform only expansion work.

The second law introduces the entropy function as the SMI of a macroscopic system at equilibrium, and states that this function never decreases for a spontaneous process occurring inside an isolated system. The differential of the entropy is given by eq. (7.26), which we report here too:

$$dS = \frac{1}{T}\,dE + \frac{P}{T}\,dV - \sum_{i=1}^{c} \frac{\mu_i}{T}\,dN_i \tag{8.81}$$

The last term on the r.h.s. of (8.81) is related to exchange of material between the system and its surroundings. When no such exchange can happen, we speak about a **closed system**. In absence of chemical reactions, in a closed system the composition $N \equiv (N_1,\ldots,N_c)$ is constant in time. In chapter 11 we will treat changes in the composition, viewed as part of an internal process within the system.

When a *spontaneous* process occurs *within* the system at constant E, V, N the entropy always increases ($dS_{E,V,N} > 0$) until the equilibrium is reached, when $dS_{E,V,N} = 0$ and S has a maximum.

It is useful to define the quantity dS_{in} as the entropy change due to any spontaneous process occurring *within* the system (the "in" in dS_{in} stands for "internal" or "interior"). Thus, instead of (8.81) we write

$$dS = \frac{1}{T}\,dE + \frac{P}{T}\,dV - \sum_{i=1}^{c} \frac{\mu_i}{T}\,dN_i + dS_{in} = dS_{ex} + dS_{in} \tag{8.82}$$

where dS_{ex} is the change in entropy of the system due to exchanges of energy (E), volume (V) or material (any of the N_i's) with the environment (the "ex" in dS_{ex} stands for "exchange"), and dS_{in} is the change of entropy in a spontaneous process at the *interior* of the system when $E, V, N_1, \ldots, N_c$ are kept constant.[10]

Combining (8.79), (8.80) and (8.82) we have

$$dS = \frac{\delta Q}{T} - \sum_{i=1}^{c} \frac{\mu_i}{T} dN_i + dS_{in} \qquad (8.83)$$

For a closed system in which all N_i are constant (no chemical reaction) the previous equation simplifies to

$$dS = \frac{\delta Q}{T} + dS_{in} \qquad (8.84)$$

As $dS_{in} \geq 0$ for a spontaneous internal process, it follows that

$$dS \geq \frac{\delta Q}{T} \qquad (8.85)$$

Note that when a small amount of heat is exchanged between the system and its surroundings, and when no other spontaneous process occurs *within* the system, we have the *equality* sign in (8.85). The inequality sign should be used whenever there exists a spontaneous process *within* the system, like for example a chemical reaction.

If the system is closed and adiabatic ($\delta Q = 0$), then it follows from (8.85) that $dS \geq 0$. In general, δQ can be either positive or negative depending on whether the heat flows into the system or out from the system.

A comment on semantics is now in order. The quantities dE, dV and dN_i in equation (8.82) represent small amounts of energy, volume and number of particles of species i, that are *exchanged* between the system and its surroundings.[11] Therefore, the corresponding change in entropy is due to exchange of energy, volume and number of particles. On the other hand, dS_{in} is the entropy change due to some spontaneous process

---

[10] We will not make much use of the distinction between dS_{in} and dS_{ex} in the rest of the book, where practical applications are addressed.

[11] In addition dN_i could be a result of a chemical reaction occurring inside the system. In this case, the exchange of particles happens between different species, rather than with the surroundings.

within the system. Although it is sometimes said that the total change in entropy is due to "flow" of entropy into or from the system, and to the creation of entropy within the system, we shall *not* do so. Instead, we shall say that the entropy can *increase* or *decrease* due to *flow* of energy, volume or material, but the entropy itself does not "flow" (more on this in section 8.10). Note also that (8.85) is only valid for a *closed* system. The more general result is (8.83), for an open system.

The equality sign in (8.85) is often said to hold for a reversible process. However, we shall call **reversible process** any process for which there is no change in the entropy. Thus, *every spontaneous process in an isolated system (with constant E, V, N) is considered to be an irreversible process in this book*. It is irreversible not in the sense that the process cannot be reversed (it can be reversed), but only in the sense that it does not occur in the reverse direction *spontaneously*. From (8.84), we see that the equality sign holds in (8.85) when $dS_{in} = 0$, that is when no internal process increases the entropy.

It is important to understand the meaning of the two terms dS_{ex} and dS_{in} defined in (8.82). The first term dS_{ex} is the entropy change due to *exchange* of either energy (E), volume (V) or material (N_i) between the system and its environment. The three terms included in dS_{ex} may be positive or negative, depending on the direction of flow of energy, volume or material between the system and the environment (or the surroundings). We can calculate dS_{ex} by virtue of either our knowledge of the function $S(E, V, N)$ for an ideal gas, or by postulating the extension of the validity of the differential (8.81) for any thermodynamic system. Specifically, we know that $S(E, V, N)$ is a monotonic increasing function of E, V and N, that is, the partial derivatives with respect to these variables are positive. The knowledge of the *properties* of the entropy function $S(E, V, N)$ enables us to calculate entropy changes for any finite quasi-static process, by integrating (8.81).

The second law tells us that, if a spontaneous process occurs in an isolated system (i.e. with constant E, V, N), then the entropy of the system must increase. This increase in entropy is dS_{in} (or ΔS_{in} for a finite change): since the system is *isolated*, $dS_{ex} = 0$ (or $\Delta S_{ex} = 0$ for a finite quasi-static process). Therefore, dS_{in} is referred to as the change in entropy due to a spontaneous process *within* the system, i.e. in the interior of the system, not due to some exchange with the surroundings.

In general, since the entropy is not defined for non-equilibrium states, we cannot follow S along the spontaneous process, therefore we can not calculate dS_{in} (nor ΔS_{in}) directly in terms of the parameters of the system. All we can say is that in a spontaneous process dS_{in} (or ΔS_{in}) is positive. This statement follows from the second law, and not from the properties of the function $S(E,V,N)$ and its derivatives with respect to the variables E,V,N.

How do we calculate the change in entropy in a spontaneous process in an isolated system?

To answer this question we exploit the fact that the entropy is a state function. This means that if a spontaneous process occurred in an isolated system, leading from state (1) to state (2), the entropy change $\Delta S = S_2 - S_1$ is *independent* of the path leading from the state (1) to state (2). Therefore, in order to calculate the change ΔS_{in} for the spontaneous process, we must devise a quasi-static process leading from state (1) to state (2), and calculate the change of entropy by integrating dS_{ex} along the path.

This sounds paradoxical: we calculate dS_{in} for a spontaneous process and in isolated system by using dS_{ex}... but dS_{ex} is only defined for a quasi-static process in a system which is not isolated! The trick for overcoming this difficulty was already expoloited in section 8.8. The general idea is that we start with an initial equilibrium state. We view the system as composed of r subsystems, all at equilibrium, such that for each one there exists a function $S^{(i)}(E^{(i)},N^{(i)},N^{(i)})$. Next, the spontaneous process in the entire combined system is viewed as a quasi-static process of *exchange between the subsystems*. Thus, what is viewed as ΔS_{in} for the *combined* system, is calculated as the sum of the ΔS_{ex} over all subsystems. In other words, the ΔS_{in} for the spontaneous process that occurs *within* the combined system, and for which we do not know how to compute the entropy change *directly*, is calculated indirectly in terms of exchanges of energy, volume or material between the subsystems, in a quasi-static process.

Although the general trick sounds abstract, the exercises below illustrate it in a simple way.

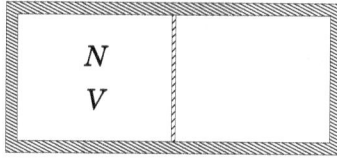

Figure 8.17: Initial state: all the N molecules occupy one half of the total volume.

Exercise 8.4. Spontaneous expansion of an ideal gas when the partition in figure 8.17 moves all the way to the right, and (2) when instead a hole is opened to let the gas diffuse and occupy the entire volume $2V$.

Note that the combined system is isolated: the values of total energy E, volume $2V$, and particles number N are constant. In order to compute the change in entropy ΔS_{in} in the combined system, we look at the two subsystems (1) and (2). Each of these performs a quasi-static process, for which we can compute the corresponding entropy variation $\Delta S_{ex}^{(i)}$. Next, we calculate $\Delta S_{in} = \Delta S_{ex}^{(1)} + \Delta S_{ex}^{(2)}$.

Exercise 8.5. Calculate ΔS_{in} for a spontaneous flow of one component from one subsystem to another.

Exercise 8.6. Calculate ΔS_{in} for a spontaneous mixing of two gases, using a quasi-static process with the help of semi-permeable partitions.

Exercise 8.7. Calculate ΔS_{in} for a finite heat flow from a hot gas to a cold gas.

Exercise 8.8. Calculate ΔS_{in} for a finite heat flow from a hot body to a cold body.

8.10 The Helmholtz and Gibbs Energies

In most applications of thermodynamics we do not keep the system isolated. In addition, it is much easier to keep the temperature and the pressure constant, rather than the energy and the volume. Hence it is useful to introduce the following two auxiliary functions[12].

The **Helmholtz energy** is defined as

$$A \equiv E - TS \tag{8.86}$$

and the **Gibbs energy** is defined as

$$G \equiv H - TS = E + PV - TS \tag{8.87}$$

where H is the enthalpy (not the SMI) defined in (8.33).

Partial derivatives of A and G with respect to the independent variables provide the value of the other thermodynamic variables. In the following equalities, all independent variables are kept constant apart the one over which the partial derivative is taken.

Let's start with a system characterized by $(T, V, N_1, \ldots, N_v)$:

$$S = -\frac{\partial A}{\partial T} \qquad P = -\frac{\partial A}{\partial V} \qquad \mu_i = \frac{\partial A}{\partial N_i} \tag{8.88}$$

The total differential of the Hemoltz energy is then

$$\begin{aligned}
\mathrm{d}A &= \frac{\partial A}{\partial T}\,\mathrm{d}T + \frac{\partial A}{\partial V}\,\mathrm{d}V + \sum_{i=1}^{c} \frac{\partial A}{\partial N_i}\,\mathrm{d}N_i \\
&= -S\,\mathrm{d}T - P\,\mathrm{d}V + \sum_{i=1}^{c} \mu_i\,\mathrm{d}N_i
\end{aligned} \tag{8.89}$$

Thus, for a system at constant T, V and composition we have

$$\mathrm{d}A_{T,V,N} = 0 \tag{8.90}$$

Hence the Helmoltz energy has an extremum. Actually, *at equilibrium the function $A(T,V,N)$ has a minimum*, as it can be shown by computing the second derivatives of A, which are all negative at the extremum.

The Helmoltz energy is an analytic function. As such, the order of differentiation over any two variables is irrelevant (Schwarz theorem). From the symmetry of second derivatives and the partial derivatives (8.88), one can derive identities that belong to the set of **Maxwell's relations**. Additional identities can be obtained by considering the second derivatives of the total energy and of the Gibbs energy.

---

[12]They were also referred to as "free energies". However, this term may be misleading and therefore was abandoned in favor of Helmholtz and Gibbs energies.

Exercise 8.9. Compute the second derivatives of $A(T,V,N)$ and prove that

$$
\left(\frac{\partial^2 A}{\partial T^2}\right)_{V,N} = -\left(\frac{\partial S}{\partial T}\right)_{V,N} = -\frac{C_V}{T}
$$

$$
\left(\frac{\partial^2 A}{\partial V^2}\right)_{T,V} = -\left(\frac{\partial P}{\partial V}\right)_{T,N}
\tag{8.91}
$$

$$
\left(\frac{\partial^2 A}{\partial N_i^2}\right)_{T,V} = \left(\frac{\partial \mu_i}{\partial N_i}\right)_{T,V}
$$

(all are negative) where C_V is the molar heat capacity at constant volume defined in eq. (8.26).

Exercise 8.10. From $\frac{\partial^2 A}{\partial T \partial V}$ prove the following Maxwell relation:

$$
\left(\frac{\partial P}{\partial T}\right)_V = \left(\frac{\partial S}{\partial V}\right)_T = \frac{\alpha}{\kappa}
\tag{8.92}
$$

where α is the coefficient of thermal expansion defined in (8.29) and κ is the isothermal compressibility defined in eq. (8.30).

Similarly, for a system characterized by $(T,P,N_1,\ldots,N_v)$:

$$
S = -\frac{\partial G}{\partial T} \qquad V = \frac{\partial G}{\partial P} \qquad \mu_i = \frac{\partial G}{\partial N_i}
\tag{8.93}
$$

and the total differential of the Gibbs energy is

$$
\mathrm{d}G = \frac{\partial G}{\partial T}\,\mathrm{d}T + \frac{\partial G}{\partial V}\,\mathrm{d}V + \sum_{i=1}^{c} \frac{\partial G}{\partial N_i}\,\mathrm{d}N_i
$$

$$
= -S\,\mathrm{d}T + V\,\mathrm{d}P + \sum_{i=1}^{c} \mu_i\,\mathrm{d}N_i
\tag{8.94}
$$

For a system at constant T,P and composition we have

$$
\mathrm{d}G_{T,P,N} = 0
\tag{8.95}
$$

In particular, *at equilibrium the function $G(T,P,N)$ has a minimum*, as it can be shown by computing the second derivatives.

Exercise 8.11. Compute the second derivatives of $G(T,P,N)$ with respect to T, P, N and prove that

$$\left(\frac{\partial^2 G}{\partial T^2}\right)_{P,N} = -\left(\frac{\partial S}{\partial T}\right)_{P,N} = -\frac{C_P}{T}$$

$$\left(\frac{\partial^2 G}{\partial P^2}\right)_{T,P} = \left(\frac{\partial V}{\partial P}\right)_{T,N} = \kappa V \qquad (8.96)$$

$$\left(\frac{\partial^2 G}{\partial N_i^2}\right)_{T,P} = \left(\frac{\partial \mu_i}{\partial N_i}\right)_{T,P}$$

(all are negative) where C_P is the molar heat capacity at constant pressure and κ the isothermal compressibility.

Exercise 8.12. Prove the following relations using the second derivatives of G with respect to T and P:

$$\frac{\partial \mu}{\partial P} = \frac{\partial V}{\partial N}$$

$$\frac{\partial \mu}{\partial T} = -\frac{\partial S}{\partial N} \qquad (8.97)$$

Exercise 8.13. From $\frac{\partial^2 G}{\partial T \partial P}$ prove the following Maxwell relation:

$$\left(\frac{\partial V}{\partial T}\right)_P = -\left(\frac{\partial S}{\partial P}\right)_T = \alpha V \qquad (8.98)$$

where α is the coefficient of thermal expansion.

Exercise 8.14. Prove the **Gibbs-Helmoltz equation**

$$\left(\frac{\partial (G/T)}{\partial T}\right)_P = -\frac{H}{T^2} \qquad (8.99)$$

In practical applications the Gibbs energy is more useful than A, as most experiments (say, chemical reactions) are performed in a laboratory when the temperature and the pressure are fixed.

8.10.1 *The Gibbs-Duhem equation*

The Gibbs energy $G(T,P,N_1,\ldots,N_c)$ is an extensive thermodynamic quantity, hence a homogeneous functions of degree 1. Thus Euler's theorem (see appendix B.2) gives the important identity

$$G(T,P,N_1,\ldots,N_c) = \sum_{i=1}^{c} N_i \mu_i \qquad (8.100)$$

Taking the total differential

$$dG = \sum_{i=1}^{c} N_i \, d\mu_i + \sum_{i=1}^{c} \mu_i \, dN_i \qquad (8.101)$$

and equating it to (8.94)

$$dG = -S \, dT + V \, dP + \sum_{i=1}^{c} \mu_i \, dN_i$$

we get the **Gibbs-Duhem equation**

$$\sum_{i=1}^{c} N_i \, d\mu_i = -S \, dT + V \, dP \qquad (8.102)$$

named after Josiah Willard Gibbs and Pierre Duhem (1861–1916). At constant T and P (8.102) gives

$$\sum_{i=1}^{c} N_i \, d\mu_i = 0 \qquad (8.103)$$

The Gibbs-Duhem equation features the differentials of the intensive variables (μ_i, P, T), multiplied by the extensive variables (N_i, V, S), and describes the changes in chemical potential for the components in a thermodynamical system. In particular, only $c - 1$ of c components have independent values for the chemical potential. The Gibbs-Duhem equation (8.102) shows that the intensive properties are not independent but related.

8.10.2 *The principle of maximum work*

Gibbs (1874) showed that, for all thermodynamic processes between the same initial and final states, the maximum work is produced by a reversible process. This is also known as the "principle of maximum work" although it is actually a theorem, which we prove here.

Consider a system interacting with a thermostat at constant temperature T, undergoing infinitesimal changes of internal energy dE and of entropy dS while performing a work $-\delta W$ on some other system[13] and

---

[13] According to our convention, the work is positive when performed *on* the system and negative when done *by* the system.

receiving heat δQ from the thermostat. From the first law (8.79) we know that $dE = \delta Q - \delta W$. Accounting for the inequality (8.85), the first law implies

$$\delta W \leq dE - T\,dS = dA_T \qquad (8.104)$$

where the last term expresses the variation in Helmholtz energy (8.86) at constant temperature. This relation says that the change of Helmholtz energy is the maximum work that can be obtained in a reversible process at constant temperature.

If we consider a system at constant T, P and write

$$\delta W = \delta W_{\text{net}} - P\,dV$$

where δW_{net} is any other work other than $P\,dV$, from the inequality (8.104) we get

$$\delta W_{\text{net}} \leq dE - T\,dS + P\,dV = dG_{T,P} \qquad (8.105)$$

This means that the maximum net work (total work minus PV-work) that can be done by a system at constant temperature and pressure is the change in Gibbs energy.

8.11 Reflections on the Meanings of Entropy and the Second Law

In this and the previous chapters we have developed the tools of thermodynamics. We have used some concepts from mechanics with which we were already familiar. We have also defined a few new concepts that were not part of mechanics. In the rest of the book we shall be concerned mainly with applications of these tools. At this point, it is appropriate to pause and ponder about the meaning and the significance of the new concepts that were introduced into our vocabulary. The central concept, the one that has stirred so many disputes, spawned many interpretations (and misinterpretations), and was enshrouded in thick layers of mystery, is undoubtedly the concept of entropy.

8.11.1 *Name and interpretation*

The story of entropy starts in 1850, when Rudolf Clausius coined the term *entropy*. Here is a quotation from the original passage from Clausius, as cited by Cooper (1969).

> I prefer going to the ancient languages for the names of important
> scientific quantities, so that they mean the same thing in all living
> tongues. I propose, accordingly, to call S the **entropy** of a body,
> after the Greek word for "**transformation**". I have designedly
> coined the word entropy to be similar to **energy**, for these two
> quantities are so analogous in their physical significance, that an
> analogy of denominations seems to be helpful.

Right after quoting Clausius' motivation for choosing the word "entropy", Cooper commented:

> By doing this, rather than extracting a name from the body of the
> current language (say: **lost heat**), he succeeded in coining a word
> that meant the same thing to everybody: **nothing**.

There have been lots of criticism regarding the appropriateness of the
term "entropy" to convey the real meaning of the concept that Clausius
had in mind. However, one should remember that the term "entropy"
was defined in terms of heat transferred and temperature, two measurable quantities, without any reference to the atoms and molecules. This
term was introduced at a time when the atomic nature of matter was far
from being universally accepted by physicists. Today we know that the
concept of entropy, as defined by Clausius, is not what the word "entropy"
means either in ancient or in contemporary Greek. Also, we know that
the two *quantities*, energy and entropy, are not "analogous in their physical significance". Therefore, the very term "entropy" chosen by Clausius
was inadequate, and perhaps even misleading. Nevertheless over a hun-
dred years this term stuck, and scientists normally use this term without
giving it a thought to its original meaning.

In this book, we shall accept the word "entropy" to describe the quantity as defined by Clausius, disregarding the *meaning* of the word in either ancient or in modern Greek. Thus, we accept the fact that there is
a quantity denoted by S which is a state function, and that for any spontaneous (natural) process in an isolated system can only increase. We
agree to call this quantity *entropy*.[14]

---

[14]Forget about the *meaning* of the *word* entropy. Thermodynamics does not even provide any
meaning to the quantity denoted by S, nor it does use the meaning of entropy. Thermodynamics is concerned only with measurements and calculations of changes in entropy, not with the
meaning of these changes.

Throughout the years people have given many qualitative interpretations, descriptions and metaphors of what entropy really means. The most popular of these is no doubt the "disorder metaphor".[15] Indeed, in many processes which occur simultaneously, and for which we have a good idea of what is happening on a molecular level, it is intuitively acceptable to associate an increase in entropy with an increase in disorder. As we have shown in a few examples, this intuitive interpretation is sometimes correct but other times it could be even misleading (a specific example is gas mixing, as we have seen in section 7.4).

Late in the 19-th century, Ludwig Boltzmann and Joshua Willard Gibbs developed the statistical mechanics. They found that a quantity, which we shall here denote by S_{SM} (for the statistical mechanical entropy), has all the properties of Clausius' entropy S. Therefore, for all practical purposes one could identify S_{SM} with S, i.e. the statistical mechanical entropy with the thermodynamic entropy.

With this identification of S, we have gained a *microscopic* description of entropy in terms of some probability distributions. The entropy defined by Gibbs had the mathematical form

$$S_{\text{SM}} = -k_{\text{B}} \sum_{i=1}^{n} p_i \ln p_i \qquad (8.106)$$

whereas Boltzmann's entropy applies to the particular uniform distribution $p_i = 1/n$, as

$$S_{\text{SM}} = k_{\text{B}} \ln n \qquad (8.107)$$

Although Boltzmann's entropy had a simple and straightforward meaning in terms of the total number of states (or the number of "complexions") n, the more general statistical mechanical expression by Gibbs lacks a simple intuitive meaning.

In 1948 Claude Shannon was interested in communication theory, a subject remote from, and totally unrelated to, thermodynamics. Shannon found that *for any probability distribution* $p_1, \ldots, p_n$ one can define a quantity H by

$$H = -\sum_{i=1}^{n} p_i \log p_i \qquad (8.108)$$

---

[15]Other less popular metaphors are the "spreading", "mixed-upness", "chaos", "information" or "missing information".

This quantity has a simple intuitive meaning as a measure of the amount of information *contained* in the distribution.[16] Clearly this quantity, which we have referred to as Shannon's measure of information (SMI), is a very general concept. It is defined for *any* distribution. It has also a very simple and intuitive meaning — so simple that anyone who has ever played the popular 20-question game could easily understand — and, of course, there is no tinge of mystery in the SMI.

It is not surprising to find that the SMI was found a very useful tool in so many fields of research, from communication theory to economics, sociology, psychology, linguistics, arts and many more. Whenever a distribution features in a theory, one can define the SMI on that distribution and study the *measure* of information contained in this distribution.

It is an unfortunate practice that most people who intend to use the *concept* of Shannon measure of information, refer to it as "entropy". The origin of this practice is found in Tribus & McIrvine (1971):

> What's in a name? In the case of Shannon's measure the naming was not accidental. In 1961 one of us (Tribus) asked Shannon what he had thought about when he had finally confirmed his famous measure. Shannon replied: "My greatest concern was what to call it. I though of calling it 'information', but the word was overly used, so I decided to call it 'uncertainty'. When I discussed it with John von Neumann, he had a better idea. Von Neumann told me, 'You should call it entropy, for two reasons. In the first place your uncertainty function has been used in statistical mechanics under that name. In the second place, and more important, no one knows what entropy really is, so in a debate you will always have the advantage'."

Shannon accepted von Neumann's suggestion and called the SMI "entropy".

On von Neumann's suggestion, Denbigh (1981) commented

> In my view von Neumann did science a disservice! (...) There are, of course, good mathematical reasons why information theory and statistical mechanics both require functions having the same formal structure. They have a common origin in probability theory, and they also need to satisfy certain common requirements such as additivity. Yet, this formal similarity does not imply that the

---

[16]One can also regard H as the amount of *missing* information, if one contemplates on the problem of retrieving the information by asking binary questions (see chapter 4).

functions necessarily signify or represent the same concepts. The term 'entropy' had already been given a well-established physical meaning in thermodynamics, and it remains to be seen under what conditions, if any, thermodynamic entropy and information are mutually inter-convertible.

Indeed, von Neumann's suggestion did science a disservice! It is not because entropy is not an SMI, but because the SMI is not entropy. In other words, the two concepts are not "mutually interconvertible".

It is true that the SMI (8.108) and the statistical mechanical entropy (8.106) have the same mathematical form (except for the constant k_B and possibly for the choice of the base of the logarithm). However, these mathematically identical forms do not imply the identity of the concepts. Thus the SMI is, in general, not the same as Clausius' entropy. However, if we agree to identify Clausius' entropy with the statistical mechanical entropy, then we can *interpret* Clausius' entropy as a special case of SMI, but not the other way around. The SMI is not entropy!

In a recent article by Sheehan & Gross (2006) we find the statement:

> Let me begin by stressing the spectacular successes of, and the foundational roles played by the entropy and the second law, not just in statistical and thermal physics, but across all the sciences and technology — and beyond.

What the authors had in mind is that the concept that has been so successful in so many fields is the SMI, although it somehow pervertedly is referred to as entropy. Clearly, an analysis of the SMI defined on the distribution of letters in English or in French has nothing to do with the thermodynamic entropy. Therefore, the reference to SMI as entropy in all the sciences, except in thermodynamics, should be discontinued.

8.11.2 *The concept*

In the traditional derivation of the second law, one usually starts with the Carnot cycle, next one obtains the Clausius inequality, which has the form $\sum \frac{Q_i}{T_i} \geq 0$ (or its continuous analog $\int \frac{\delta Q}{T} \geq 0$), then one introduces the *entropy* by defining its infinitesimal change for a "reversible" process as

$$dS = \frac{\delta Q}{T} \qquad \text{(reversible)} \qquad (8.109)$$

Finally one formulates the *second law* as

$$dS > \frac{\delta Q}{T} \quad \text{(irreversible)} \tag{8.110}$$

The first equation (8.109) introduces the function of state S (called entropy) in terms of heat transfer divided by the absolute temperature of the system. The second inequality (8.110) is valid for all irreversible changes in a closed system.

Note that above we used the traditional terms "reversible" and "irreversible" in (8.109) and (8.110), although in this book we use a different terminology: the equality (8.109) is valid for a quasi-static heat exchange between a *closed* system and its surroundings, while the inequality (8.110) is valid for a spontaneous process *within* a closed system, when there is also a quasi-static exchange of heat with its surroundings. If the system is adiabatic, $\delta Q = 0$, and the inequality in (8.110) becomes $dS > 0$.

The two statements (8.109) and (8.110) may be combined by introducing a quantity $\delta Q'$, called the *uncompensated heat* by Clausius, which is always positive and is *defined* by

$$\delta Q' \equiv T\,dS - \delta Q > 0 \tag{8.111}$$

Denoting

$$dS_{ex} \equiv \frac{\delta Q}{T} \quad \text{and} \quad dS_{in} \equiv \frac{\delta Q'}{T} \tag{8.112}$$

one has

$$dS = dS_{ex} + dS_{in} \tag{8.113}$$

which is the same as eq. (8.82).

Here are a few statements which are typical in the traditional introduction of the second law. Prigogine & Defay (1954) say

> The entropy of the system can vary for two reasons and for two reasons only; either by *transport of entropy* to or from the surroundings through the boundary surface of the system, or by *creation of entropy* inside the system.

Thus, dS_{ex} is referred to as the entropy that is "transported", or "exchanged" between the system and its surroundings, while $dS_{in} = \delta Q'/T > 0$ means that entropy can only be created in an irreversible process. They argue then that the definition of dS_{in} in (8.112) "gives the uncompensated heat a physical significance". However, we have already explained that entropy is not created, and most importantly it cannot "flow".

The traditional derivation of the second law is not only lengthy and tedious, formal and abstract, but above all leaves the student who first encounters the entropy and the second law baffled and bewildered. All efforts in attempting to understand the meaning of entropy and the second law are doomed to fail. Frustration gradually turns into a feeling of desperation, leaving the impression that entropy is a mystery and its understanding will forever be hopeless. This is true for students of physics and chemistry who will apply the second law in their studies and later in their researches. It is a fortiori true for students of all other branches of sciences, who may keep only a faint memory of something they have once learned, called entropy, something that "have resisted complete explanation", something that "taps into the deepest unresolved mysteries in modern physics" (Greene, 2004).

Let us see where the mystery sprang from and whether we can do anything about it.

First, δQ is a small amount of heat, that is some form of energy that flows across the walls of the system. The temperature T is also understood as something that has a constant value at any point in the system at equilibrium. Since dS_{ex} is *defined* as $\delta Q/T$, it is natural to conclude that, if δQ *flows* into (or from) the system, then also dS_{ex} is also something that *flows* into (or from) the system. But what is the thing that is "smuggled along" with the heat, when heat crosses the wall of the system? Thermodynamics does not provide any answer to this question.

Second, as dS_{in} is defined in terms of the "uncompensated heat" $\delta Q'$, which is not even heat, it leaves the impression that it is another kind of entropy spirited along with the uncompensated heat. So what is this entropy? Is it created? Created from what? Perhaps from the uncompensated heat (which is not heat), or from nothing?

In our approach to the second law we do not provide answers to any of these questions: those questions simply do not arise. The mystery is not removed, it simply does not appear.

In our approach, the entropy is derived from a simple, well defined and mystery-free concept, Shannon's measure of information (SMI). As we have seen in chapter 4, this concept can be made even more familiar by relating it to the average number of binary questions that one needs to ask in the familiar 20-question game.

Once we adopt this point of view, then it becomes awkward to talk about "entropy flow" or "entropy crossing a wall", as the entropy is a special kind of SMI, and the SMI can increase or decrease, but does not flow.[17] The entropy is defined as a measure on certain distributions, characterizing a macroscopic system at thermodinamic equilibrium. As such, this measure can *increase* or *decrease*, but it does not *move* from one system to another. Of course, when heat *moves* into the system, it causes a change in some distribution. This change in the distribution in turn causes a change in SMI, but the SMI does not (and can not) flow into or from the system. Saying that the entropy flows into the system makes no sense, as much as saying that the number of states flows into the system.

A lady taking vitamins to improve her health, and perhaps to look more attractive, clearly causes the vitamins to *diffuse into* her body, as a result of which she might become healthier and more attractive. Would anyone claim that "health" or "beauty" have *flown* into her body? The vitamins have *flown* into her body, and her "health" or her "beauty" have changed as a result of the flow of the vitamins, but they have not flown into her body!

Similarly, when heat, volume, or material are exchanged between the system and its surroundings, the entropy of the system can increase or decrease. The change of entropy is a result of a change in some *distribution within* the system, not a result of being *exchanged* with the surrounding.

The same is true when a spontaneous process occurs within the system: the entropy of the system changes because some distribution has changed *within* the system. To say that entropy is *created* in the system implies that something substantial is created. Although this is less awkward than saying that "entropy flows", it should be avoided too.

Actually, although today we continue saying that "heat flows" following the tradition started by assuming that the heat is some kind of fluid,

---

[17]The reader might argue that using the informational interpretation of entropy permit us to talk about "information that flows into or from the system". That is true. Information can flow into or from the system, but the SMI is not information, but a particular *measure* of information.

in reality nothing really flows. Rather, kinetic energy is exchanged between molecules and the walls of the system, and this energy is transmitted from the walls to the molecules of another system. We can continue to call this energy transfer "heat flow", provided that we keep firm in mind that, when heat flows into the system, it affects the velocity distribution of the molecules in the system.[18] This change in the distribution causes a change in the entropy. This is true for any spontaneous process occurring within the system: some distribution is changed and, as a result, the entropy changes.

In this chapter, we introduced the concept of entropy as a generalization of the entropy of an ideal gas derived in chapter 6. Of course there are many systems for which the explicit function for the entropy cannot be calculated, neither by thermodynamics (where it is not needed), nor by statistical mechanics (where it is needed, but it is usually extremely difficult to calculate). Therefore, we are lucky that we have one system (an ideal gas) for which the entropy can be calculated exactly, and the meaning of which is crystal clear and mystery-free. Having that function in mind, we proceeded with the assignment of an entropy function for any thermodynamic system, even when we do not know its explicit form, maintaining its meaning and its significance.

8.11.3 *The role of time*

Finally, we recall that entropy, as well as all other derived quantities, is defined *only* for equilibrium states. As such, strictctly speaking *the entropy is not a function of time*. This is in contrast to the statement that the second law is the "classical statement that the entropy of an isolated system increases with time" (Prigogine & Defay, 1954). Actually, since thermodynamics deals only with equilibrium states, and the entropy is defined only for an equilibrium state, there is no need to talk about entropy as a function of time.

As we have seen in the previous sections, if we remove an internal constraint in an isolated system at equilibrium, the system evolves spontaneously to a new equilibrium state. The entropy of the new equilibrium

---

[18]This is strictly true only for an ideal gas. For the general case, the heat flowing into the system will cause a change in the distribution over the energy levels of the system.

state is higher than the total initial entropy. The entropy of the system is *not defined* in any intermediate state, between the initial and the final equilibrium states. Therefore, we can not speak of entropy as a function of time. The entropy is only a *function of the state of a thermodynamic system at equilibrium*.

It is true that the entropy always changes in the same direction, as the time elapses. This fact does not imply that there exists a function $S(t)$ that describes the change in entropy with the time. The entropy is only defined at the initial and at the final equilibrium states. It is not defined at any intermediate stage.

When we remove a constraint in the system, the system moves from an initial equilibrium state (i) to a final equilibrium state (f). If this process occurs spontaneously in an isolated system, then $S_f > S_i$. Since S_f is attained at a *later* time than S_i, it is correct to say that the direction of change of S is the same as the direction of change of time. However, this does not imply the existence of a function $S(t)$. All we know is that $S(\xi_f) > S(\xi_i)$ for two equilibrium states ξ_i and ξ_f occurring at times t_f and t_i such that $t_f > t_i$.

If we perform the process quasi-statically, we actually follow the system at some finite sequence of discrete steps $\xi_1 = S_i, \xi_2, \ldots, \xi_n = S_f$. Correspondingly, we have a finite (non continuous) sequence of entropy values $S(\xi_1), \ldots, S(\xi_n)$. Since $S(\xi_{i+1})$ is larger than $S(\xi_i)$, and since the state ξ_{i+1} occurs at a later time than ξ_i, we might be tempted to infer that the entropy is a function of time. However, we should realize that this "function" of time is only a series of values of the entropy pertaining to the specific sequence of steps $\xi_1, \ldots, \xi_n$. There are infinite sequences of steps connecting the initial and final states, along which this process can actually occur, and the time intervals between the steps can also be changed in arbitrary ways. For each of these sequences there is a *different* sequence of entropy values. Therefore, there is no single function that we can denote as $S(t)$.

There have been many attempts to extend the definition of entropy to non-equilibrium states. In fact, Boltzmann himself has done that in his famous H-theorem. In all of these attempts, including Boltzmann's one, the quantity which is defined on any arbitrary (non-equilibrium) distribution is actually what we call the SMI, not the entropy, which is defined in thermodynamics only for equilibrium states. Thus, when the state of the system evolves between two equilibrium states, there is a corresponding evolution of some distribution, and a corresponding evolution of the SMI. However, the entropy does not evolve with time: it changes its value from one equilibrium state to another.

Of course one can define the SMI H on states other than equilibrium states. For example, if we start with a system of particles in one compartment and open a hole in the partition, such that particle can flow from one compartment to another (as in section 8.8.3), we can define a *distribution*, and a corresponding SMI at any time. Therefore, we have a function $H(t)$. Note however that this function is not unique for the expansion of the gas from volume V to volume $2V$: different hole sizes will cause different flow rates of particles, hence different rates of change for the distribution of the particles, hence different functions $H(t)$. For each specific hole, or more generally for each well defined process, we shall have a different function $H(t)$.

The situation is more complicated for the entropy of the system. We can open the hole for a short period of time Δt, wait until the system reaches equilibrium, open again the hole for another short time Δt, and so on. We shall obtain a sequence of equilibrium states $\xi_1, \ldots, \xi_n$. Corresponding to this sequence of states, we have a sequence of entropies $S(\xi_1), \ldots, S(\xi_n)$. Also we know that the *order* of these numbers is the same as the order of the times, as $t_1 < t_2 < \ldots < t_n$ and $S(\xi_1) < S(\xi_2) < \ldots < S(\xi_n)$, but this does not mean that there exist a function $S(t)$. We can change the sequence of opening and closing of the hole, and we can change the intervals Δt of the times between the steps: in each case we shall get a different sequence of $S(\xi_1), \ldots, S(\xi_n)$.

Thus, we cannot talk of entropy as a function of time even for the simplest process of expansion of gas. Can we talk about the change of entropy in a living system with time? Perhaps one day we would be able to do so, but that time is very far from now.

8.12 Summary

In this chapter, we learned the first law of thermodynamics and the related concepts of internal energy (a state function), heat and work (which are not state functions). By focusing on the mechanical work performed in an expansion process of a gas (or its reverse, the gas compression), we considered different types of quasi-static processes: isothermal expansion, adiabatic expansion, and isochoric transformations. Idealized cyclic processes like the Carnot and Stirling cycles have been reviewed, noticing that their efficiency represents the maximum achievable value.

We also reviewed the basic postulates that are used to build thermodynamics in this book. The most important assumption is the existence of an equilibrium state. After having postulated the existence of an additive state function with similar properties as the entropy function for an ideal gas, which corresponds to the SMI for the equilibrium distribution, we have formulated the second law as the principle of maximum entropy over all possible constrained equilibrium distributions. If some constraint is removed, the system will evolve toward a new equilibrium state with larger entropy, because this corresponds to a much more probable macroscopic state. By further postulating that the total differential of the entropy function is the same as for an ideal gas, eq. (8.57), the evolution after removal of some constraint can be also understood and quantitatively described.

In this chapter we have also presented the main tools with which we shall be working with in the rest of this book. They are very general and may be applied to any thermodynamic system at equilibrium. The tools were derived from the first and the second laws.

It is important for the reader to understand what these laws imply with respect to the feasibility or unfeasibility of certain processes. We have introduced the concept of a quasi-static transformation, to help us calculating work and heat exchanges between the system and its surroundings, in some specific processes. An example that used to be the cornerstone of thermodynamics is the Carnot cycle, an idealized heat engine working with two characteristic temperatures, which is able to convert the maximum amount of heat into work.

We have seen the implications of the properties of the entropy function $S(E, V, N)$ regarding the conditions for the equilibrium state. The most important result of this kind is the condition of equilibrium with respect to flow of matter, either from one location to another, or from one species to another. In both cases, matter flows from a high to a low chemical potential. In the rest of the book, we shall see that the chemical potential is the most important tool in the application of thermodynamics to real systems.

In most applications of thermodynamics one does not need to know the meaning of entropy. However, we have built up the concept of entropy from the simple, well defined and easily interpreted Shannon's measure of information (SMI). This removes any sense of mystery related to entropy and the second law.

PART 2

Applications

Chapter 9

The Phase Rule and Phase Diagrams

In this chapter we consider different systems characterized by different phases, and learn how to describe a thermodynamic system graphically. We shall first present the *phase rule*, which was derived by Gibbs. This rule is essentially an extension of the experimental fact that a single phase of a pure component can be characterized thermodynamically by two intensive parameters. Its molecular interpretation is far from being trivial.

Next we consider the coexistence curves of a pure substance in different phases (solid, liquid, gas) and several examples of mixtures of two components, as a function of their composition and temperature.

9.1   States of Matter and Phase Transitions

A *phase*, or *state of matter*, is a form of matter that is uniform throughout in chemical composition and physical state. In other words, a phase is a region of a thermodynamic system, throughout which all intensive quantities of a substance are essentially uniform. Typical phases are *solid* (shape and volume are well defined), *liquid* (fixed volume, with shape driven by the container), and *gas* (both shape and volume determined by the container), but there are also other phases.

For example, a gas at very low pressure and high temperature may be fully or partially ionized, in the sense that electrons are free to move, whereas in normal conditions they are confined inside some atomic or molecular volume. This ionized gas is called a *plasma*, and is found in several astrophyisical conditions, from the van Allen radiation belts in the Earth magnetosphere to the solar corona, from stars to intergalactic regions. Indeed, plasma is by far the most abundant phase in the Universe.

When considering quantum effects, other exotic states of matter appear, as we mention briefly in section 9.1.1. However, even at laboratory conditions one finds more variety. For example, it is possible to find elements that, in a single phase, may show further differentiation in terms of atomic aggregation, i.e. different **allotropes** of the same substance. For example, the allotropes of carbon include diamond (where the carbon atoms are bonded together in a tetrahedral lattice arrangement), graphite (where the carbon atoms are bonded together in sheets of a hexagonal lattice), graphene (single sheets of graphite), and fullerenes (where the carbon atoms are bonded together in spherical, tubular, or ellipsoidal formations). They have different physical properties.

At molecular or crystalline level, it is also possible to have different forms of atomic aggregation. The ability of a solid material to exist in more than a single form of crystal structure is called **polymorphism** (the term "allotropy" only refers to pure chemical elements). For example, calcite and aragonite are both forms of calcium carbonate. Another example is silicon dioxide ($Si\,O_2$), also known as silica, which is known to form many polymorphs, like α-quartz, β-quartz, tridymite, cristobalite, coesite, stishovite, etc.

There are also molecules whose 3-dimensional structure admits two mirror images that are not superposable, called **enantiomers**. They have the same chemical and physical properties, apart from their effect on light polarization: they rotate the polarization in opposite directions.

Whether these different forms of the same substance have to be considered as different phases depends on the problem under consideration. For example, unless the observable of interest is the polarization of light, the enantiomers of the same molecule belong to the same phase.[1]

A **phase transition** is the conversion of one phase into another phase, which occurs at a characteristic temperature for a given pressure. During a phase transition of a given medium, certain properties of that substance change, often discontinuously, as a result of the change of some external condition, such as temperature, pressure, or others.

---

[1] Sometimes enantiomers can also form separate solid phases.

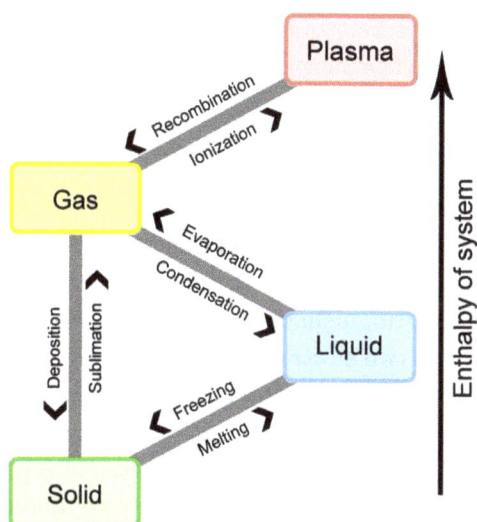

Figure 9.1: Phase changes.

For example, when increasing the temperature of a liquid at constant pressure, when the medium reaches its ***boiling point*** the volume changes abruptly, as the result of molecules passing from the liquid phase to the vapour phase of the substance. Similarly, the ***freezing point*** may be reached by lowering the temperature. As moleculs move from the liquid to the solid phase, the density of the medium changes in a discontinuous way.

The passage from solid to liquid phases is called ***melting***, while ***sublimation*** is the solid-gas transition. The reverse phase transitions are called ***freezing*** (liquid-solid) and ***deposition*** (gas-solid) (figure 9.1).

The passage from liquid to gas phases is called ***boiling*** or ***evaporation***. A ***vapor*** (USA spelling) or ***vapour*** (UK spelling) is a substance in its gas phase when the temperature is below the critical point (see section 9.3.3 below). The reverse phase transitions is called ***condensation***.

The gas-plasma phase transition is called ***ionization***, and happens for example during lightning, the powerful sudden flow of electricity accompanied by thunder that occurs during an electric storm. The result of the electric discharge is that electrons are stripped off their atoms. Then the electrostatic force between these negative electrons and the positively charged ions acts to restore neutrality, and **recombination** happens with the emission of the visible light.

9.1.1   *Non-classical states of matter*

The four phases solid, liquid, gas, and plasma, can be described in the framework of classical physics, but other states of matter emerge when considering quantum mechanics. At microscopic level, particles of the same species are indistinguishable. For example, there is no way of distinguish between any two electrons. For example, when they collide one cannot tell which particle came from the left and which from the right side.

The **spin** is the intrinsic angular momentum of a particle, independent of its motion across the space. There are two fundamental families of particles, called **fermions** and **bosons**, characterized by semi-integer (1/2, 3/2, 5/2, etc.) and integer (0, 1, 2, etc.) spin, respectively, in units of the **reduced Planck constant** $\hbar \equiv h/(2\pi)$.

The quantum description of the system is the same after swapping two identical bosons, but changes sign when exchanging two identical fermions. As a consequence, fermions and bosons obey different quantum-mechanical rules, following the so-called Fermi-Dirac[2] and Bose-Einstein[3] statistics. The main difference is that two identical fermions cannot occupy the same quantum state, whereas any number of identical bosons can occupy the same quantum state.

An extreme example of the latter situation is when a gas of bosons is cooled down to almost $0\ \mathrm{K} = -273.16°\mathrm{C}$. In this case, a large fraction of bosons occupies the quantum state with the lowest energy, obtaining a macroscopic quantum system, called **Bose-Einstein condensate**, with very peculiar behavior.

Electrons are fermions, and this explains the structure of atoms. The electrostatic force that, classically, would bring the negatively charged electrons inside the positively charged nucleus (made of protons, with charge +1, and neutrons, with charge 0) would cause an atom to collapse in a very short time (of order of 10^{-9} s). Despite this, atoms are stable.

---

[2]Named after Enrico Fermi and Paul Adrien Maurice Dirac (1902–1984).
[3]Named after Satyendra Nath Bose (1894–1974) and Albert Einstein.

The electrons move in a much larger volume than the nucleus (which occupies only 10^{-4} of the atomic volume), and are characterized by quantum numbers that fix their energy levels and orbital angular momentum. Once these quantum numbers are fixed, this defines a unique quantum state. Because electrons are fermions, only one electron can occupy each state (the so-called ***Pauli's exclusion principle***). In addition, at atomic level energy is discrete: electrons can move across energy levels by making finite "jumps".

A more exotic form of matter, which manifests itself only at extremely high densities, is the neutron-degenerate matter that constitutes the core of neutron stars. One speaks of ***degenerate matter*** when the density is so high that the dominant contribution to the pressure comes from Pauli's exclusion principle.

9.2   The Phase Rule

The phase rule is essentially a straightforward extension of the experimental fact that the thermodynamic state of a system consisting of a pure component in a single phase can be described by two intensive variables. Thus, we only need a pair of variables like T and μ, or T and P, or T and $\rho = N/V$ (the number density), etc. Given the values for this pair of variables, all other intensive variables are determined. For example, if we choose P and T to describe the system, the chemical potential is *determined* by T and P, i.e. $\mu = \mu(T,P)$.[4]

The reader can easily realize that for a system composed of some 10^{23} particles we would need an enormous number of parameters to describe its microscopic state or ***microstate***, in short. This description would require the specification of the position (3 values) and momentum (3 additional values) for each particle. Hence it is a remarkable fact that the thermodynamic state of such a system, which is a macroscopic state or ***macrostate***, may be described by only two parameters. Of course, the thermodynamic state is not a detailed description of the location and momentum of each particle. Nevertheless, it is far from obvious that only

---

[4]Here, we describe the *state* of the system by intensive variables. If we need to describe also the *size* of the system, we need to specify at least one more extensive variables, such as the volume, the number of molecules (or moles), etc.

two parameters are sufficient for the thermodynamic description of the state of a pure single-component system.

In all discussions in this chapter we shall assume that the system is very large, and that surface effects are negligible. Also, we assume that there are no external fields acting on the system, or that they have negligible effects on the properties of the system (as it is the case for the gravitational field in a laboratory). We always discuss systems at equilibrium.

Having a single phase of one component at equilibrium, we can say that we have two *degrees of freedom*, meaning that one could change the two intensive variables arbitrarily, as long as we maintain a single phase.

Now suppose that we have a single phase, which contains c components. At this stage, we also assume that there are no chemical reactions between these components. The composition of the system may be specified by the c mole fractions $x_1, x_2, \ldots, x_c$, which satisfy the condition $\sum_{i=1}^{c} x_i = 1$. Hence only $c - 1$ of these c parameters are actually independent, while we change the composition of the system. When we include also the two intensive parameters needed to specify a unique macrostate for a single component, we find that a total of $c + 1$ independent parameters is required to specify the thermodynamic state of the whole system. Thus, *a single phase with c components has $c + 1$ degrees of freedom.*

Next, suppose we have p phases (e.g. solid, liquid, gas) at equilibrium. Intuitively, one would think that $p(c + 1)$ parameters are needed to define a unique state, but this is not true: less parameters are actually sufficient. If each phase is isolated from all the others, then of course we must describe the system by $p(c + 1)$ parameters. However, this is not the case when the phases are at thermal, mechanical and material equilibrium. For instance, if we have just two components and two phases, then the first phase is described by $c + 1 = 3$ parameters, say P, T and x_1, but for the entire system we do not need twice as many parameters, since the system is at equilibrium (thermal, mechanical and material): the temperature, the pressure and the chemical potential of all components will be uniform throughout the system.

Thus, in general for p phases and c components in thermodynamic equilibrium, the following conditions hold[5]:

$$
\begin{cases}
T = T^{(1)} = T^{(2)} = \cdots = T^{(p)} & \text{(thermal equilibrium)} \\
P = P^{(1)} = P^{(2)} = \cdots = P^{(p)} & \text{(mechanical equilibrium)} \\
\mu_1 = \mu_1^{(1)} = \mu_1^{(2)} = \cdots = \mu_1^{(p)} & \text{(material equil. w.r.t. species 1)} \\
\mu_2 = \mu_2^{(1)} = \mu_2^{(2)} = \cdots = \mu_2^{(p)} & \text{(material equil. w.r.t. species 2)} \\
\vdots & \\
\mu_c = \mu_c^{(1)} = \mu_c^{(2)} = \cdots = \mu_c^{(p)} & \text{(material equil. w.r.t. species } c)
\end{cases}
$$

Altogether, we have $(p-1)$ equalities in each row and we have $(c+2)$ rows, therefore, we have $(p-1)(c+2)$ conditions that must hold at equilibrium. We have to subtract this number from the $p(c+1)$ parameters describing p isolated systems in a single phase with c components, in order to obtain the number d of degrees of freedom of a system in which all these phases are in thermodynamic equilibrium. Thus, we obtain the **Gibbs' phase rule**

$$
\boxed{d = c - p + 2} \tag{9.1}
$$

Thus, d is the number of intensive parameters that we can change independently while p phases are maintained in a system with c components at equilibrium. For a system with a single species ($c = 1$) and typical laboratory conditions[6], there are three possible situations compatible with equation (9.1):

- $p = 1$ (one phase, like solid, liquid or gas) implies $d = 2$. The system must have at least one phase, in which case we can change independently two intensive parameters. We can represent this on a plane, as in figure 9.2, showing the water phase diagram;
- $p = 2$ (two phases) implies $d = 1$. When there are two phases at equilibrium, we can change only one parameter: we move along a two-phase equilibrium line in the phase diagram. In figure 9.2 they are called SL (solid-liquid separation), LV (liquid-vapour), and SV (solid-vapour);

---

[5]"w.r.t." = with respect to.
[6]We do not consider exotic phases like a plasma or the core of a neutron star.

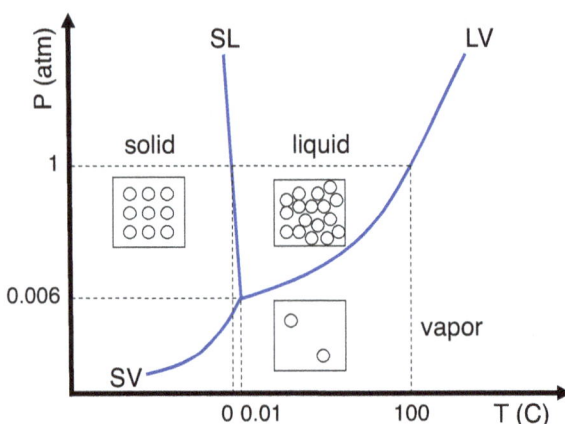

Figure 9.2: Phase diagram of water. Melting and boiling temperatures at the standard 1 atm = 101 325 Pa pressure are (by definition) 0°C = 273.15 K and 100°C = 373.15 K, respectively. The triple point has pressure of 611.73 Pa = 0.006037 atm and temperature 273.16 K = 0.01°C.

- $p = 3$ (three phases) implies $d = 0$. If solid, liquid and vapour phases coexist, there is no degree of freedom: we are at a well defined equilibrium point, called the **triple point** because it connects the solid-liquid equilibrium line with the liquid-vapour and the solid-vapour equilibrium lines. The triple point is obtained with a single set of values for the intensive quantities.

So far we have derived the phase rule for a system of c *non-reacting* components. We shall see in chapter 11 that, if there is a chemical reaction between some of the components, then we have one additional condition that we call **chemical equilibrium**. In this case, we have to take into account also the chemical equilibrium when computing the number of degrees of freedom.

For instance, a gas mixture containing hydrogen, nitrogen and ammonia at very low temperature may be considered as a three-component system, and the number of degrees of freedom is

$$d = c + 2 - p = 3 + 2 - 1 = 4 \tag{9.2}$$

However, if the temperature is high or if we add a catalyst so that a chemical equilibrium exists between the three components, then

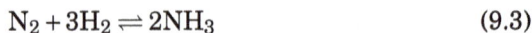

$$N_2 + 3H_2 \rightleftharpoons 2NH_3 \tag{9.3}$$

In this case, the condition of chemical equilibrium is

$$\mu_{N_2} + 3\mu_{H_2} = 2\mu_{NH_3} \tag{9.4}$$

(we will discuss it in more details in chapter 11). This condition effectively reduces the number of degrees of freedom by one: instead of (9.2), we have $d = 3$.

Another way of viewing the change in d is to recognize that, although we have three species, we have only two components whose concentration can be changed arbitrarily: the third concentration is determined by the condition of the chemical equilibrium. We must count only the number of *independent* components.

In general, if we have c species undergoing r chemical reactions, we have $c - r$ independent components.

Note that there is some arbitrariness in counting the number of components even when there are no chemical reactions taking place in the system. For instance, two isotopes of argon in the gas phase would have three degrees of freedom. However, if we are not interested in changing the isotopic composition of such a system, we can ignore the isotopic composition, and treat the system as a one-component system for which $d = 2$. Similarly, if we have water with negligible amounts of salt, we can ignore these impurities and treat the system as a pure one-component system.

9.3 One-Component Systems

A simple phase diagram is one where there exists only one kind of solid state. In such a case the phase diagram consists essentially of three regions: vapour, solid, and liquid.

Figure 9.3 shows a typical simple phase diagram of a one component system (CO_2 in this example), where we choose the pressure P and the temperature T to describe the state of the system.[7] There are three regions (vapour, liquid and solid) in which we have only one phase ($p = 1$), therefore two degrees of freedom ($d = 2$). This means that within these

---

[7]As usual we do not care to specify the *size* of the system. We just assume that the system is large enough that surface effects can be neglected.

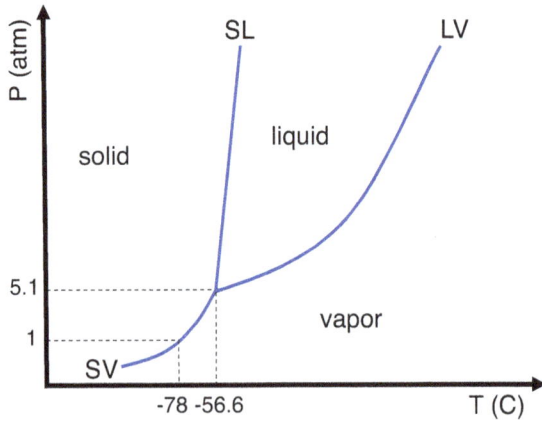

Figure 9.3: Phase diagram of carbon dioxide (CO_2).

regions we can change P and T as we wish, as long as we do not hit one of the boundary lines denoted by LV, SL and SV.

When changing both P and T inside the same region, we continue having a single phase. However, when we reach one of those boundary lines another phase appears and there is only one degree of freedom: at equilibrium we can not independently change P and T any more. Instead, we have to follow that line in one of the two possible directions (that is, P and T are related by some function). What is the shape of this curve? In general, we cannot write the equation of the curve explicitly, but measurements have been performed on many systems.

9.3.1 *Coexistence of two phases of the same component*

We know that if there are two phases, each of the phases may be described by two intensive variables. As we have chosen P and T, any other intensive variable must be determined by the values of P and T. In particular, the chemical potentials of each phase is determined when pressure and temperature are given. Let us consider liquid and vapour phases: then $\mu_{\text{vap}} = \mu_{\text{vap}}(P, T)$ and $\mu_{\text{liq}} = \mu_{\text{liq}}(P, T)$. When the two phases are at equilibrium we must have the condition

$$\mu_{\text{vap}}(P, T) = \mu_{\text{liq}}(P, T) \qquad (9.5)$$

Thus, although we do not know the *explicit* functions $\mu_{\text{vap}}(P, T)$ and

$\mu_{\text{liq}}(P, T)$, we know that when the two phases are at equilibrium their chemical potentials must have the same value. Equation (9.5) is an *implicit* relationship between P and T, which means that if we knew the explicit functions $\mu_{\text{vap}}(P, T)$ and $\mu_{\text{liq}}(P, T)$ we could solve for P in terms of T (or vice versa) and obtain a function

$$P = f_{\text{LV}}(T) \qquad (9.6)$$

For a single substance in a closed system with liquid and vapor phases at equilibrium, this defines the value of the **vapor pressure** at the given temperature.

As a hypothetical example suppose we knew that the chemical potentials follow two explicit functions

$$\mu_{\text{vap}}(P, T) = c_1 + c_2 T \log P \qquad (9.7)$$
$$\mu_{\text{liq}}(P, T) = c_3 + c_4 T \log P \qquad (9.8)$$

where c_1, c_2, c_3 and c_4 are some numerical constants. When the two phases are in equilibrium, we can apply the condition (9.5) to obtain

$$c_1 + c_2 T \log P = c_3 + c_4 T \log P \qquad (9.9)$$

and solving for T we get

$$T = \frac{c_3 - c_1}{(c_2 - c_4) \log P} = f_{\text{LV}}(P) \qquad (9.10)$$

which is an explicit function of the pressure.

Although in general we do not know the explicit form of the curves in the phase diagram, we can say something important about the *slopes* of these curves.

Consider the phase transition $\alpha \rightarrow \beta$, for example an evaporation with $\alpha =$ liquid and $\beta =$ vapor, and define the change in chemical potential as

$$\Delta\mu(\alpha \rightarrow \beta) \equiv \mu_\beta(P, T) - \mu_\alpha(P, T) \qquad (9.11)$$

On the coexistence curve, the condition of equilibrium (9.5) implies

$$\Delta\mu(\text{eq}) = 0$$

Let's take the differential of $\Delta\mu$ along the equilibrium line (this simply means that we move along a very small segment on the two-phase equilibrium line):

$$d(\Delta\mu)_{\text{eq}} = \left(\frac{\partial \Delta\mu}{\partial T}\right)_P dT + \left(\frac{\partial \Delta\mu}{\partial P}\right)_T dP = 0 \qquad (9.12)$$

From (9.11) we immediately get

$$\left(\frac{\partial \Delta \mu}{\partial T}\right)_P = \left(\frac{\partial \mu_\beta}{\partial T}\right)_P - \left(\frac{\partial \mu_\alpha}{\partial T}\right)_P = -S_m(\beta) + S_m(\alpha) = -\Delta S_m$$

$$\left(\frac{\partial \Delta \mu}{\partial P}\right)_T = \left(\frac{\partial \mu_\beta}{\partial P}\right)_T - \left(\frac{\partial \mu_\alpha}{\partial P}\right)_T = V_m(\beta) - V_m(\alpha) = \Delta V_m$$

where the last terms in each equation are the changes in **molar entropy** (i.e. the entropy for one mole in a single phase)

$$S_m \equiv -\left(\frac{\partial \mu}{\partial T}\right)_P \tag{9.13}$$

and **molar volume**

$$V_m \equiv \left(\frac{\partial \mu}{\partial P}\right)_T \tag{9.14}$$

(check also Exercise 8.12 on page 224).

Thus we can rewrite (9.12) as

$$d(\Delta \mu)_{eq} = -\Delta S_m \, dT + \Delta V_m \, dP = 0 \tag{9.15}$$

or equivalently

$$\boxed{\left(\frac{dP}{dT}\right)_{eq} = \frac{\Delta S_m(\alpha \to \beta)}{\Delta V_m(\alpha \to \beta)} = \frac{\Delta H_m(\alpha \to \beta)}{T \Delta V_m(\alpha \to \beta)}} \tag{9.16}$$

where $\Delta H_m \equiv H_m(\beta) - H_m(\alpha)$ is the change in molar entalpy. The first equality is known as the **Clausius-Clapeyron equation**.[8] The last equality in (9.16) follows from the condition of equilibrium $\Delta \mu = \Delta H - T\Delta S = 0$. In general, ΔH and ΔV are functions of P and T.

The Clausius-Clapeyron equation is remarkable and deserves to be examined carefully. First, note that this is an exact equation, derived from the condition of equilibrium (9.5). Second, note that it connects experimental quantities which can be measured: the slope of the coexistence line in the (T, P) plane, the temperature, the molar entalpy change, and the molar volume change. These four apparently independent measurable quantities must fulfill equation (9.16). Therefore, we can also use this equation to assess the accuracy of their measured values.

However, equation (9.16) is most useful for a qualitative reasoning about the sign of the slopes of the equilibrium lines in the phase diagram.

---

[8]Note the full (in contrast to partial) derivative in this equation. This means that a change in T determines a change in P, and viceversa.

First, let us consider the solid-liquid equilibrium line SL:

$$\left(\frac{dP}{dT}\right)_{SL,eq} = \frac{\Delta S_{SL}}{\Delta V_{SL}} = \frac{\Delta H_{SL}}{T \Delta V_{SL}} \tag{9.17}$$

where $\Delta H_{SL} \equiv H_{liq} - H_{sol}$ and $\Delta V_{SL} \equiv V_{liq} - V_{sol}$.

For most substances ΔV_{SL} is positive (water is an exception, as we will see in chapter 12; see also figure 9.4). In this case, melting of a solid is associated with an *increase* in the molar volume: the solid *expands* on melting.

In addition, for all substances $\Delta H_{SL} > 0$. This means that the molar enthalpy is positive: one must *invest* energy to transform one mole of a solid into liquid. The reason is simple: the solid is characterized by a regular packing of the molecules. In this arrangement the molecules maximize their interactions and the whole system minimizes the potential energy. Upon melting, the regular structure breaks down, the average distance between the molecules increases, and the extent of the intermolecular interactions is reduced. Therefore, energy must be supplied to melt a solid.

From these two experimental observations one can conclude that *for most substances the solid-liquid equilibrium line has a positive slope.*

An important exception is the case of water, which has negative slope. This exception reveals to us an important aspect of water which is sometimes considered as an anomaly (see chapter 12).

It should be noted that sometimes the positive sign of the slope of the solid-liquid curve is deduced from the first equality on the r.h.s. of (9.17) rather than the second equality, i.e. one uses the entropy change in melting instead of the enthalpy of melting. Thus, $\Delta S_{SL} \equiv S_{liq} - S_{sol}$ is argued to be positive. Why? Because the solid is more "ordered" or more "structured" than the liquid. Therefore, we expect that the change in entropy in melting of a solid will be positive. This argument and the following conclusion are correct for this case, and seem to support the traditional association of entropy with order or disorder. However, one should be careful in reaching such conclusions based on order-disorder for the entropy change. It is safer, whenever possible, to *use arguments based on energy changes* rather than entropy changes. The argument we presented above was based on the enthalpy change rather than the entropy change. This is generally easier to interpret and should be preferred.

Next, we turn to the liquid-vapour and solid-vapour curves. For both of them we can simplify the expression for the slope as follows. In any transition from a condensed phase (solid or liquid) to a vapour phase, the molar volume of the vapour is much larger than the molar volume of either the solid or the liquid phase. Therefore, we can approximate $\Delta V_{LV} \equiv V_{vap} - V_{liq} \approx V_{vap}$ and $\Delta V_{SV} \equiv V_{vap} - V_{sol} \approx V_{vap}$. Furthermore, assuming that the vapour is approximately an ideal vapour, we can write, $V_{vap} \approx RT/P$. Hence, we can rewrite (9.16) as

$$\left(\frac{dP}{dT}\right)_{LV,eq} \approx \frac{\Delta H_{LV}}{RT^2/P}$$

or equivalently

$$\boxed{\left(\frac{d\ln P}{dT}\right)_{LV,eq} \approx \frac{\Delta H_{LV}}{RT^2}} \tag{9.18}$$

And similarly for the solid-vapour curve

$$\boxed{\left(\frac{d\ln P}{dT}\right)_{SV,eq} \approx \frac{\Delta H_{SV}}{RT^2}} \tag{9.19}$$

These equations are known as the **Clapeyron equations**. From (9.18) and (9.19) we can conclude that *the slope of the solid-vapour and liquid-vapour curves is always positive*. No exceptions. In order to transfer a molecule (or a mole of molecules) from the solid or liquid phase into the vapour phase, one needs to overcome the "binding forces" (or intermolecular interactions), i.e., one must invest energy in this process. Hence, $\Delta H > 0$ and the r.h.s. of (9.18) and (9.19) is always positive.

Furthermore, the slopes of the SV and LV curves are of similar order of magnitude and, in general, very different from the slope of the solid-liquid coexistence curve.

An approximate formula for the liquid-vapour or solid-vapour curve may be obtained if one assumes that ΔH_{LV} and ΔH_{SV} are nearly independent of the temperature. In this approximation one can integrate the Clapeyron equations [either (9.18) or (9.19)]

$$\ln P \approx C - \frac{\Delta H}{RT} \tag{9.20}$$

where C is an integration constant.

This equation is useful in calculating the change in pressure associated with the change in temperature when moving along the coexistence curves.

Exercise 9.1. The density of ice at $0°C$ and 1 atm is 0.917 gr/cm^3. The density of liquid water at the same T and P is 1 gr/cm^3. ΔH of melting water is 1440 cal/mol. Assuming that ΔV and ΔH of melting are independent of temperature, calculate the melting temperature of ice at $P = 400\,atm$.

9.3.2 Coexistence of three phases of the same component

Having discussed the curves of two phases at equilibrium, we next turn to the case of three phases at equilibrium. From the phase rule (9.1) we know that the number of degrees of freedom is zero ($d = 1 + 1 - 2 = 0$). This means that we cannot change any of the intensive variables while keeping the three phases at equilibrium. Thus, the three phases can be at equilibrium on single points in the phase diagram. In the simplest case we have only one kind of solid, hence there exists only one point in the phase diagram at which the three phases can be at equilibrium. Such point is referred to as the **triple point**, and is found by solving the two equations

$$\mu_{liq}(P,T) = \mu_{vap}(P,T) = \mu_{sol}(P,T) \tag{9.21}$$

As we have seen, each of these equations determines a line in the phase diagram. Therefore, the simultaneous coexistence of three phases may be obtained by the intersection of two equilibrium lines (either LV and SV, or SL and SV, or LV and SL). In other words the the triple point (T_{TP}, P_{TP}) is the single point that satisfies the two equalities in (9.21), and we could find it analytically if we had the explicit functions $\mu_{liq}(P,T)$, $\mu_{vap}(P,T)$ and $\mu_{sol}(P,T)$. This applies to any three phases at equilibrium. In chapter 12 we shall see several triple points in the phase diagram of water at high pressures.

9.3.3 The critical point

There is one more important and unique point in the phase diagram that is called the **critical point**, at which no phase boundary exist. When we move along the liquid-vapour equilibrium line by increasing the temperature as well as the pressure, we find experimentally that the density of the vapour phase increases while the density of the liquid decreases.

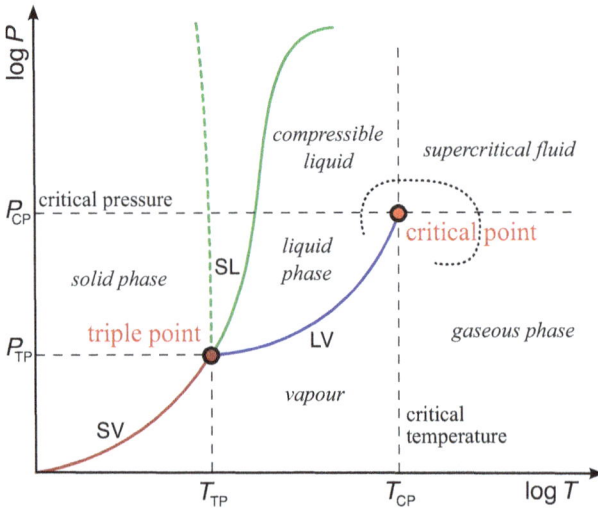

Figure 9.4: Phase diagram with critical point. The thick dashed line is the solid-liquid coexistence line of water. The dotted line shows a possible transformation connecting a point in the liquid phase to a point in the gaseous phase without encountering discontinuities.

In other words, the thermal expansion of the liquid phase overcomes the compression due to the increase of pressure, whereas the compression overcomes the thermal expansion of the vapour. Thus, as we move upward along the LV equilibrium line, the densities of the two phases become more and more similar, until at some point they become equal. Once this critical point is reached, the two phases become identical: there is only one phase referred to as **supercritical fluid** (figure 9.4). Hence the critical point is the end of the liquid-vapour equilibrium curve.

For example, the critical point of carbon dioxide (CO_2) is at $T = 31.04°C$, $P = 72.8$ atm and the critical point of water is at $T = 373.946°C$, $P = 217.7$ atm.

The existence of the critical point tells us an important message: the liquid and the gaseous phases are not different in any fundamental attribute. They are both random, disordered phases, mainly differing in their densities. These densities can be made very close to each other, and at the critical point there exists no distinction between the two phases. This also means that we can move on the P-T plane from any point in the gaseous phase to any point in the liquid phase without encountering a phase transition, just by changing (T, P) values such that we turn around the critical point. One possible transformation is shown by the dotted path in figure 9.4.

The critical point is a unique point, and it "belongs" to the liquid-vapour coexisting curve. There is no similar point along the solid-liquid curve. This fact tells us that, unlike the liquid and gaseous phases, there is a fundamental difference between a solid and a liquid: a solid has a well defined structure, whereas in a liquid the arrangement of the molecules is random or nearly random. When we move along the solid-liquid equilibrium line, the difference between the two phases is not merely a difference in density (or molar volume): one phase is structured, while the other is not. There is no intermediate state between a structured and a non-structured phase. In other words, one cannot move continuously on the P-T plane from a structured phase into a random phase; there is no way of avoiding a discontinuity.

Thus, one can follow the solid-liquid curve without encountering a new critical point. At most, one finds a new triple point when a new solid phase appears (see examples below).

It should be noted that water is considered to be a structured liquid. Indeed, in some sense it retains some degree of local structure similar to ice. When the temperature decreases, the degree of structure of the liquid increases but this structure is only local. Whatever the temperature is, there is no long range structure that is the fingerprint of the solid state, in this case the structure of ice. We shall further discuss water and aqueous solutions in chapter 12.

Another graphical way of representing the critical point is the P-V diagram, in which isotherms (curves at constant T) appear as broken lines below the critical point (figure 9.5). At very high temperatures the system behaves like an ideal gas, or nearly so: the pressure is inversely proportional to the volume. This means that (much) above the critical point, isotherms are hyperbolic curves. As we lower the temperature, these curves deviate more and more from the ideal gas and below the critical temperature the isotherms show three distinct regions: at very large volume and low pressures the curve is similar to an ideal gas curve; at very high pressures and small volume the curve is very steep (as a small change in volume gives a large pressure variation); and in an intermediate region the curve is horizontal.

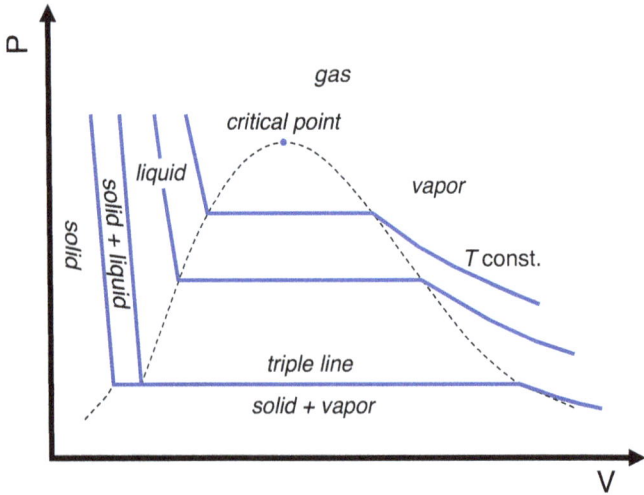

Figure 9.5: Phase diagram P-V.

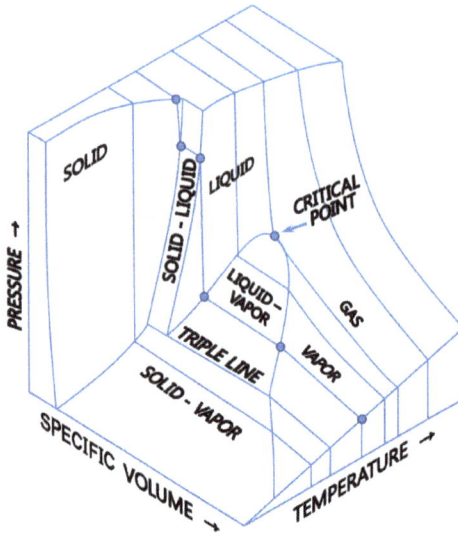

Figure 9.6: Phase diagram P-V-T.

At the triple point, pressure and temperature are fixed: if we change the container volume, we move along a horizontal line in the P-V diagram (the ***triple line***). Showing 3 dimensions as in figure 9.6,[9] is also useful to get a more complete picture of all phase transitions.

Let V_{liq} and V_{vap} denote the molar volumes of the liquid and gas at specific pressure (and temperature). They correspond to the points at which the slope of isotherms below the critical temperature have a discontinuity. For increasing T, the molar volumes V_{liq} and V_{vap} approach each other, and the length of the horizontal segment becomes increasingly shorter until the two points V_{liq} and V_{vap} coincide. This point is the critical point.

9.3.4 *Allotropy*

Some chemical elements exhibit two or more solid forms differing in physical properties but having almost the same chemical nature. This phenomenon is known as ***allotropy***, the name for ***polymorphism*** for pure chemical elements. Carbon and sulfur have many allotropes, but also phosphorus and oxygen exist in different forms. Among the metals, boron, silicon, arsenic, germanium, and antimony have several allotropes.

There are 3 types of allotropy:

- ***Enantiotropy***: two forms of a solid substance exist together in equilibrium with each other at a particular temperature under normal pressure. For example, at normal pressure and temperature between 368.6 K and 285 K, sulfur exist in two solid forms, called rhombic sulfur (S_r) and monoclinic sulfur (S_m), in equilibrium with each other.
- ***Monotropy***: only one allotrope is stable under normal conditions, the other being unstable. Examples are diamond and graphite, or molecular oxygen (O_2) and ozone (O_3).
- ***Dynamictropy***: there is a true equilibrium between the two allotropes, one changing into the other at exactly the same rate as the reverse occurs. Both allotropes are stable over a wide temperature range. For example, liquid sulfur exist in two forms, the pale yellow mobile form called S_λ and a dark viscous form called

---

[9]From https://en.wikipedia.org/wiki/Phase_diagram

S_μ, in equilibrium with each other: $S_\lambda \rightleftharpoons S_\mu$. With increasing temperature the proportion of S_μ increases, whereas with decreasing temperature it is the fraction of S_λ that increases. Thus sulfur shows both enantiotropy and dynamictropy.

In general the allotropy among solid substances is due to the difference in crystalline structure. It may also be due to the presence of different number of atoms, like oxygen and ozone (O_2 and O_3), or the sulphur rings like S_8 and S_{12} (cyclic sulphus molecules exist with 6, 7, 8, 9–15, 18, and 20 atoms).

9.3.5 *The phase diagram of sulfur*

Sulfur forms more than 30 different allotropes. The best known form is α-octasulfur S_8, with a molecule formed by a ring of 8 S atoms which stay alternatively higher and lower with respect to the plane containing the barycenter. Octasulfur is a soft, bright-yellow solid with only a faint odor, which easily breaks and form small debris or powder.

By slow heating, at 95.6°C α-octasulfur starts transforming into β-octasulfur, which has different intermolecular interactions. As they have different melting temperatures (113°C for S_α and 119°C for S_β, which are relevant for fast heating), the phase diagram can be viewed as the superposition of two sets of coexisting lines. Above 119°C sulfur is liquid, hence

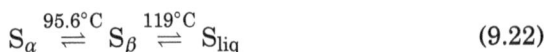

$$S_\alpha \overset{95.6°C}{\rightleftharpoons} S_\beta \overset{119°C}{\rightleftharpoons} S_{\text{liq}} \tag{9.22}$$

Figure 9.7 shows the solid-liquid and solid-vapour equilibrium lines for S_α as dashed red curves, whereas continuous blue curves refer to S_β. In the solid phase, sulfur cristals may exhibit rhombic and monoclinic unit cells. The line connecting points A and D in figure 9.7 separates them. At point A (at 95.6°C and 10^{-5} atm), the differentiation between S_α and S_β begins. Point C is the triple point of S_α, whereas point B (119°C, 10^{-4} atm) is the triple point of S_β. Point D is at 154°C and 10^3 atm.

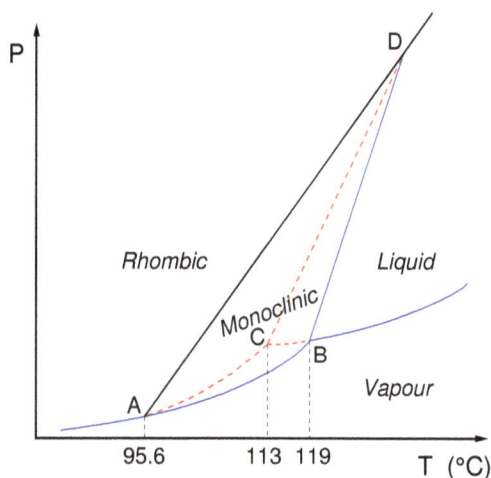

Figure 9.7: Phase diagram of sulfur, adapted with changes from Goh & Chia (1983). Point C is the triple point of S_α and point B is the triple point of S_β. Dashed lines represent coexistence curves of monoclinic sulfur.

9.3.6 *The phase diagram of phosphorous*

Phosphorus has several allotropes with different physical properties. The two most common allotropes are white phosphorus (with tetrahedral molecule P_4) and red phosphorus. The phase diagram of phosphorus is shown in figure 9.8, where dashed red lines refer to white phosphorus and continuous blue lines to red phosphorus.

White phosphorus is the least stable, the most reactive (highly inflammable: it ignites at $30°C$), the most volatile, the least dense, and the most toxic of the allotropes, and is present in solid, liquid and gas phases up to $800°C$, when it starts decomposing into P_2 molecules (diphosphorus, not stable as liquid or solid). When solid, two forms are possible: α-phosphorus is stable at higher temperatures (including room temperature), and β-phosphorus is stable at lower temperatures (prepared from α form at about $-80°C$). Light and heat accelerate its transformation into red phosphorus, which is typically present in non-pure white phosphorus samples (which then appear more and more yellow while aging). Under standard pressure, the melting point of white phosphorus is at $44.15°C$, and the boiling point is at $280.5°C$.

Red phosphorus has polymeric structure, looking like a chain of tetrahedra. It may be formed by heating white phosphorus to $250°C$ or by

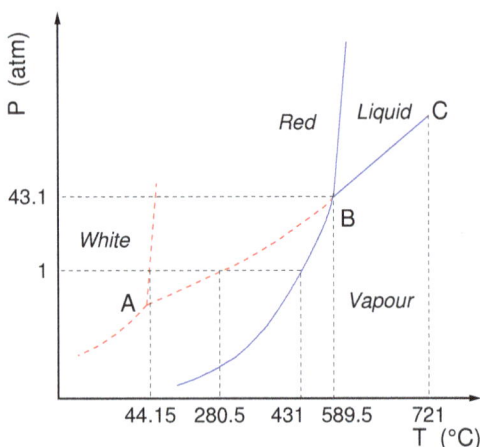

Figure 9.8: Phase diagram of phosphorus, adapted with changes from Goh & Chia (1983). A is the triple point of white phosphorus, B is the triple point of red phosphorus, and C is the critical point. Dashed red lines refer to white phosphorus and continuous blue lines to red phosphorus.

exposing white phosphorus to sunlight. After prolonged heating its color darkens and it becomes less reactive. A long annealing of red phosphorus above 550°C produces violet phosphorus, which is another crystalline form. Red phosphorus sublimes at 431°C, with triple point at 589.5°C under 43.1 atm (Haynes *et al.*, 2015).

Black phosphorus is the least reactive allotrope and the thermodynamically stable form below 550°C. It is also known as β-metallic phosphorus and has a structure somewhat resembling that of graphite. Very high pressures (~ 12000 atm) are usually required to produce black phosphorus, but it can also be produced at ambient conditions using metal salts as catalysts.

Exercise 9.2. The point A in the phase diagram of phosphorous (figure 9.9) is the triple point of white phosphorus: vapour, liquid and white phosphorous coexist at $P_A = 0.17$ mmHg $= 2.2 \times 10^{-4}$ atm and $T_A = 44.1$°C. The heat of vaporization of liquid white phosphorous is $\Delta H_{\text{liq}} = 12.5$ kcal/mol. The vapour pressure of red phosphorous at $T_D = 308.5$°C is $P_D = 0.07$ atm and the heat of sublimation of red phosphorous is $\Delta H_{\text{r}} = 25.8$ kcal/mol. Assuming that ΔH_{r} and ΔH_{liq} are independent of temperature, find the triple point of the red phosphorous (point B in the phase diagram).

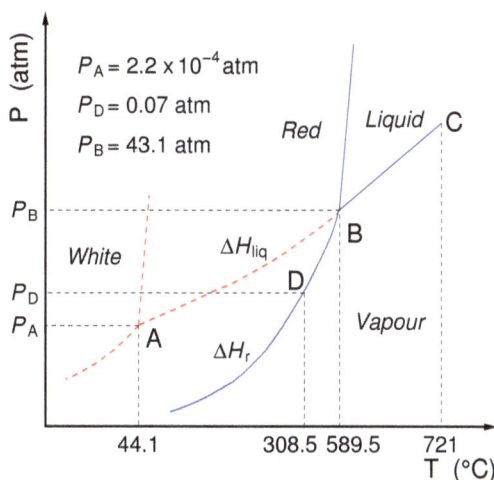

Figure 9.9: Phase diagram of phosphorus, with values for exercise 9.2.

9.3.7 *The phase diagram of carbon*

The phase diagram of carbon is shown in the left panel of figure 9.10 (Bundy, 1989), where region A is for catalytic transformation of graphite to diamond; B is the region of spontaneous fast graphite-to-diamond transformation; C of spontaneous fast diamond-to-graphite transformation; regiond D is for spontaneous slow diffusionless transformation of hex graphite to hex diamond.

A *diffusionless transformation* is a phase change that occurs without long-range diffusion of atoms but rather by some form of homogeneous movement of many atoms that results in a change in crystal structure. These movements are small, usually less than the interatomic distances, and the atoms maintain their relative relationships.

Huge pressures are necessary to observe phase transitions between graphite and diamond, and very high temperatures are required to observe the gas and liquid phases. The reason is that the binding energy between C atoms is very high. The triple point is around $5000°C$ and 10^5 atm. Liquid carbon is an insulator below 200 atm but becomes a good conductor above 10^4 atm. The transition between insulating and "metallic" liquid carbon would happen just above region E in figure 9.10.

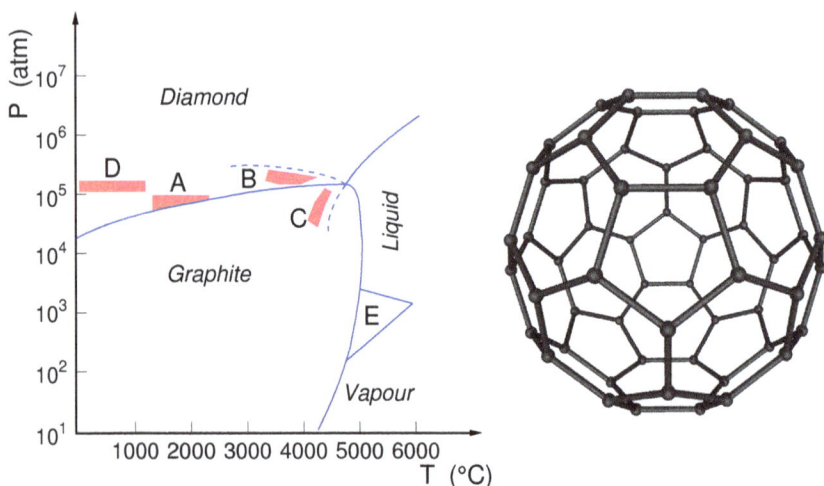

Figure 9.10: Phase diagram of carbon (left; adapted with changes from Bundy, 1989) and buckminsterfullerene C_{60} (right).

Several other allotropes of carbon have been discovered, called fullerenes. The buckminsterfullerene C_{60} was the first fullerene to be discovered (in 1985), named after Richard Buckminster Fuller, an architectural modeler who popularized the geodesic dome. Its atoms form a spherical structure which resembles a soccer ball (figure 9.10, right). For this reason, spherical fullerenes are also called buckyballs. Different sizes are also known, from the smallest C_{20} to C_{60}, followed by C_{70}, C_{76}, C_{82}, and C_{84}, up to the icosahedral fullerene C_{540}.

Fullerenes may also have different forms, like the carbon nanotubes which look like "wraps" made by a single graphite layer (with C atoms grouped in exagonal mesh). These tubes have interesting properties, with diameter of few nanometres but length spanning from micrometers to millimeter size. Their extraordinary macroscopic properties include high tensile strength, high electrical conductivity, high ductility, high heat conductivity, and very low chemical activity.

9.4 Two-Component Systems

For systems with two independent components ($c = 2$) we have four possible numbers of degrees of freedom. These are obtained as follows. The smallest number of degrees of freedom is $d = 0$, hence from the phase rule (9.1) we get $p = 4$, i.e. four coexistent phases. Recall that in a one-component system we have at most three coexistent phases at equilibrium. For $d = 1$, we find from the phase rule that $p = 3$; for $d = 2$ we find $p = 2$; finally for $d = 3$, we have a single phase ($p = 1$). Thus, we have altogether four cases.

The fact that we have at most $d = 3$ degrees of freedom suggests to use three-dimensional phase diagrams, as in figure 9.6. Indeed, such phase diagrams are used mainly by chemical engineers. Another possible a phase diagram has as independent variables the intensive quantities P, T and x = mole fraction of one of the components.

The ***mole fraction*** or ***molar fraction*** of component i is defined as the ratio between the number n_i of moles of i and the total amount of components, expressed in moles: $x_i \equiv n_i/n_{\text{tot}}$, where $n_{\text{tot}} = \sum_{j=1}^{c} n_j$ (note that $n_j \in \mathbb{R}^+$ for all j and that $\sum_{i=1}^{c} x_i = 1$).

If $c = 2$ one typically defines $x = x_1$, such that $x_2 = 1 - x$ is the molar fraction of the other component. In addition, with $c = 2$ we call the diluted component (whose x is smaller) the ***solute*** and the other component (with molar fraction $1 - x > x$) the ***solvent***.

In this chapter we shall fix either P, T, or x. By fixing one degree of freedom, say setting $P = 1$ atm, we are left with a maximum of two degrees of freedom that we can describe in a two-dimensional phase diagram. In most cases fixing the pressure is not a serious restriction, because the phase diagram does not change much when we change the pressure within reasonable limits around the atmospheric pressure. Of course, if we are interested in exploring a wide pressure range, we have to study either a three-dimensional phase diagram, or work with several two-dimensional diagrams, each taken at constant temperature or mole fraction.

9.4.1 *Two liquid phases at equilibrium*

Any mixture of two liquids has some vapour pressure. The vapour pressure varies with the composition of the system. In this section, we fix the pressure of the system at a value higher than the largest vapour pressure of the mixture. With such a choice we have only two liquid phases at equilibrium, and no presence of a gaseous phase. We shall revert back to discuss a system with a gaseous phase in section 9.4.5.

We can classify the various mixtures of the two components in terms of the extent of the mutual miscibility. Two extreme cases are: two liquids which mix with each other in the entire range of composition x, and two liquids which *practically* do not mix at all. Examples of the former are water and ethanol, and examples of the latter are water and mercury.

It should be noted that if two liquids A and B are in contact and are at material equilibrium (i.e. there is no impermeable partition between the two liquids), the chemical potentials of the two components A and B must be the same in the two phases α and β:

$$\begin{cases} \mu_A^{(\alpha)} = \mu_A^{(\beta)} \\ \mu_B^{(\alpha)} = \mu_B^{(\beta)} \end{cases} \tag{9.23}$$

We shall see in chapter 10 that, whenever the concentration of one component is very low in the second component, it is experimentally found that the chemical potentials of both components have the following dependence on the mole fractions, when A is the solute[10] and B is the solvent:

$$\mu_A = \mu_A^0 + RT \ln x_A \quad \text{(valid for } x_A \approx 0) \tag{9.24}$$

$$\mu_B = \mu_B^p + RT \ln x_B \quad \text{(valid for } x_B \approx 1) \tag{9.25}$$

The constant parameters μ_A^0 ('0' reminds us of the limit to zero concentration of A) and μ_B^p ('p' stands for "pure", i.e. practically without solute) are functions of P and T but not of the concentrations x_A and x_B. These two constants are very different in their "physics content": μ_A^0 is determined by the average interactions between one molecule of A and the surroundings of pure B; μ_B^p is determined by the average interactions between one molecule of B and surroundings of pure solvent B (see illustration in figure 10.2).

Formally, these two constants are defined by

$$\mu_A^0 \equiv \lim_{x_A \to 0} (\mu_A + RT \ln x_A) \tag{9.26}$$

$$\mu_B^p \equiv \lim_{x_B \to 1} (\mu_B + RT \ln x_B) \tag{9.27}$$

---

[10]A should not be a ionic solute.

The existence of the first limit is based on experimental facts as well as theoretical support from statistical mechanics (see also chapter 10). The second constant is simply obtained by substituting $x_B = 1$ in (9.25) to obtain the chemical potential μ_B of pure B.

When the two liquids A and B are in contact and in material equilibrium, μ_A^0 and μ_B^p are *finite* quantities. On the other hand, if A has *zero* concentration in the phase β of pure B, then $\mu_A^{0,(\beta)} = -\infty$, which means there is no material equilibrium with respect to A in the two phases, as $\mu_A^{(\alpha)} > \mu_A^{0,(\beta)} = -\infty$, hence the first condition in (9.23) does not hold. Therefore, *theoretically* there will always be some solubility of A in "pure" B, and some solubility if B in "pure" A. However, in some cases the solubility is so small that we can practically treat the two phases as pure A and pure B.

Total miscibility is rare, and total immiscibility is even rarer. The most common case is when we have partial miscibility of A in B, and B in A, and the extent of miscibility varies with temperature (as well as with pressure, but in this section we keep P fixed; the effect of pressure is usually less pronounced).

Consider the example of water and phenol (C_6H_6O, with density $1.07\,g/cm^3$). We keep the pressure sufficiently high, so that only two liquid phases exists at equilibrium. We denote by α the phase richer in water (occupying the upper part of the container), and β the phase richer in phenol (heavier than water, hence at the bottom of the container).

Figure 9.11 shows the (x_p, T) diagram of a mixture of water and phenol, where x_p is the total mole fraction of the phenol. We draw the diagram for a range of temperature where there are only liquid phases. Note that from the phase rule $d = 4 - 2 = 2$ but, since we have used one degree of freedom by fixing the pressure, we are left with only one degree of freedom. Thus fixing the temperature, say at T_1, determines the compositions of the two phases α and β at equilibrium $x_p^{(\alpha)}(T_1)$ and $x_p^{(\beta)}(T_1)$, where $x_p^{(\beta)}(T_1) > x_p^{(\alpha)}(T_1)$, i.e., the mole fraction of phenol (x_p) corresponding to the temperature T_1 in phase β is larger than the mole fraction of phenol in phase α.

The values of $x_p^{(\alpha)}$ and $x_p^{(\beta)}$ are determined by the conditions of equilibrium:

$$\begin{cases} \mu_w^{(\alpha)}(P,T,x_p^{(\alpha)}) = \mu_w^{(\beta)}(P,T,x_p^{(\beta)}) \\ \mu_p^{(\alpha)}(P,T,x_p^{(\alpha)}) = \mu_p^{(\beta)}(P,T,x_p^{(\beta)}) \end{cases} \tag{9.28}$$

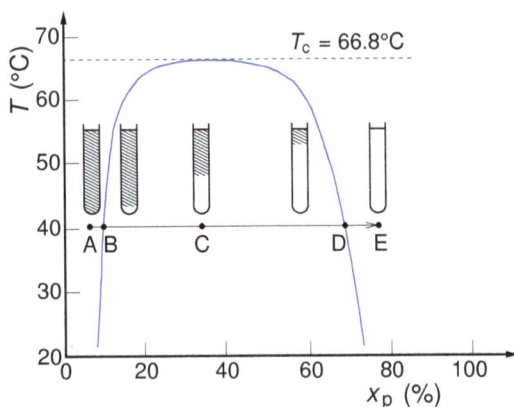

Figure 9.11: Phase diagram of mixture of liquid water and phenol.

Fixing T and P leaves two equations for $x_p^{(\alpha)}$ and $x_p^{(\beta)}$, whose values can then be uniquely determined. These two solutions are called conjugated solutions.

As we increase the temperature, the solubility of phenol in water increases and the solubility of water in phenol increases too. Therefore, the two compositions $x_p^{(\alpha)}$ and $x_p^{(\beta)}$ approach each other as the temperature increases. At a certain temperature ($T_c = 66.8°C$ in the case of water and phenol) the composition of the two phases becomes identical and we have only a single phase. The temperature T_c at which this occurs is known as the ***critical solution temperature***, and the corresponding composition of the single phase is known as the ***critical composition***.

The approach to the critical composition for increasing solution temperature is reminiscent of the approach to the critical point (section 9.3.3). At the critical point the (pure) liquid and gas phases become identical in density. Here, the two phases become identical in composition.

Above the critical solution temperature, we have only one liquid phase. From the phase rule we find that $d = 2 + 2 - 1 = 3$. Since we fixed the pressure we are left with two degrees of freedom: we can change both $T > T_c$ and x_p in an arbitrary way, as long as we maintain a single phase.

Suppose that we start at $T_1 = 40°C$ from pure water (point A in figure 9.11), and add some phenol. Initially, the fraction of phenol $x_p^{(\alpha)}$ in the "pure" water phase is so small that it is completely dissolved in water, and we have a single phase. This means that, keeping P fixed, there are two degrees of freedom and we can adjust both T and $x_p^{(\alpha)}$. If we continue to add phenol at constant temperature, at point B the second liquid phase appears: we have reached the blue curve in figure 9.11. The value of $x_p^{(\alpha)}$ at which this happens is the **solubility** of phenol in water at the temperature T_1.

Once we have two phases, we lose one degree of freedom. Therefore determining T and P also determines the compositions $x_p^{(\alpha)}$ and $x_p^{(\beta)}$ of the two phases. As we add more and more phenol, the total mole fraction x_p of the entire system increases, but the compositions of the two phases α and β do not change: the concentrations $x_p^{(\alpha)}$ and $x_p^{(\beta)}$ are determined by the condition of the material equilibrium for the two components. When we are inside the blue curve (for example at point C), changing the total mole fraction x_p induces a change in the proportion between the two phases. At some point we reach again the curve, at point D, and the water phase disappears. At point E we have again two degrees of freedom, and a single phase: pure phenol.

Exercise 9.3. Let N_{tot} be the total quantity of material (either the total number of moles or the total weight) of the system with total composition $x_p^{(tot)}$ (mole fraction or weight fraction). Show that the ratio between the quantities of the two phases α and β is equal to the ratio of the segments $x_p^{(tot)} - x_p^{(\alpha)}$ and $x_p^{(\beta)} - x_p^{(tot)}$:

$$\frac{N^{(\beta)}}{N^{(\alpha)}} = \frac{x_p^{(tot)} - x_p^{(\alpha)}}{x_p^{(\beta)} - x_p^{(tot)}} \qquad (9.29)$$

Thus, when we move along the BD line in figure 9.11, the ratio of the quantities $N^{(\beta)}/N^{(\alpha)}$ increases since $x_p^{(tot)}$ increases, while $x_p^{(\alpha)}$ and $x_p^{(\beta)}$ remain constant, until we reach the point D. At that point, adding more phenol to the system "absorbs" all the water in the system: then we have only one phase which may be considered as a solution of water in phenol. Adding more phenol moves the system along the line DE, where we have a single phase system, therefore two degrees of freedom.

9.4.2   LCST and UCST

The **upper critical solution temperature** (UCST) or **upper consolute temperature** is an upper bound to the critical temperature above which the components of a mixture are miscible in all proportions. Thus the UCST is the hottest point at which cooling will immediately induce a phase transition.

The UCST is in general dependent on pressure, and the phase separation below the UCST happens because the interactions between components favor a partially demixed state. Any infinitesimal fluctuation in composition or density will then trigger the spontaneous phase transition.

For example, hexane-nitrobenzene mixtures have a UCST of 19°C, aniline-water system at 168°C (at pressures high enough for liquid water to exist at that temperature), and lead-zinc system at 798°C (a temperature where both metals are liquid).

The limit of local stability with respect to small fluctuations is defined by the condition that the second derivative of Gibbs free energy is zero. The locus of these points in a (x, T) plot is known as the **spinodal curve**. For compositions within this curve, phase separation is a spontaneous process, called **spinodal decomposition**. Outside of the curve, a single phase system may be obtained (possibly via some careful process).

The local points of coexisting compositions form the **coexistence curve** (also known as **binodal curve**), which denotes the minimum-energy equilibrium state of the system.

In a phase diagram of the mixture components, the UCST is the shared maximum of the (concave down) spinodal and coexistence curves. Thus, the binodal and spinodal curves meet at their extremum, the **liquid-liquid critical** (or **consolute**) **point** defining the critical temperature (the highest temperature at which two phases are present) and the corresponding critical composition. For the mixture of phenol and water, the critical temperature is 66.8°C, which is also the UCST.

Two types of liquid-liquid critical points exist. In addition to the UCST, there may be also the **lower critical solution temperature** (LCST), which is the coldest point at which heating will immediately

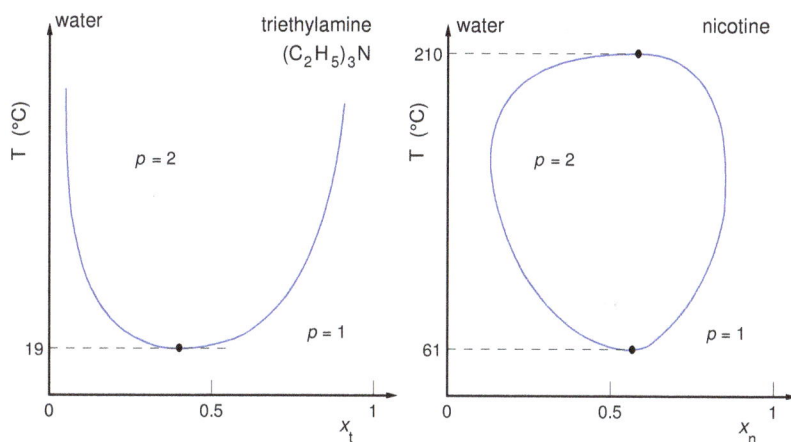

Figure 9.12: Phase diagram of mixture of liquid water with triethylamine (left) or nicotine (right).

induce phase separation. The LCST is the lower bound to the critical temperature below which the components of a mixture are miscible for all compositions.

In the phase diagram of the mixture components, the LCST is the shared minimum of the (concave up) spinodal and coexistence curves. It is in general pressure dependent, increasing as a function of increasing pressure.

For small molecules, the existence of a LCST is much less common than the existence of the UCST. For example, the system triethylamine-water has a LCST of 19°C (figure 9.12, left).

The nicotine-water system has a LCST of 61°C and also a UCST of 210°C, at pressures high enough for liquid water to exist at that temperature (figure 9.12, right). The components are therefore miscible in all proportions below 61°C and above 210°C (at high pressure), and partially miscible in the interval from 61 to 210°C.

Partially miscible polymer solutions often exhibit both UCST and LCST, which both depend on the molar mass and the pressure. Some polymer solutions have a LCST at temperatures higher than the UCST, which means that there is a temperature interval of complete miscibility, with partial miscibility at both higher and lower temperatures.

9.4.3 *System with one liquid phase in equilibrium with solid phases*

We consider a two-component system and assume that the two components A and B are miscible in the liquid state, but not in the solid phase. Keeping the pressure constant, we have from phase rule $d = 3 - p$, which has three possible types of solution.

When a single phase exists ($p = 1$) we have $d = 2$ degrees of freedom and can change both T and x_A independently. The single phase may be liquid (region L in figure 9.13) or solid (regions S_A, S_B, and S_{A+B} in figure 9.13).

When we have liquid and solid phases in equilibrium, $p = 2$ and $d = 1$: there is a single degree of freedom and we move along a coexistence curve $T = T(x_A)$. The liquid phase may be in equilibrium with pure solid A (region $L + S_A$ in figure 9.13) or B (region $L + S_B$).

In addition, there exists a certain composition $x_A = x_e$ at which the liquid phase is obtained when the whole solid A + B melts at the unique temperature T_e, the **eutectic temperature**. The values of x_e and T_e define the **eutectic point**, at which we have $p = 3$ and $d = 0$: the eutectic point is a single point in the phase diagram (at constant pressure), which is the third type of solution of the phase rule.

At the eutectic composition, the atoms or molecules of A and B form a lattice that melts at the same temperature, whereas at any other com-

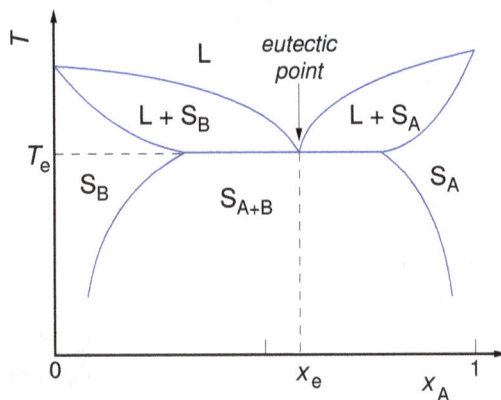

Figure 9.13: At the eutectic composition x_e the solid melts as a whole when the temperature reaches T_e, the eutectic temperature.

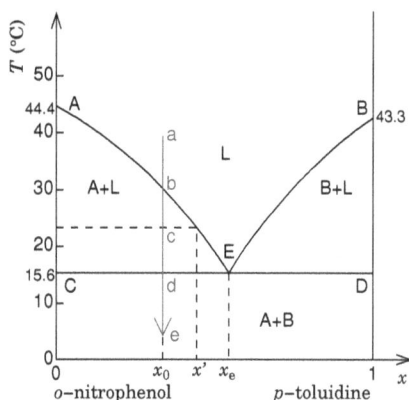

Figure 9.14: Mixture of p-toluidine and o-nitrophenol. The arrow connecting points a–e is discussed in Exercise 9.4.

position the lattice of A or the lattice of B will melt first. When the percentage of A is very small, its atoms or molecules are too far to be able to create a lattice (their intermolecular interactions are too weak) and one has a solid of B with some impurities of A (region S_B in figure 9.13). The situation is reversed when $x_A \approx 1$ (region S_A).

The eutectic point is similar to the melting point of a pure substance: heat is exchanged with the system but the temperature remains constant, as long as there are three phases at equilibrium (pure A, pure B and solution of composition x_e). The number of degrees of freedom is $d = 1$ but, since we have fixed the pressure, we are left with zero degrees of freedom.

An example of such a system is p-toluidine and o-nitrophenol (figure 9.14). This system forms a nearly *symmetric ideal solution* in the liquid phase. We shall discuss symmetric solutions in section 10.3.5 but for now it is sufficient to say that for them the chemical potentials of each component has the same form

$$\mu_A = \mu_A^p + RT \ln x_A$$
$$\mu_B = \mu_B^p + RT \ln x_B$$

for the entire range of compositions, where x_A and x_B are the mole fractions of the two components, and $\mu_A^p = \mu_A(x_A = 1)$ and $\mu_B^p = \mu_B(x_B = 1)$ are the chemical potentials of pure A and B, respectively.

For this system, this means that the coexistence curves liquid-solid A and liquid-solid B are given by

$$\mu_A^{p,sol} = \mu_A^{p,liq} + RT \ln x_A \tag{9.30}$$

$$\mu_B^{p,sol} = \mu_B^{p,liq} + RT \ln x_B \tag{9.31}$$

where $\mu_A^{p,sol}$ and $\mu_A^{p,liq}$ are the chemical potentials of the pure solid A and the pure liquid A, respectively, at the same T and P. These quantities are independent of the composition, therefore equation (9.30) is an implicit equation $T = T(x_A)$ for the liquid-solid A curve in figure 9.13. Similarly, equation (9.31) is an implicit equation for the liquid-solid B curve. We can rewrite them as

$$R \ln x_A = \frac{\mu_A^{p,sol} - \mu_A^{p,liq}}{T} \tag{9.32}$$

$$R \ln x_B = \frac{\mu_B^{p,sol} - \mu_B^{p,liq}}{T} \tag{9.33}$$

An approximate explicit expression for these curves may be obtained after differentiating by T (at constant P), by assuming that the difference in the molar enthalpy of the pure solid and pure liquid A, $\Delta H_A \equiv H_A^{p,sol} - H_A^{p,liq}$, is independent of temperature.

By differentiation of (9.32) and using the Gibbs-Helmoltz equation (8.99) we get

$$R \left(\frac{\partial \ln x_A}{\partial T} \right)_P = \frac{\partial}{\partial T} \left(\frac{\mu_A^{p,sol} - \mu_A^{p,liq}}{T} \right)_P = -\frac{\Delta H_A}{T^2} \tag{9.34}$$

which, if ΔH_A, is independent of T, can be easily integrated to obtain the approximate relationship

$$R \ln x_A = \frac{\Delta H_A}{T} + c \tag{9.35}$$

The integration constant c can be determined by imposing the melting temperature T_A of pure A: for $x_A = 1$ eq. (9.35) gives $c = -\Delta H_A/T_A$, such that

$$\ln x_A = \frac{\Delta H_A}{R} \left(\frac{1}{T} - \frac{1}{T_A} \right) \tag{9.36}$$

Thus, by determining the heat of melting of pure A (at constant pressure) and the melting temperature of pure A (at the same pressure), we

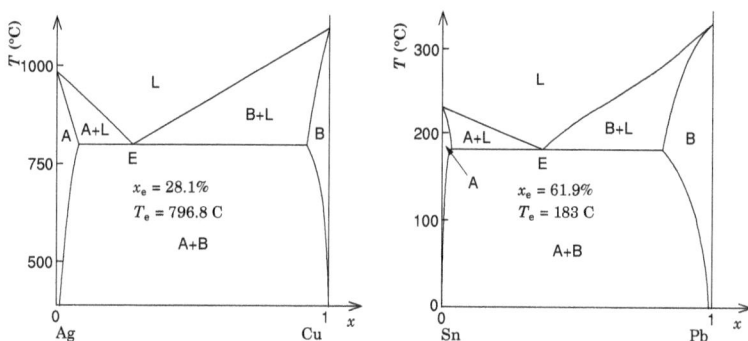

Figure 9.15: Copper-silver (left) and tin-lead (right) phase diagrams. In both cases, the composition is expressed as mass fraction.

have an explicit expression for the x_A as a function of T. A similar expression may be obtained for the liquid-solid B curve.

Two other systems with one eutectic point are silver (Ag) with copper (Cu), and tin (Sn) with lead (Pb). Their phase diagrams are shown in figure 9.15.

Exercise 9.4. Describe what happens when we cool a solution of p-toluidine (A) and o-nitrophenol (B) along the line abcde in figure 9.14.

9.4.4 *Systems with a congruent melting point*

If a solid compound melts into a liquid with the same composition of the solitd, one speaks of **congruent melting**. When the phase transition brings to a liquid with different composition, one has **incongruent melting**.

Let A and B be the two components, and call AB the solid compound formed by their chemical combination. The phase diagram (figure 9.16) has different regions and has many similarities with the system discussed in section 9.4.3. Indeed, the phase diagram of this system may be viewed as the combination of two phase diagrams of the type shown in figure 9.13: one for the two "pure" solids B and AB on the left side, and another for the solids AB and A on the right side. All we have said in section 9.4.3 applies for each sub-diagram. Note that this system has

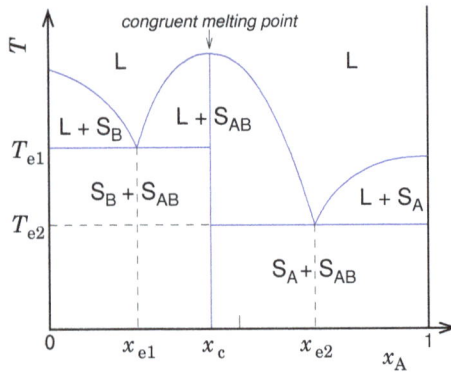

Figure 9.16: Phase diagram of formation of a compound with congruent melting point.

two eutectic points (x_{e1}, T_{e1}) and (x_{e2}, T_{e2}). At the first one, the solids B and AB precipitate; at the second eutectic point the two solids A and AB precipitate.

There are 3 solid phases: pure A (denoted as S_A in figure 9.16), pure B (S_B), and the crystal AB (S_{AB}). This implies the existence of the 3 liquid-solid coexistence curves at the top of figure 9.16.

The solid phases are always pure, and mixing happens only in the liquid phase. Above some temperature, there is a single liquid phase. However, there are also regions in which the liquid coexist with some deposition of solid A or solid B.

The new feature of this system is the occurrence of a congruent melting point. The **congruent melting point** is the point of the phase diagram where the liquid and solid phases have the same composition x_c: for $x_A = x_c$ there is only one component AB in the phase diagram, with two phases (solid and liquid).

If we cool down the liquid with composition x_c, at the **congruent freezing point** the solution precipitates are pure AB, without track of solid A or solid B which characterize the freezing at any other mole fraction.

At the **congruent composition**, there is a single phase and a single component. Therefore, at fixed pressure we remain with one degree of freedom that, in this case, is the temperature. This gives the vertical coexistence line in figure 9.16.

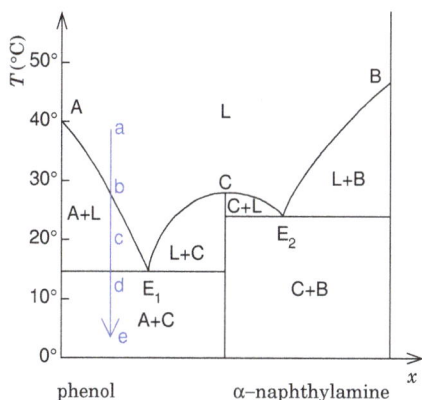

Figure 9.17: Phase diagram of a solution of phenol and naphthylamine.

Exercise 9.5. Describe what happens when we cool a solution of phenol (A) and naphthylamine (B) along the line abcde in figure 9.17.

9.4.5 *Two miscible liquids at equilibrium with a vapour phase*

We discuss here the simplest mixture of two components miscible in the entire range of composition. We also assume that both the liquid and the gas phases are ideal, in the sense that the liquid is a symmetric ideal solution (alread mentioned in section 9.4.3 and to be discussed in details in section 10.3.5) and the gas phase is a mixture of ideal gases.

In this system we have $p = 2$ phases and $c = 2$ components, therefore the number of degrees of freedom is $d = 2$.

We shall call y_1 and y_2 the mole fractions of the two components in the gaseous phase, with x_1 and x_2 being their mole fractions in the liquid phase.

At equilibrium between the two phases we must have the equalities

$$\begin{cases} T = T^{\text{gas}} = T^{\text{liq}} \\ P = P^{\text{gas}} = P^{\text{liq}} \\ \mu_1^{\text{gas}} = \mu_1^{\text{liq}} \\ \mu_2^{\text{gas}} = \mu_2^{\text{liq}} \end{cases} \tag{9.37}$$

From the assumption of ideality we can rewrite the last two conditions as

$$\mu_1^{gas} = \mu_1^{p,gas} + RT\ln y_1 = \mu_1^{p,liq} + RT\ln x_1 = \mu_1^{liq} \tag{9.38}$$

$$\mu_2^{gas} = \mu_2^{p,gas} + RT\ln y_2 = \mu_2^{p,liq} + RT\ln x_2 = \mu_2^{liq} \tag{9.39}$$

where the superscript 'p' stands for "pure" component. We obtain

$$\Delta\mu_1^p \equiv \mu_1^{p,gas} - \mu_1^{p,liq} = RT\ln\frac{x_1}{y_1} \tag{9.40}$$

$$\Delta\mu_2^p \equiv \mu_2^{p,gas} - \mu_2^{p,liq} = RT\ln\frac{x_2}{y_2} \tag{9.41}$$

We now wish to obtain the explicit functions $x_1(T)$ and $y_1(T)$. To do this, we further make the approximation that the enthalpy of evaporation of the two components is independent of T.

We use the identity (9.32) in the form

$$\frac{\partial}{\partial T}\left(\frac{\Delta\mu}{T}\right) = -\frac{\Delta H}{T^2}$$

with the last assumption to obtain

$$\frac{\Delta\mu_1^p}{T} = \frac{\Delta H_1^p}{T} + c_1 \tag{9.42}$$

The integration constant c_1 can be determined by imposing $\Delta\mu_1^p = 0$ at the boiling temperature T_1^b of the first component:

$$c_1 = -\frac{\Delta H_1^p}{T_1^b} \tag{9.43}$$

Therefore, from (9.40), (9.42) and (9.43) we obtain

$$\frac{\Delta\mu_1^p}{RT} = \frac{\Delta H_1^p}{R}\left(\frac{1}{T} - \frac{1}{T_1^b}\right) = \ln\frac{x_1}{y_1} \tag{9.44}$$

Similarly, for the other component we obtain

$$\frac{\Delta\mu_2^p}{RT} = \frac{\Delta H_2^p}{R}\left(\frac{1}{T} - \frac{1}{T_2^b}\right) = \ln\frac{x_2}{y_2} \tag{9.45}$$

From (9.44) and (9.45) we can solve for $x_1(T)$ and $y_1(T)$

$$x_1(T) = \frac{\exp[f_1(T) + f_2(T)] - \exp[f_1(T)]}{\exp[f_2(T)] - \exp[f_1(T)]} \tag{9.46}$$

$$y_1(T) = \frac{\exp[f_2(T)] - 1}{\exp[f_2(T)] - \exp[f_1(T)]} \tag{9.47}$$

with the help of the auxiliary function

$$f_i(T) \equiv \frac{\Delta H_i^{\mathrm{p}}}{R}\left(\frac{1}{T} - \frac{1}{T_i^{\mathrm{b}}}\right) \tag{9.48}$$

Thus, if we know the boiling temperature and the heat of vaporization of the two components, we have the explicit functions of the compositions x_1 and y_1 in the two phases.

Exercise 9.6. Find T_i^{b} and ΔH_i^{p} for the water-ethanol mixture, and draw the two functions (9.46) and (9.48).

When the components have different boiling temperatures, they can be obtained from their liquid mixture by **distillation**, which exploits selective evaporation and condensation to get an almost complete separation, or an increased concentration of the selected component. When the liquid and vapour phases have the same composition, further distillation is ineffective. Such a liquid mixture is called an **azeotrope**, and cannot be altered by simple distillation.

Each azeotrope has a characteristic boiling point, which can be either less than the boiling point temperatures of any other ratio of its constituents (a **positive azeotrope**), or greater than the boiling point of any other ratio of its constituents (a **negative azeotrope**).

A positive azeotrope is 95.63% ethanol (C_2H_6O) and 4.37% water by weight. At atmospheric pressure ethanol boils at 78.4°C and water boils at 100°C, but the azeotrope boils at 78.2°C, which is lower than either of its constituents.

An example of a negative azeotrope is hydrochloric acid (HCl) at a concentration of 20.2% and 79.8% water by weight. HCl boils at −84°C and the azeotrope boils at 110°C, which is higher than either of its constituents.

Chapter 10

Mixtures and Solutions

Traditionally, one calls a **solution** a mixture where one component, the **solute**, is diluted in the **solvent**, which represents the bigger fraction of molecules. The term "mixture" is considered to be more general, applicable to any kind of composition, although we shall use these two terms as synonyms. We shall discuss mainly systems at some fixed temperature T and pressure P, as these are the most common variables under which solution thermodynamics is studied.

We offer in this chapter a brief presentation of the pair correlation functions, although they do not feature in thermodynamics and are only used in statistical mechanics. The reason is that some qualitative familiarity with the meaning and properties of these functions is essential for a deeper understanding of the molecular origin of the various types of ideal solution, as well as the deviations from the ideal behavior.

10.1 Partial Molar Quantities

Let's consider an extensive property of a mixture, like the volume. The **partial molar volume** of species i is defined as

$$\overline{V}_i \equiv \left(\frac{\partial V}{\partial n_i} \right)_{T,P,n_{j \neq i}} \tag{10.1}$$

$\overline{V}_i$ can be measured as the amount by which V changes when one mole of species i is added to the mixture, while keeping constant temperature, pressure and amounts of all other species. The additivity rule for the volume is then expressed as

$$V = \sum_i n_i \overline{V}_i \tag{10.2}$$

A partial molar quantity is an intensive thermodynamic quantity, which in general depends on the temperature, pressure, and composition of the mixture.

Similarly, one can define other partial molar quantities. The chemical potential of species i is defined as the **partial molar Gibbs energy**

$$\mu_i \equiv \left(\frac{\partial G}{\partial n_i} \right)_{T,P,n_{j \neq i}} \tag{10.3}$$

Inserting the relation $G = H - TS$ into (10.3) one immediately gets

$$\mu_i = \overline{H}_i - T\overline{S}_i \tag{10.4}$$

where the **partial molar enthalpy** and **partial molar entropy** are defined as

$$\overline{H}_i \equiv \left(\frac{\partial H}{\partial n_i} \right)_{T,P,n_{j \neq i}} \tag{10.5}$$

$$\overline{S}_i \equiv \left(\frac{\partial S}{\partial n_i} \right)_{T,P,n_{j \neq i}} \tag{10.6}$$

Relation (10.4) can be compared with the relationship between molar quantities

$$\mu = G/n = H_{\mathrm{m}} - TS_{\mathrm{m}} \tag{10.7}$$

which holds for pure substances.

10.2   Pair Correlation Function and Kirkwood-Buff Integrals

The concept of *pair correlation function* (PCF) takes a central place in the modern theory of liquids and liquid mixtures. However, in this book we shall not need the PCF itself, but integrals over the PCF, which are called the *Kirkwood-Buff integrals* (KBI). The latter are very important quantities in the characterization of the *local* properties of liquid mixtures.

In this section we shall present a qualitative description of the PCF in pure liquid and in mixtures, and describe some of its basic properties. Once we are familiar with the notion of PCF, we shall define the Kirkwood-Buff integrals, illustrate briefly their meaning, and mention some applications. Finally, we will quote one result due to Kirkwood and Buff, which we shall use in several sections in this chapter. More details can be found on Ben-Naim (2006b).

10.2.1 *The pair correlation function for simple one component liquid*

Consider a system of N simple particles in a volume V at some temperature T. By simple particles we mean particles that are spherically symmetric, like liquid argon and neon or, to first approximation, methane. Basically, the interaction between two simple particles is described by a pair potential $U(r)$ that is a function of their distance r only.

A simple model that approximates the interaction between a pair of neutral atoms or molecules is the ***Lennard-Jones potential***, shown in figure 10.1 and defined by

$$U(r) = 4\varepsilon \left[\left(\frac{\sigma}{r} \right)^{12} - \left(\frac{\sigma}{r} \right)^{6} \right] \tag{10.8}$$

where ε and σ are parameters with the dimensions of an energy and a distance, respectively. The energy ε is the depth of the ***potential well***, i.e. the negative potential peak. The potential is zero at distance σ and at infinity, with a repulsive force for $r < \sigma$, which becomes very intense for approaching particles, which behave then as "hard spheres" with radius σ. Thus σ is a characteristic distance that can be interpeted as the size of the particle. The force is attractive at larger distances with decreasing intensity beyond the potential peak at $r = 2^{1/6}\sigma$. Hence this is the typical distance between two particles, and ε is the amount of work required to separate them.

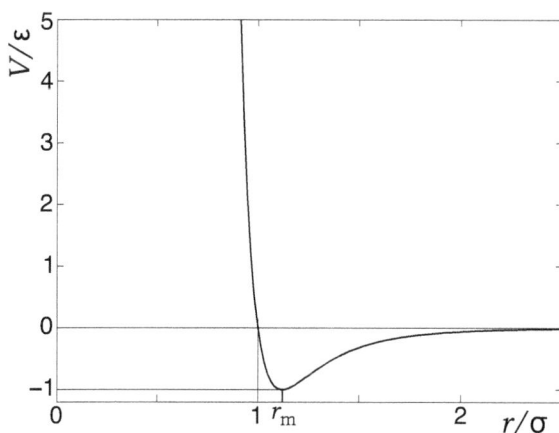

Figure 10.1: Lennard-Jones potential.

Actually, we shall not need the explicit form of the pair potential $U(r)$. All we need to know at this stage is that the **interaction energy**, i.e. the energy change for bringing two particles from infinite separation to the final distance r, depends only on the distance r and not on the relative orientations of the two particles. In chapter 12 we shall discuss an important liquid, water, which is not a simple liquid: its pair potential strongly depends on the relative orientations of the two molecules.

Now we want to distinguish between the **bulk density** of the particles in any system (not necessarily of simple particles), defined by

$$\rho_b \equiv \frac{N}{V} \tag{10.9}$$

and the **local density** $\rho(r)$ at some location $r \equiv (x, y, z)$ in the liquid, measured from some origin. In order to define $\rho(r)$, we select an infinitesimal element of volume $dr \equiv dx\,dy\,dz$, take the average number $\overline{N}$ of particles in such infinitesimal volume, and define

$$\rho(r) \equiv \frac{\overline{N}(r)}{dr} \tag{10.10}$$

Note that in our notation, although r is a vector, dr is a *scalar*: it is the the volume of the cube whose edges are long dx, dy, dz in the three directions, centered on r. Often one writes $dV \equiv dx\,dy\,dz$ or d^3r instead, but here we prefer the notation dr to remind ourselves that the volume element dV is at position r.

It is intuitively clear that, if the system is not subjected to any external field, say gravitational, then the local density at each point r is constant, i.e. $\rho(r) = \rho$, independent of the location. Hence the total number of particles in the system is

$$N = \int_V \rho(r)dr = \rho V \tag{10.11}$$

which implies that $\rho_b = \rho$, i.e. the local and bulk densities are the same, at any point r in the fluid.

Next, we introduce the concept of **conditional density**, defined as the density $\rho_c(r)$ as seen from a specific location. Suppose that we choose the origin of the coordinate system at a center of one specific particle, and let $\rho_c(r)$ be the local density at the position r relative to such origin. Clearly, fixing the coordinate system at the center of a specific particle makes the *conditional density* $\rho_c(r)$ dependent on the *relative* location

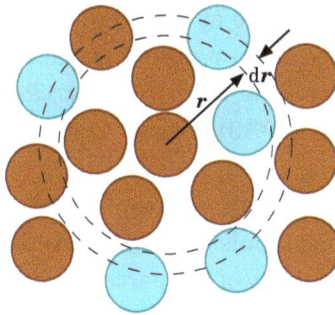

Figure 10.2: Spherical shell centered on a specific simple particle.

r. However, because the particles are simple, the conditional density around a specific particle will depend only on the scalar distance $r = |r|$, and not on the orientation of the vector r. In other words the conditional density $\rho_c = \rho_c(r)$, i.e. it is spherically symmetric about the origin chosen at the center of a simple particle. This implies that the local density is the same at each point inside a spherical shell of radius r around the origin (figure 10.2). The corresponding element of volume is $4\pi r^2 \, dr$.

It is intuitively clear that, in general, the conditional density $\rho_c(r)$ is different from the bulk density ρ_b. The field generated by the particle chosen as the origin of the coordinates can not be neglected in its neighbors: it acts as an external field.

We now define the ***radial distribution function*** $g(r)$ by the ratio

$$g(r) \equiv \frac{\rho_c(r)}{\rho_b} \tag{10.12}$$

such that $\rho_c(r) = \rho_b \, g(r)$. Since here $g(r)$ depends only on the scalar distance r, it coincides with the ***pair correlation function*** (or PCF), which in general may depend also on the angles. The PCF $g(r)$ quantifies the deviation of the local density at r with respect to the bulk density ρ_b.

Two properties of $g(r)$ can be deduced immediately from its definition. First, when r is very large, i.e. the distance between the two particles is very large, the local density becomes equal to ρ_b and

$$\lim_{r \to \infty} g(r) = 1 \tag{10.13}$$

This is a consequence of the fact that the pair potential $U(r)$ has a relatively short range, a few molecular diameters at most. When we go very far from the origin, we will not notice any more the effects of the field of force generated by the particle at the origin. Therefore the conditional density $\rho_c(r)$ will be the same as the bulk density ρ_b. Hence we can say that *at large distances there exists no correlation between the particles*. Note that the limit (10.13) is exact in an *open* system (Ben-Naim, 2006b).

Second, when r is very small, say less than the size σ of the particles, there exists a very strong repulsive force (figure 10.1). At such short distances the conditional density of the particles will be nearly zero, simply because particles cannot penetrate into each other at distances $r \le \sigma$, hence $g(r \le \sigma) \approx 0$.

Thus, we have a qualitative idea of the behavior of the function $g(r)$ at very large, and at very small distances. At intermediate distances, we get information on the behavior of $g(r)$ from experiments (diffraction of X-rays) or from theoretical calculation (by solving integral equations or performing simulations). Typically, the PCF is not a monotonic function of r, but it resembles the amplitude of a damped oscillator, although not as regular (Ploetz *et al.*, 2010), reflecting the fact that the local density of particles $\rho_c(r)$ changes significantly in the immediate neighborhood of the origin. By thinking in terms of "hard balls" and looking at figure 10.2, one easily understands why: the closest particles tend to be located in the innermost shell at radial distances of about one particle diameter, then inside a shell roughly twice as far. However, a liquid has no regular structure and the particles are free to move. Hence for increasing r the density becomes gradually more uniform, reaching $g(r) \approx 1$ already at distances of few diameters. In the simple case of a Lennard-Jones potential, the PCF is shown in figure 10.3.

As the pressure of the liquid affects its density, the PCF also depends on P. At very low density, one can show that $g(r)$ is related to the pair potential $U(r)$ in a simple way:

$$\lim_{\rho \to 0} g(r) = \exp\left[-\frac{U(r)}{k_B T} \right] \tag{10.14}$$

k_B is the Boltzmann constant and T the absolute temperature.[1]

---

[1] A subtle point should be pointed out: the relation (10.14) is strictly correct for an *open* system. If the system is *closed*, there exists a correction term that is, however, negligibly small (Ben-Naim, 2006b).

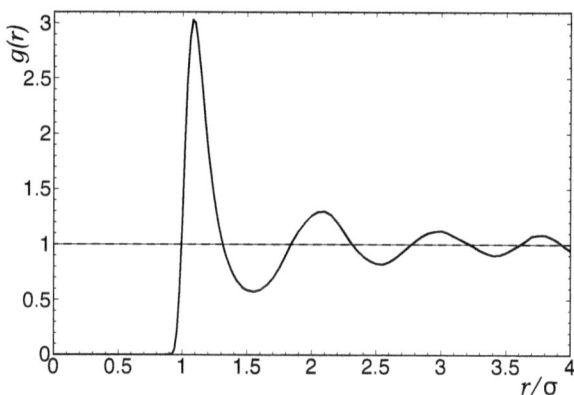

Figure 10.3: Lennard-Jones PCF with $\varepsilon = 0.71 k_B T$ and $\rho = 0.844$.

Note carefully that, even when the *local* density $\rho(r)$ at each point r is the same as the bulk density $\rho_b = N/V$, if you were sitting at the center of one specific particle, while observing the density around you, you would see that at distances $r < \sigma$ there are no other particles except from the one you are sitting on. At very large distances, you would measure the same density as the bulk density. However, because of the attractive interaction at distances $r \geq \sigma$, very close to your position you would observe a conditional density $\rho_c(r)$, which is larger than the bulk density. Other maxima would be seen at $r \approx 2\sigma, 3\sigma, \ldots$, alternated with density minima as shown in figure 10.3.

10.2.2 *The pair correlation function for multi component systems of simple particles*

The extension of the definition of the pair correlation functions for multi-component systems is straightforward. Suppose we have a two-component system of N_A A-particles and N_B B-particles. The bulk densities are

$$\rho_A = \frac{N_A}{V} \quad \text{and} \quad \rho_B = \frac{N_B}{V} \tag{10.15}$$

Note that the subscripts A and B refer to the species A and B: both densities are *bulk* densities.

If we choose an A-particle at the origin of the coordinate system, then we can ask about the conditional density of either A-particles or of B-particles around it. Similarly, if we choose to place a B-particle at the origin of the coordinate system, then again we can ask about the conditional density of either A-particles or B-particles. Thus, in this system we can define four pair correlation functions: $g_{AA}(r)$, $g_{AB}(r)$, $g_{BA}(r)$ and $g_{BB}(r)$. However, because of the spherical symmetry around both A and B particles $g_{AB}(r) = g_{BA}(r)$, therefore there are only three independent pair correlation functions in this system.

10.2.3 The Kirkwood-Buff integrals

Again, we consider a two-component system of A and B species, with bulk densities ρ_A and ρ_B as in (10.15). The conditional density of A-particles at a distance r from a specific B-particle is $\rho_A g_{AB}(r)$. The conditional average number of A-particles in an element of volume $d\mathbf{r}$ at a distance r from the B-particle at the origin is $\rho_A g_{AB}(r) d\mathbf{r}$.

We now choose the element of volume $d\mathbf{r}$ to be a spherical shell of radius r and width dr centered on the given B-particle (figure 10.2). Since $g_{AB}(r)$ depends only on the scalar distance r, the average number of A-particles in the spherical shell, given a B-particle at the center, is $\rho_A g_{AB}(r) 4\pi r^2 dr$ and the average number of A-particles in a sphere of radius r_M centered on the same B-particle is

$$\overline{N}(A|B; r_M) = \int_0^{r_M} \rho_A g_{AB}(r) 4\pi r^2 dr \tag{10.16}$$

Now suppose we randomly choose the center of the sphere of radius r_M within the liquid. Since the bulk density of A is ρ_A and the volume of the sphere is $4\pi r^3/3$, the average number of A-particles in this sphere is

$$\overline{N}(A; r_M) = \int_0^{r_M} \rho_A 4\pi r^2 dr = \rho_A \frac{4\pi r^3}{3} \tag{10.17}$$

Note carefully the difference between the two quantities in (10.15) and (10.16). The first is a *conditional* average, the condition being the B-particle chosen at the center of the sphere. The second is the average of A-particles in a sphere of radius r_M, the center of which is chosen at any point in the liquid. Therefore the difference

$$\Delta \overline{N} \equiv \overline{N}(A|B; r_M) - \overline{N}(A; r_M)$$
$$= \rho_A \int_0^{r_M} [g_{AB}(r) - 1] 4\pi r^2 dr \tag{10.18}$$

is simply the *change* in the average number of A-particles in a sphere of radius r_M, caused by the choice of looking at the system from the center of a B-particle.

All we have said above is valid for any radius r_M, provided that the sphere does not reach the boundary of the liquid. As the system dimensions are very large compared to the typical molecular distances, neglecting the boundaries is a very good approximation and we will always assume that we are far enough from them. For each chosen r_M we can define the integral that appears on the r.h.s. of eq. (10.18) and we expect $g_{AB}(r)$ to become unity for r_M much larger than the typical molecular distances, for example when $r_M \gg \sigma$ in the case of the Lennard-Jones potential (10.8). This is also supported by the experimental results. Hence the integrand in (10.18) will be null far enough from the center of the chosen B-particle, and we can formally extend the integration to infinity without any complication.

Of course, "infinity" actually means very large compared to the typical molecular distances, but still smaller than the distance to the nearest boundary. Thus we always think about a very large sphere (compared, say, to σ) that is fully contained inside the liquid. This sphere has volume $4\pi r_M^3/3$ and is an open system to both A and B, as all molecules are free to move across the fictitious surface at distance r_M from the chosen B-particle.[2]

It should be noted that for most systems of interest, the distance for which the PCF is practically one is of the order of a few molecular diameters (figure 10.3). This is not true for perfect solids, nor for systems near the critical point, but we shall not be concerned with these exceptions here.

The formal extension of the integration to infinity gives us the *Kirkwood-Buff integral* (KBI)

$$G_{AB} \equiv \int_0^\infty [g_{AB}(r) - 1]\, 4\pi r^2 \, dr \qquad (10.19)$$

---

[2]There are some subtle points in the long range behavior of $g_{AB}(r)$ in open and closed systems. For more details see Ben-Naim (2006b).

In a similar way we define the other two KBI's

$$G_{AA} \equiv \int_0^\infty [g_{AA}(r) - 1]\, 4\pi r^2\, dr \qquad (10.20)$$

$$G_{BB} \equiv \int_0^\infty [g_{BB}(r) - 1]\, 4\pi r^2\, dr \qquad (10.21)$$

In summary, in a two-component system we have three different PCF's and three different KBI's, G_{AA}, $G_{AB} = G_{BA}$ and G_{BB}.

The best way to grasp the meaning of $G_{\alpha\beta}$ is to look first at $\rho_\alpha G_{\alpha\beta}(r_M)$ for a *finite* r_M, then choose a radius r_M beyond which the integrand $g_{\alpha\beta}(r) - 1$ is practically zero, then formally extend the integration to infinity, assuming that for $r > r_M$ the integrand is strictly zero.

The quantity $g_{\alpha\beta}(r) - 1$ is sometimes referred to as the **total correlation** between the two species α and β. This is a correlation in the probabilistic sense.[3] The KBI $G_{\alpha\beta}$ is thus a measure of the overall correlation between α and β.

10.2.4   *An exact expression from the Kirkwood-Buff theory of solutions*

Here is a simple, beautiful and information loaded formula:

$$\boxed{\left(\frac{\partial \mu_A}{\partial x_A}\right)_{P,T} = \frac{k_B T}{x_A(1 + \rho_{tot}\, x_A x_B\, \Delta_{AB})}} \qquad (10.22)$$

where μ_A is the chemical potential of A, x_A is the mole fraction of A, $\rho_{tot} = \rho_A + \rho_B$ is the total (bulk) density of the particles in the system, and

$$\Delta_{AB} \equiv G_{AA} + G_{BB} - 2G_{AB} \qquad (10.23)$$

where $G_{\alpha\beta}$ are the KBI's defined in (10.19), (10.20) and (10.21).

The formula (10.22) was first derived by Kirkwood & Buff (1951) but remained dormant for over 20 years, together with the Kirkwood-Buff theory of solution. Starting from the mid 1970's a surge of applications of the Kirwood-Buff theory were published (for a review see Ben-Naim

---

[3]Here, we have derived the KBI using the concept of conditional densities, but we could also use the probabilistic notions of conditional probabilities [Ben-Naim (2006b)].

2006b). The derivation of this formula is admittedly not easy. Fortunately, you do not need to derive this equation.[4] Nevertheless, we feel that presenting it in this book is helpful at least for three reasons.

First, we shall encounter several applications of this formula in the forthcoming sections. In each case, we shall discuss both the conventional approach and the "modern" approach based on this formula. The conventional approach is based on some experimental data which is not part of the thermodynamic formalism, and it might leave the reader with an uneasy feeling that the whole approach, as well as the results, are only empirical, lacking a firm theoretical support. On the other hand, the Kirkwood-Buff approach does provide such a firm theoretical support.

Second, besides the theoretical support, the Kirkwood-Buff formula provides molecular conditions under which various "ideal solution" behaviors are manifested. These conditions are not accessible experimentally.

Finally, we hope that after comparing the experimental approach with the theoretical approach based on the Kirkwood-Buff theory, you will be encouraged to study the Kirkwood-Buff theory, and hopefully will enjoy its beauty, generality, and its interpretative power, together with its wide range of applicability.

We shall discuss a few specific applications of (10.22) in the forthcoming sections. Here, we shall add a few comments regarding the quantity Δ_{AB} which is defined in (10.23).

The KBI, $G_{\alpha\beta}$ measures the "overall" correlation between the species α and β. As in general the PCF differs from unity only at distances of a few molecular diameters, $G_{\alpha\beta}$ may be considered as a *local* quantity. It is local in the sense that its value depends only on the local environment of single molecules. The KB theory provides many relationships between *local* and *global* quantities. Equations (10.19) to (10.21) are among these.

Clearly, when the two molecules α and β are very similar, say two isotopes of argon, we also expect that the form of the functions $g_{\alpha\beta}(r)$ does not depend on the indices α and β. Therefore, we also expect the value

---

[4]We do encourage however the curious reader to study the Kirkwood-Buff theory of solutions and the result (10.22). See Kirkwood & Buff (1951) and Ben-Naim (2006b).

of $G_{\alpha\beta}$ to be independent of α and β. If all the $G_{\alpha\beta}$ functions have nearly the same value, then the quantity Δ_{AB} will be nearly zero. However, Δ_{AB} can be nearly zero even when this condition is not met: it is sufficient that G_{AB} is very close to the arithmetic average of G_{AA} and G_{BB}, that is

$$G_{AB} = \frac{1}{2}(G_{AA} + G_{BB}) \qquad (10.24)$$

Clearly, two species that are not very similar can fulfill the condition (10.24), hence making Δ_{AB} equal to zero. We shall see the importance of the value of Δ_{AB} in section 10.7 in the context of solution stability.

10.3   The Three Reference *Ideal* Solutions

Thermodynamics does not provide us with explicit functions of the dependence of the chemical potential on the concentrations of the various components in the mixtures. Nevertheless, it is useful to use additional information (i.e. not coming from thermodynamics) to obtain an explicit expression for the dependence of the chemical potential on the concentration in some special cases. We shall discuss now three important cases.

10.3.1   *Ideal gas mixture. Experimental approach*

An ideal gas of pure component is defined experimentally through the equation of state

$$PV = nRT \qquad (10.25)$$

Inserting V/n from (10.25) in the thermodynamic relationship (9.14) and integrating it, we obtain[5]

$$\mu = C + RT\ln P$$
$$= \mu(1\,\mathrm{atm}, T) + RT\ln P \qquad (10.26)$$

where the integration constant $C = \mu(1\,\mathrm{atm}, T)$ in general depends on T. Thus, the explicit dependence of the chemical potential on the pressure is based on the experimental knowledge of the equation of state of an ideal gas.

---

[5]Remember that, whenever we write $\ln P$, the pressure P has to be intended without physical units. In other words, if we are measuring P in atmospheres, we have to understand $\ln P$ as $\ln(P/1\,\mathrm{atm})$. Of course, the same is also true for any other physical quantity: the argument of each mathematical function must be dimensionless.

An ***ideal gas mixture*** is an ideal mixture for which each component behaves as if it were the only ideal gas filling the entire volume of the system. In other words, different species do not interact, they do not "see" each other.

We define the ***partial pressure*** of the i-th component by

$$P_i = y_i P \tag{10.27}$$

where $y_i \equiv n_i/n_{tot}$ is the mole fraction,[6] so that the equation of state of a mixture of c components is

$$PV = \sum_{i=1}^{c} n_i RT \tag{10.28}$$

Equations (10.27) and (10.28) are also referred to as ***Dalton's law of partial pressure***.

Now, suppose we have a mixture of gases at equilibrium with pure component i. The chemical potentials in the two phases must be equal. Therefore, the chemical potential of the i-th component in the mixture must have the same functional form as (10.26), i.e., as if the the i-th component were filling the entire space by itself with pressure P_i. Thus, using (10.27) we obtain

$$
\begin{aligned}
\mu_i &= \mu_i(P_i = 1, T) + RT \ln P_i \\
&= \mu_i(P_i = 1, T) + RT \ln P + RT \ln y_i \\
&= \mu_i^{p}(P, T) + RT \ln y_i
\end{aligned} \tag{10.29}
$$

where $\mu_i^{p}(P, T)$ is the chemical potential of the pure component i at the same pressure and temperature of the mixture.

Equation (10.29) is sometimes used to define an ideal gas mixture, sometimes also referred to as ***perfect mixture of gases***. However, a more usual definition of an ideal gas mixture is as a mixture for which the mole fraction equals the ratio between the partial and total pressures and the ratio between partial and total volumes:

$$y_i = \frac{n_i}{n_{tot}} = \frac{P_i}{P} = \frac{V_i}{V} \tag{10.30}$$

The last equality is basically the ***law of partial volumes*** formulated by Émile Hilaire Amagat (1841–1915) in 1880, which states that in an

---

[6]We adopt a notation in which we use y_i for mole fractions in a gaseous phase and x_i for mole fractions in a liquid or a solid phase.

ideal gas mixture the volume V is equal to the sum of all volumes that the pure components would have, at the same pressure and temperature. In other words, the law of partial volumes implies that the volume is an extensive quantity, as it adds up for non-interacting systems.

As an example of ideal gas mixture, we mention the air: a gas mixture mostly made by nitrogen and oxygen, with mole fractions of about 21% O_2 and 79% N_2 at 25°C. The air behaves like an ideal gas mixture when the pressure is smaller than about 3 atm.

Before passing to the theoretical approach, the reader should note carefully that the derivation of (10.29) is based on the thermodynamic relation (9.14) supplemented by the equation of state of an ideal gas and Dalton's law. The latter are experimental facts that are not derivable from within thermodynamics.

10.3.2   *Ideal gas mixture. Theoretical approach*

Consider a mixture of two components A and B, with x_A being the mole fraction of A in the liquid phase.[7] Using the Kirkwood-Buff formula (10.22) we find that, when $\rho_{tot} \to 0$, which is equivalent to $P \to 0$, we get

$$\lim_{P \to 0} \left(\frac{\partial \mu_A}{\partial x_A} \right)_{P,T} = \frac{k_B T}{x_A} \tag{10.31}$$

Integrating (10.31) and applying the result (9.38) for one mole of the A component we have

$$\lim_{P \to 0} \mu_A = \mu_A^{ig,0} + RT \ln x_A = \mu_A^{liq,p} + RT \ln y_A \tag{10.32}$$

where $\mu_A^{ig,0}$ is the chemical potential of an ideal gas of pure A component at the specified pressure P and temperature T.

Equation (10.32) is essentially the same as the result (10.29), but now we have derived this relationship from the Kirkwood-Buff theory, for the special case of a two-component mixture. Thus, although we have not derived the Kirkwood-Buff formula (10.22), we ask the reader to trust that the result (10.32) is based on an exact theoretical foundation, which is different from the experimental-based derivation of (10.29).

---

[7]Whenever we refer to a gaseous phase we shall use y_A instead of x_A for the mole fraction.

10.3.3 *Dilute ideal solutions. Experimental approach*

It is known experimentally that, when the concentration of one component in the mixture is very low, then its partial pressure in the vapor phase is proportional to its mole fraction in the solution. This is known as **Henry's law**, formulated in 1803 by the English chemist William Henry (1774–1836):

$$P_i = k_{H,i} x_i \qquad \text{(valid for } x_i \approx 0\text{)} \qquad (10.33)$$

where $k_{H,i}$ is a parameter that may be dependent on T but is independent of x_i.

At equilibrium between the liquid mixture and its vapor we must have the equality

$$\mu_i^{\text{liq}} = \mu_i^{\text{gas}} = \mu_i^{\text{ig},0} + RT \ln P_i \qquad (10.34)$$

By inserting Henry's Law (10.33) we get for $x_i \approx 0$

$$\mu_i^{\text{liq}} = \mu_i^{\text{gas}} = \mu_i^{\text{ig},0} + RT \ln k_{H,i} + RT \ln x_i$$
$$= \mu_i^{0,x} + RT \ln x_i \qquad (10.35)$$

Thus, we have an explicit dependence of the chemical potential of the i-th component on its mole fraction x_i in the vapor phase.[8] The parameter $\mu_i^{0,x}$ is, in general, a function of P, T and of the other components, but does not depend on x_i.

Remember that, if there are c components, only $c - 1$ mole fractions are independent. Fixing x_i leaves only $c - 2$ independent mole fractions. For 2-component systems the composition is determined only by one mole fraction, say x_A, and $\mu_A^{0,x}$ is independent of the composition. Note also the superscript x in $\mu_i^{0,x}$, used to distinguish this standard chemical potential from other standard chemical potentials.

Whe have met again the relationship between the chemical potential of a component and its concentration, which we had encountered in section 9.4.1. There we considered the case of a mixture of two components A (a very diluted non-ionic solute) and B (solvent), in which case one has the identities (9.24) and (9.25):

$$\mu_A = \mu_A^0 + RT \ln x_A \qquad \text{(valid for } x_A \approx 0\text{)} \qquad (10.36)$$

$$\mu_B = \mu_B^p + RT \ln x_B \qquad \text{(valid for } x_B \approx 1\text{)} \qquad (10.37)$$

---

[8]The ideal solution behavior discussed here is for non-ionic solutes.

Actually, (10.37) is a consequence of the experimental result (10.36) and of Henry's law.

Exercise 10.1. Show that (10.37) is obtained from (10.36) and (10.33) when the solution and gaseous phase are at equilibrium.

10.3.4 *Dilute ideal solutions. Theoretical approach*

Expanding the r.h.s. of the Kirkwood-Buff formula (10.22) about $x_A = 0$ we have the leading term

$$\left(\frac{\partial \mu_A}{\partial x_A}\right)_{P,T} = \frac{k_B T}{x_A} \qquad \text{(valid for } x_A \to 0) \qquad (10.38)$$

By integrating this relation for one mole of the A component we get for $x_i \to 0$

$$\mu_A = RT \ln x_A + \text{const.} \qquad (10.39)$$

As we get the same dependence of μ_A on x_A of the experimental approach, we can identify the integration constant with $\mu_i^{0,x}$ defined in (10.35), for $i = A$.

From (10.38) we can recover Henry's law. Assuming that there exists equilibrium between a dilute liquid solution and its vapor, and that the vapor above the solution is an ideal gas mixture, we must have the equality of the chemical potentials of A in the two phases:

$$RT \ln x_A + \mu_A^{0,x} = \mu_A = \mu_A^{gas} = \mu_A^{ig,0} + RT \ln P_A \qquad (10.40)$$

This implies that P_A and x_A must be proportional:

$$P_A = k_{H,A} x_A \qquad (10.41)$$

Thus, we have recovered Henry's Law from the Kirkwood-Buff theory of solutions.

Although the functional dependence of μ_A on x_A is the same in (10.32) and (10.39), with a term $RT \ln x_A$ in both cases, it should be stressed that the integration constants are different. In (10.32) we assumed that P is very small, but the equation is valid for any value of x_A. Therefore, we could identify the constant $\mu_A^{ig,0}$ with the chemical potential of a *pure* (i.e., $x_A = 1$) ideal gas at the same P and T. On the other hand, (10.39) is valid for only very small x_A (at the limit $x_A \to 0$, i.e. when A is very diluted in a solvent). Therefore, we can *not* use (10.39) for $x_A = 1$. The integration constant in (10.39) is not specified a priori, although it is related to the experimentally determinable Henry's constant k_H defined in (10.41), i.e. by (10.33) with $i = A$.

Again, we stress that the experimental approach is based on some experimental result, like Henry's law, which is not part of thermodynamics. On the other hand, the theoretical approach provides a firm theoretical basis for Henry's Law, and the typical behavior of dilute ideal solutions.

10.3.5 *Symmetric ideal solution. Experimental approach*

It is found experimentally that certain mixtures obey a relationship similar to Henry's Law (10.33), but valid for all components at all compositions. Specifically, for a two component system of A and B we find

$$P_i = k_{H,i} x_i \tag{10.42}$$

for $i = A, B$ and $0 \le x_i \le 1$.

Clearly, in this case we can substitute $x_i = 1$ in (10.42) and identify the constant $k_{H,i}$ as the vapor pressure P_i^p of pure component i at the same temperature, i.e.

$$P_i = P_i^p x_i \tag{10.43}$$

The relationship (10.43) is known as the **Raoult's law**.

Using essentially the same argument as in the dilute-ideal case, we assume that the mixture is at equilibrium with an ideal gas mixture. Therefore, the chemical potential of, say, A must be the same in the two phases, hence

$$
\begin{aligned}
\mu_A^{liq} = \mu_A^{gas} &= \mu_A^{ig,0} + RT \ln P_A \\
&= [\mu_A^{ig,0} + RT \ln P_A^p] + RT \ln x_A \\
&= \mu_A^p(P,T) + RT \ln x_A
\end{aligned}
\tag{10.44}
$$

where we used eq. (10.34) in the first line and we have identified the constant $\mu_A^p(P,T) = \mu_A^{ig,0} + RT \ln P_A^p$ as the chemical potential of pure A (i.e. when $x_A = 1$) at the same values of P and T of the solution, thanks to the fact that (10.44) is valid for any concentration value $0 \le x_A \le 1$. Note also that the dependence of μ_A^{liq} on the x_A is the same as in (10.32) and (10.39), but the meaning of the integration constant is different.

Exercise 10.2. Show that if (10.44) is valid for component A, then also

$$\mu_B^{\text{liq}} = \mu_B^{p}(P,T) + RT\ln x_B$$

valid for $0 \le x_B \le 1$.

10.3.6 Symmetric ideal solution. Theoretical approach

For any mixture of two components A and B, if $\Delta_{AB} = 0$ in the entire range of compositions we get from the Kirkwood-Buff formula (10.22)

$$\left(\frac{\partial \mu_A}{\partial x_A}\right)_{P,T} = \frac{k_B T}{x_A} \qquad (0 \le x_A \le 1) \tag{10.45}$$

Integrating (10.45) we obtain the chemical potential per mole of A as

$$\mu_A(P,T,x_A) = \mu_A^{p}(P,T) + RT\ln x_A \tag{10.46}$$

with $0 \le x_A \le 1$ and $\mu_A^{p}(P,T)$ identified as the chemical potential of pure A at the same P,T as in the mixture. This is the same as the result (10.44), but now instead of relying on the experimental Raoult's law, we started from the Kirkwood-Buff theory of solutions.

Furthermore, in the theoretical approach we have a clear requirement which defines the condition for obtaining the symmetrical ideal behavior, namely $\Delta_{AB} = 0$. Here, we only showed that $\Delta_{AB} = 0$ is a sufficient condition for the behavior of a symmetric ideal solution, but it is possible to prove that $\Delta_{AB} = 0$ is both a necessary and sufficient condition (Ben-Naim, 2006b).

10.4 Examples and Applications

10.4.1 Lowering the freezing temperature

Suppose we have a solid, say ice, in equilibrium with a solution, say water (W) and ethanol (E). If the solution is dilute ideal with respect to E we have, for $\mu_E \to 0$

$$\mu_E = \mu_E^{0} + RT\ln x_E \tag{10.47}$$

and, for $x_W \approx 1$

$$\mu_W^{\text{liq}} = \mu_W^{p} + RT\ln x_W \tag{10.48}$$

Note that μ_W^p is the chemical potential of pure water. We assume that the solid phase at equilibrium with the solution is pure ice, therefore

$$\mu_W^{sol} = \mu_W^{liq} = \mu_W^p + RT\ln x_W \tag{10.49}$$

By using the identity (9.34) one obtains

$$R\frac{\partial \ln x_W}{\partial T} = \frac{\Delta H_W^{(m)}}{T^2} \tag{10.50}$$

where $\Delta H_W^{(m)} \equiv H_W^{liq} - H_W^{sol}$ is the enthalpy of melting of pure water at the freezing point at say, one atmospheric pressure. Again, we assume that $\Delta H_W^{(m)}$ is approximately independent of temperatures, in a small range of temperatures about the freezing point. Hence, integrating (10.50) we get

$$\ln x_W = \frac{-\Delta H_W^{(m)}}{RT} + \text{const} \tag{10.51}$$

Since we know that at $x_W = 1$ the temperature is the freezing temperature of pure water, denoted T_f, we can find the integration constant in (10.51) and write

$$\ln x_W = \frac{-\Delta H_W^{(m)}}{R}\left(\frac{1}{T} - \frac{1}{T_f}\right) = \frac{\Delta H_W^{(m)}(T - T_f)}{RTT_f} \tag{10.52}$$

From this result we see that, since $x_W \leq 1$, it must be that $T - T_f < 0$, which implies that the addition of a small quantity of a solute to a pure liquid always lowers its freezing temperature.

A very common example is spreading salt on the roads during the winter, to prevent the formation of ice which would be extremely dangerous for all vehicles.

If the very small mole fraction of the generic solute A is x_A we can rewrite (10.52) as

$$\ln x_W = \ln(1 - x_A) \approx -x_A = \frac{\Delta H_W^{(m)}(T - T_f)}{RT_f^2}$$

which gives

$$\Delta T \equiv T - T_f \approx -\frac{RT_f^2}{\Delta H_W^{(m)}}x_A \tag{10.53}$$

Thus, for very low concentrations ($x_A \approx 0$) the lowering of the freezing temperature is proportional to the mole fraction of the solute.

10.4.2   *Elevation of the boiling temperature*

For the same system as discussed in section 10.4.1, we assume that the solution is at equilibrium with the vapor phase. If the partial pressure of the water is P_W, then at equilibrium we have

$$
\begin{aligned}
\mu_W^{\text{liq}} &= \mu_W^{\text{p,liq}} + RT\ln x_W = \mu_W^{\text{gas}} \\
&= \mu_W^{\text{p,gas}} + RT\ln y_W = \mu_W^{\text{gas}}(1\,\text{atm}) + RT\ln P_W
\end{aligned}
\tag{10.54}
$$

Following similar steps as in the previous section we get

$$
\Delta T \equiv T - T_b \approx \frac{RT_b^2}{\Delta H_W^{(b)}} x_A
\tag{10.55}
$$

where T_b is the boiling of the pure liquid water, and $\Delta H_W^{(b)} \equiv H_W^{\text{vap}} - H_W^{\text{liq}}$ is the boiling enthalpy of pure liquid water.

10.4.3   *Osmotic pressure*

The process of **osmosis** is the spontaneous flow of solvent molecules through a semi-permeable membrane (impermeable to the solute) into a region of higher solute concentration, in the direction that tends to minimize the difference in solute concentration on the two sides.

The **osmotic pressure** is defined as the excess pressure that one has to exert on the solution to prevent the flow of pure solvent into the solution through the semi-permeable membrane.

Figure 10.4 shows an example in which sugar ($C_6H_{12}O_6$) is added to pure water (H_2O) on one side of a U-shaped tube in which a membrane permeable to water but impermeable for sugar is placed. The level differences reaches the equilibrium value h at which $\rho g h$ equals the osmotic pressure (ρ is the mass density of the solution and g is the gravity acceleration).

Osmosis is of vital importance in biology as the cell membrane is selective toward many of the solutes found in living organisms.

Let's consider again the case of dilute solution of a solute A in a solvent, say water (W). The chemical potential of the solvent is

$$
\mu_W^{\text{liq}}(P,T,x_W) = \mu_W^{\text{p}}(P,T) + RT\ln x_W
\tag{10.56}
$$

where $\mu_W^{\text{p}}(P,T)$ is the chemical potential of pure liquid water at the same temperature and pressure as in the solution. Since $x_W \leq 1$ always, we

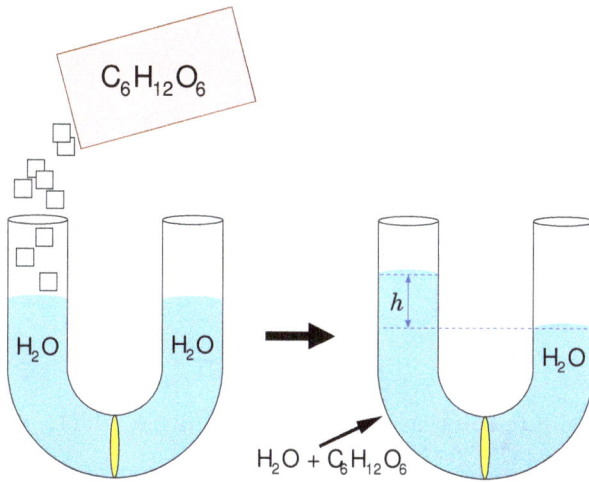

Figure 10.4: Osmosis in a U-shaped tube through a dialysis membrane when sugar is added to pure water on one side of the membrane. The fluid level rises on the side to which the sugar has been added and drops on the other side.

have

$$\mu_W^{\text{liq}}(P,T,x_W) \leq \mu_W^{\text{p}}(P,T) \qquad (10.57)$$

which means that if we have a solution at equilibrium with a pure liquid through a partition permeable only to the solvent, pure solvent will always flow *into* the solution.

We now use the thermodynamic identity relating the pressure variation of the chemical potential at constant temperature with the molar volume

$$\left(\frac{\partial \mu_W}{\partial P}\right)_T = \overline{V}_W \approx V_W^{\text{p}} > 0 \qquad (10.58)$$

(the molar volume of an ideal solution is practically the same as the molar volume of the pure solvent) to show that an increase of pressure on the solution will increase the chemical potential of the solvent.[9]

In practice, the osmotic pressure is the additional pressure Π that we need to exert on the solution, such that the inequality in (10.57) will turn into an equality. This overpressure is a function of the mole fraction x_A

---

[9]Note that the partial molar volume can also be negative but here the partial molar volume is nearly the same as the molar volume, which is always positive.

of the solute, so that (10.57) becomes the equality

$$\mu_W^{liq}(P + \Pi, T, x_A) = \mu_W^{p}(P, T) \tag{10.59}$$

that is the indirect definition of the osmotic pressure Π.

Assuming that that molar volume is independent of the pressure (which is a very good approximation), we can integrate (10.58) to obtain

$$\mu_W^{liq}(P + \Pi, T, x_A) - \mu_W^{liq}(P, T, x_A) = \int_P^{P+\Pi} V_W^p \, dP \simeq V_W^p \Pi \tag{10.60}$$

Hence from (10.56), (10.59) and (10.60) we get

$$-RT \ln x_W \approx V_W^p \Pi \tag{10.61}$$

For very dilute solutions we can expand $\ln x_W = \ln(1 - x_A) \approx -x_A$ in (10.61) and obtain

$$\Pi \approx x_A RT/V_W^p \tag{10.62}$$

Thus, the osmotic pressure Π is proportional to the mole fraction of the solute that, for dilute solutions, is $x_A = n_A/(n_A + n_W) \approx n_A/n_W$. Hence (10.61) may be rewritten as

$$\Pi \approx \frac{n_A RT}{n_W V_W^p} = \frac{n_A RT}{V} \tag{10.63}$$

where $V = n_W V_W^p$ is the total volume of the solution. Thus we have just derived the ***van 't Hoff equation*** for the osmotic pressure

$$\Pi V \approx n_A RT \tag{10.64}$$

that looks very similar to the equation of state of an ideal gas (2.6). Equation (10.64) is named after Jacobus Henricus van 't Hoff, Jr. (1852–1911), the first winner of the Nobel prize in chemistry.

The similarity between the equation of state of an ideal gas (2.6) and the van Hoff equation (10.64) should not lead to misinterpretations of the molecular origin of the osmotic pressure. In an ideal gas, the pressure P is a result of the "bombardment" of the particles on the wall of the container. However, the osmotic pressure Π has a quite different source. Because of the inequality (10.57), there is a tendency for the water (i.e. the solvent) to flow from the pure phase into the solution. The osmotic pressure is the *additional* pressure we have to exert on the solution to stop the flow of the solvent into the solution.

10.5 Deviations from Ideal Behavior

We have discussed three kinds of ideal mixtures: ideal gases (IG), dilute ideal (DI), and symmetrical ideal (SI) solutions. In all these cases one can write the chemical potential of component A as

$$\mu_A = RT \ln x_A + \text{const.} \tag{10.65}$$

where *the integration constant is different for each case*. This last statement is very important, in order to avoid misinterpretations (a very frequent mistake). It is only when the ideal case has been specified that it becomes possible to interpret the integration constant: there is no unique meaning.

In addition most mixtures and solutions are not ideal, although their behavior could be described in first approximation with relationships that are strictly valid only for ideal cases. Therefore, when discussing *deviations* from ideal behavior, it is of utmost importance to specify which type of ideality we are referring to. In the literature one usually expresses deviations from ideality in terms of various activity coefficients. This practice may be extremely confusing, not only because there are three ideal reference states, but also because, even within each type of ideality, one can discuss different kinds of deviations. We shall discuss below a few examples.

10.5.1 *Small deviations from ideal gas behavior*

A theoretical ideal gas (pure or mixture) is defined as a system of non-interacting particles. An experimental ideal gas (pure or mixture) can be obtained by letting the pressure P or, equivalently, the total density $\rho_{tot} = \sum_i \rho_i$ tend to zero. Because the intermolecular interactions between the particles act on a range of a few molecular diameters, taking the limit $\rho_{tot} \to 0$ makes the average intermolecular distance so large that the total interaction energy approaches zero, i.e. the system approaches the behavior of an ideal gas.

When the system is ideal we write the chemical potential of species i as

$$\mu_i = \mu_i^p + RT \ln x_i \tag{10.66}$$

where μ_i^p is the chemical potential of a pure component of ideal gas i at the same pressure and temperature as the mixture.

If the system deviates from ideal-gas behavior, we modify equation (10.66) by adding a correction term

$$\mu_i = \mu_i^p + RT\ln x_i + RT\ln\gamma_i^{IG} \tag{10.67}$$

where the ***activity coefficient*** (for IG) γ_i^{IG} accounts for the intermolecular interactions and is equal to 1 for an ideal gas.

Note that the quantity μ_i^p in (10.67) must be the same as in (10.66), and it retains its meaning as the chemical potential of pure i at the same P, T (without intermolecular interactions). Such a quantity may be calculated theoretically in statistical mechanics, but may not be accessible to experimental determination.

Substituting $x_i = 1$ in (10.67) gives

$$\mu_i^{\text{pure real}}(P,T) = \mu_i^{\text{pure ideal}}(P,T) + RT\ln\gamma_i^{IG} \tag{10.68}$$

This means that the activity coefficient γ_i^{IG} is connected to the intermolecular interactions between i-molecules in a pure i component. Note also that in general γ_i^{IG} depends on the composition of the mixture. Therefore, its value in (10.67) is not the same as the value in (10.68).

10.5.2 *Small deviations from dilute ideal solutions*

Consider a two-component system of A and B where the total density is not very low, so that intermolecular interactions are not negligible. When $x_A \to 0$, we have a dilute ideal solution of A in B and we can write

$$\mu_A(P,T,x_A) = \mu_A^{0,x}(P,T) + RT\ln x_A \tag{10.69}$$

which is valid only in the limit $x_A \to 0$. Here $\mu_A^{0,x}$ is the ***standard chemical potential*** defined as

$$\mu_A^{0,x}(P,T) \equiv \lim_{x_A \to 0} \mu_A(P,T,x_A) - RT\ln x_A \tag{10.70}$$

where the limit is taken at constant T,P values.

When the system is not dilute-ideal one formally adds a correction term to (10.69) and writes the chemical potential as

$$\mu_A(P,T,x_A) = \mu_A^{0,x}(P,T) + RT\ln x_A + RT\ln\gamma_A^{DI}(P,T,x_A) \tag{10.71}$$

where $\mu_A^{0,x}(P,T)$ is the same in (10.69) and (10.71) and $\gamma_A^{DI}(P,T,x_A)$ is another **activity coefficient**[10] (for DI).

Remember that the correction term in (10.71) is due to deviations from dilute ideal solutions, and this should be distinguished from corrections from ideal-gas behavior (section 10.5.1) and correction from symmetrical ideal solution (section 10.5.3).

When discussing dilute solutions, it is sometimes more convenient to express the concentration of A in other scales such as molality or molarity. The **molality** (b_A) of a solution is defined as the amount of substance (in mol) of solute A divided by the mass (in kg) of the solvent B (McNaught & Wilkinson, 1997):

$$b_A \equiv \frac{n_A}{m_B} \tag{10.72}$$

The **molar concentration** or **molarity** (c_A) is defined as the amount of substance (in mol) of solute A divided by the volume (in litres) of the solvent B (McNaught & Wilkinson, 1997):

$$c_A \equiv \frac{n_A}{V_B} \tag{10.73}$$

The use of molar concentration is often not convenient, because the volume of most solutions slightly depends on temperature due to thermal expansion. In these cases one may prefer the temperature-independent measure of concentration provided by the molality.

When molality or molarity are used as concentration measures, instead of (10.69) one writes either

$$\mu_A(P,T,b_A) = \mu_A^{0,b}(P,T) + RT \ln b_A \tag{10.74}$$

or

$$\mu_A(P,T,c_A) = \mu_A^{0,c}(P,T) + RT \ln c_A \tag{10.75}$$

where $\mu_A^{0,b}$ and $\mu_A^{0,c}$ are defined similarly to (10.70), with limits taken as $b_A \to 0$ and $c_A \to 0$, respectively.

When the system deviates from dilute ideal solutions the correction terms are different for each case: (10.74) and (10.75) become

$$\mu_A(P,T,b_A) = \mu_A^{0,b}(P,T) + RT \ln b_A + RT \ln \gamma_A^{DI}(P,T,b_A) \tag{10.76}$$

---

[10]In the literature it is often abbreviated as γ_A but this only creates confusion with the activity coefficients related to different ideality conditions.

and

$$\mu_A(P,T,c_A) = \mu_A^{0,c}(P,T) + RT \ln c_A + RT \ln \gamma_A^{DI}(P,T,c_A) \qquad (10.77)$$

respectively.

In the limit of dilute-ideal solution, the A-component is extremely dilute in the solvent (here, pure B, but in general the solvent could be a mixture of any number of components, other than A). In this limit, there are no solute-solute interactions: the solute molecules do not "see" each other. Theoretically, such system may be studied by considering only one A-molecule immersed in a "sea" of solvent molecules. Deviations from the dilute ideal behavior occur when the concentration of A increases, such that one cannot ignore any more the interactions between two A-molecules, three A-molecules, and so on. These interactions determine the value of the activity coefficient. But note carefully that one must specify precisely how we increase the concentration of A. In most cases we hold P and T constant and simply add A-particles into the system. However, in some cases it is found more convenient to keep c_B and T constant (in systems with constant volume), or b_B and T constant (when studying osmotic systems). In each of these cases the extent of deviation from DI behavior is different. That is the reason why we have used different notations for the corresponding activity coefficients.

Finally, we note also that Kirkwood-Buff theory of solutions provides us with exact expressions for the small deviations from DI behavior in terms of the Kirkwood-Buff integrals introduced in section 10.2.3 (Ben-Naim, 2006b).

For instance, if we keep μ_B and T constant, the first order deviation from an ideal dilute solution is

$$\mu_A(T,\mu_B,\rho_A) = \mu_A^{0,\rho}(T,\mu_B) + RT \ln \rho_A - RT G_{AA}^0 \rho_A + \cdots \qquad (10.78)$$

where G_{AA}^0 is the KBI at $\rho_A \to 0$. If we keep P and T constant, we have

$$\mu_A(P,T,\rho_A) = \mu_A^{0,\rho}(P,T) + RT \ln \rho_A - RT [G_{AA}^0 - G_{AB}^0] \rho_A + \cdots \qquad (10.79)$$

In this case we see that the first order deviation involves both G_{AA}^0 and G_{AB}^0. For more details, the reader is referred to chapter 6 of Ben-Naim (2006b).

10.5.3 *Small deviations from symmetrical ideal behavior*

One starts with SI solutions, for which the chemical potential of each component for any concentration $0 \le x_A \le 1$ has the form

$$\mu_A(P,T,x_A) = \mu_A^P(P,T) + RT \ln x_A \qquad (10.80)$$

When a system deviates from SI behavior we add a correction term to (10.80), and write

$$\mu_A(P,T,x_A) = \mu_A^P(P,T) + RT \ln x_A + RT \ln \gamma_A^{SI} \qquad (10.81)$$

In (10.81) we denoted the *activity coefficient* for SI with the short-hand notaion $\gamma_A^{SI} = \gamma_A^{SI}(P,T,x_A)$. The reason is simple: whereas in the case of dilute solutions it is very common to study deviations from ID with respect to different concentration scales and different thermodynamic variables, like (P,T), (μ_B,T) or (ρ_B,T), when studying deviations from SI behavior it is always assumed that we characterize the composition of the system by the mole fractions, and use P and T as the thermodynamic variables. Therefore, we chose the simpler notation. Note however, that the superscript SI is essential. In (10.81), γ_A^{SI} measures the extent of the deviations from "similarity" between the two components A and B. This is very different from the quantities γ_A^{IG} and γ_A^{ID} that have been discussed earlier.

Within thermodynamics, γ_A^{SI} in (10.81) is a well defined quantity that may be determined experimentally by measuring deviations from Raoult's law (10.43).

On the other hand, statistical thermodynamics provides more detailed expressions for the deviations from SI behavior. Specifically, by expanding (10.22) about $\Delta_{AB} = 0$ to first order in Δ_{AB}, we get

$$\left(\frac{\partial \mu_A}{\partial x_A}\right)_{P,T} = \frac{k_B T}{x_A}(1 - \rho_{tot} x_A x_B \Delta_{AB} + \cdots) \qquad (10.82)$$

Assuming further that $\rho_{tot}\Delta_{AB}$ is very small in the entire range of composition, so that we can neglect the variation of $\rho_{tot}\Delta_{AB}$ with x_A, we can integrate (10.82) to obtain the following approximate expression for the chemical potential:

$$\mu_A \simeq \mu_A^P(P,T) + RT \ln x_A + RT \rho_{tot} \Delta_{AB} x_B^2/2 \qquad (10.83)$$

Note that the constant of integration in (10.83) is fixed by substituting $x_A = 1$ and $x_B = 0$, and noting that the result is simply the chemical potential of pure A at the same P and T of the mixture. Thus, for small deviations from SI behavior we have an explicit expression for the corresponding activity coefficient:

$$\ln\gamma_A^{SI} = \rho_{tot}\Delta_{AB}\,x_B^2/2 \qquad (10.84)$$

Exercise 10.3. Show that for a two-component system, if (10.84) is valid for γ_A^{SI}, then a similar expression is also valid for γ_B^{SI}, i.e.

$$\ln\gamma_B^{SI} = \rho_{tot}\Delta_{AB}\,x_A^2/2$$

Exercise 10.4. Show that positive values of Δ_{AB} correspond to positive deviations from Raoult's Law (10.43).

The dependence of γ_A^{SI} on x_B^2 is typical for the deviations from SI behavior in mixtures of non-electrolytes. It was first found for solid solutions using simple lattice theory, for which one obtains (see chapter 6 of Ben-Naim 2006b)

$$RT\ln\gamma_A^{SI} = -z(E_{AA}+E_{BB}-2E_{AB})x_B^2/2 \qquad (10.85)$$

where z is the number of nearest neighbors to each site and $E_{\alpha\beta}$ is the interaction energy for the pair of species α and β occupying adjacent sites. This is used in theoretical discussion of solid mixtures.

10.6   Global and Local Characterization of Mixtures

In section 10.5 we have discussed deviations from ideal behavior, with the help of various activity coefficients assigned to each species. In the study of mixtures, it was traditionally found useful to study deviations from SI solutions in the *entire range* of composition and for the *entire mixture*, and not for each component separately.

For any extensive quantity, for example the energy E, one can define an *excess* with respect to the value expected in case of an ideal behavior:

$$E^{ex} \equiv E^{real} - E^{ideal} \qquad (10.86)$$

where E^{real} is the energy value for the entire solution at some given P, T and composition, and E^{ideal} is the value that the energy would have if the mixture were a symmetric ideal solution at the same P, T and composition.

By denoting with $\boldsymbol{x}$ the vector listing the composition components $(x_1, x_2, \ldots, x_c)$, one defines the excess of the most important thermodynamic quantities as

$$G^{\text{ex}} \equiv G^{\text{real}} - G^{\text{ideal}} = G(P, T, \boldsymbol{x}) - \sum_{i=1}^{c} N_i (\mu_i^{\text{p}} - RT \ln x_i) \qquad (10.87)$$

$$S^{\text{ex}} \equiv S^{\text{real}} - S^{\text{ideal}} = S(P, T, \boldsymbol{x}) - \sum_{i=1}^{c} N_i (S_i^{\text{p}} - R \ln x_i) \qquad (10.88)$$

$$H^{\text{ex}} \equiv H^{\text{real}} - H^{\text{ideal}} = H(P, T, \boldsymbol{x}) - \sum_{i=1}^{c} N_i H_i^{\text{p}} \qquad (10.89)$$

$$V^{\text{ex}} \equiv V^{\text{real}} - V^{\text{ideal}} = V(P, T, \boldsymbol{x}) - \sum_{i=1}^{c} N_i V_i^{\text{p}} \qquad (10.90)$$

Note that in each case we subtract the ideal part of all the components in the mixture. Hence the excess quantities measure *global deviations* from SI behavior of the *entire* mixture.

On the other hand, one may also focus on *local* deviations, referring to the close environment of each of the molecules. The fundamental quantities in this case are the KBIs and solvation thermodynamic quantities that will be introduced in section 10.8 below (we shall also discuss solvation quantities in chapter 12, in connection with solvation of simple solutes in aqueous solutions). The KBIs themselves convey information on the local densities of one component around a molecule in a mixture. Furthermore, by connecting them to measurable quantities, one can estimate the local composition around each molecule.

The difference between local and bulk composition is referred to as **preferential solvation** and can also be expressed in terms of the KBIs. Finally, knowing all the $G_{\alpha\beta}$ in a mixture, one can characterize the deviation from SI behavior by the quantity Δ_{AB}. All these quantities can be derived from the global properties, but the derivation is based on the inversion of the Kirkwood-Buff theory of solutions, which is beyond the scope of the present book. For details, see Ben-Naim (2006b).

10.7 Large Deviations from SI Solutions and Stability Condition

Traditionally, deviations from SI solutions are measured in terms of deviations from Raoult's law (10.43). For the component A in the mixture, one writes

$$\frac{P_A}{P_A^0} = x_A \gamma_A^{SI} \qquad (10.91)$$

where P_A is the partial pressure of A in the mixture and P_A^0 is the vapor pressure of pure A at the same temperature T of the solutions. When $\gamma_A^{SI} = 1$ we have Raoult's law (10.43). When $\gamma_A^{SI} > 1$ we say that we have *positive* deviations from Raoult's law, and when $\gamma_A^{SI} < 1$ we say that we have *negative* deviations.

However, when the interest is on the stability of solutions, one must pay careful attention to the fact that the value of P_A/P_A^0 does not determine uniquely the stability condition,[11] which is given by the requirement of a minimum in the Gibbs energy.

This condition can be formulated in terms Gibbs energy per particle as follows. For a solution of A in B with Gibbs energy $G = N_A \mu_A + N_B \mu_B$ the *Gibbs energy per particle* is

$$g \equiv \frac{G}{N_A + N_B} \qquad (10.92)$$

and the stability condition requires that the second derivative of g with respect to the concentration is positive, i.e. that the function $g(x_A)$ is upward concave:

$$\left(\frac{\partial^2 g}{\partial x_A^2}\right)_{P,T} = \frac{1}{x_B}\left(\frac{\partial \mu_A}{\partial x_A}\right)_{P,T} > 0 \qquad (10.93)$$

which is equivalent to requiring

$$\frac{1}{1-x_B}\left(\frac{\partial \mu_B}{\partial x_B}\right)_{P,T} > 0 \qquad (10.94)$$

In (10.93) and (10.94) the derivatives are taken at constant pressure and temperature.

Thermodynamics does not offer any molecular conditions for stability, but the Kirkwood-Buff theory does. With the help of eq. (10.22), we can rewrite condition (10.93) as

$$\left(\frac{\partial^2 g}{\partial x_A^2}\right)_{P,T} = \frac{k_B T}{x_A x_B (1 + x_A x_B \rho_{tot} \Delta_{AB})} > 0 \qquad (10.95)$$

---

[11]For more details, see appendix P of Ben-Naim (2006b).

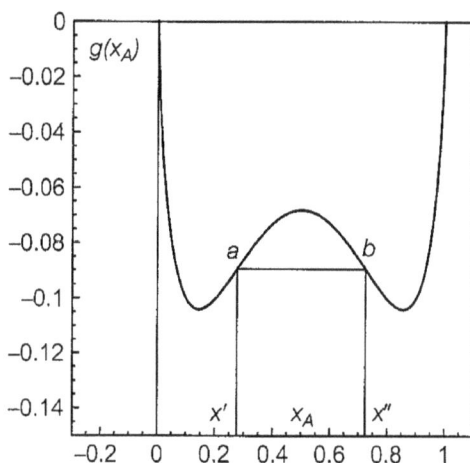

Figure 10.5: Example of unstable solution.

Clearly, whatever the value of $\rho_{tot}\Delta_{AB}$ is, when either x_A or x_B is very small we find

$$\left(\frac{\partial^2 g}{\partial x_A^2}\right)_{P,T} = \frac{k_B T}{x_A x_B} > 0$$

Therefore, *a dilute solution will always be stable*. In addition, the system will always be stable when $\Delta_{AB} \geq 0$.

When $\Delta_{AB} < 0$, stability requires that $|x_A x_B \rho_{tot}\Delta_{AB}| < 1$. Otherwise, if we have both $\Delta_{AB} < 0$ and $|x_A x_B \rho_{tot}\Delta_{AB}| > 1$, then we shall have instability.

Note that these are *exact results* based on the KB theory, whereas any discussion of stability based on first-order theories is risky, because it relies on the assumption that $\rho_{tot}\Delta_{AB}$ is very small (a more detailed analysis is presented in Ben-Naim 2006b).

An example of unstable solution is provided in figure 10.5, showing a function $g(x_A)$ with two minima, separated by a region with downward concavity in some region $x' \leq x \leq x''$ (we use the shorthand notation x for x_A here). As any point on the curve $g(x)$ has higher Gibbs energy than the pair of mixtures with compositions x' and x'', the system will spontaneously split into two phases.

From $N_A = N'_A + N''_A$ and $N_B = N'_B + N''_B$ we obtain the mole fractions $x' = N'_A/(N'_A + N'_B)$ and $x'' = N''_A/(N''_A + N''_B)$. Then it is easy to show that the overall composition x may be expressed in terms of the composition of the two phases x' and x'' as follows

$$x = \alpha x' + (1-\alpha)x'' \quad \text{with} \quad \alpha \equiv \frac{N'_A + N'_B}{N_A + N_B} \tag{10.96}$$

Let's look once again at the condition for stability (10.95) from the KB theory. The most interesting situations are found when Δ_{AB} defined in (10.23) is negative, that is when

$$G_{AB} > \frac{1}{2}(G_{AA} + G_{BB}) \tag{10.97}$$

Only when this condition is satisfied, i.e. when the KB integral for different species is larger than the average KB integral of pure substances, the system may have instabilities.

More precisely, for values $-4 < \rho_{tot}\Delta_{AB} < 0$ the r.h.s. of (10.95) is positive and the system is always stable. At $\rho_{tot}\Delta_{AB} = -4$ the second derivative diverges at $x_A = x_B = \frac{1}{2}$, but there is stability otherwise (e.g. for $x_A \approx 0$ and $x_A \approx 1$). Thus, in this case only at the centre of the composition range there is instability.

Finally, for $\rho_{tot}\Delta_{AB} < -4$ the system is unstable in the "inner" region of $x_A \approx x_B$ but is stable in the "outer" regions $x_A \approx 0$ and $x_A \approx 1$. As $\rho_{tot}\Delta_{AB}$ becomes more negative, the region of instability expands and the two regions of stability are pushed toward the edges.

In this discussion, we assumed $\rho_{tot}\Delta_{AB}$ to be independent of the composition. However, in real systems this may be not true, such that the details of the regions of stability may differ, although the qualitative picture will remain the same. Finally, the KB theory is valid only for stable systems, hence it cannot be used to study systems that are unstable with respect to the composition.

10.8   Solvation Thermodynamics

In this section, we briefly present a huge field of research in solution chemistry. We start with a qualitative discussion of what solvation quantities are, and where they are used. We then define the process of solvation, and the corresponding solvation thermodynamical quantities. We shall also discuss how solvation quantities may be obtained from experimentally measurable quantities. Specific examples will be presented in chapter 12, as solvation thermodynamics has been studied more in aqueous solutions than in any other liquid.

10.8.1 *Why do we need solvation thermodynamics?*

There are essentially three reasons for studying solvation thermodynamics. The first and simplest one is that solvation determines solubility, which is of paramount importance in many applications. For example, solubility is a fundamental property when studying micelles, detergents, drugs and slow release of drugs, not to speak about applications that will be mentioned later, as desalination and solubility of proteins in water.

The **solubility** of a pure solid into a liquid is defined as the concentration ρ_A of the solid A in the liquid B for which the following equilibrium condition holds:

$$\mu_A^p = \mu_A^B(\rho_A) \tag{10.98}$$

where μ_A^p is the chemical potential of pure A, at some pressure P and temperature T, and $\mu_A^B(\rho_A)$ is the chemical potential of A in the liquid B as a function of the concentration ρ_A. Thus, the value of ρ_A that satisfies (10.98) is called the solubility of A in B.

In most cases we do not have pure A, but two coexisting phases, say α and β, with different concentrantion. For instance, mixtures of water (A) and butanol (B) in some range of temperature and pressure form two phases at equilibrium, each phase containing both A and B, with equilibrium conditions:

$$\mu_A^\alpha(\rho_A^\alpha) = \mu_A^B(\rho_A^\beta) \quad \text{and} \quad \mu_B^\alpha(\rho_B^\alpha) = \mu_B^B(\rho_B^\beta) \tag{10.99}$$

These relationships, although they determine unique values for ρ_A^α and ρ_A^β at the given P and T, cannot be solved explicitly. However, in some specific cases, one can calculate the distribution of A (and B) in the two phases. We shall discuss important examples in chapter 12.

Another important application of solvation quantities is in the study of equilibrium constants in a solvent. In chapter 11 we shall discuss chemical equilibrium in an ideal gas phase. This is an important application of thermodynamics to chemical relations at equilibrium. More important, however, are chemical reactions that occur in a liquid phase; in particular, in aqueous solutions, a medium for most biochemical processes. We shall see in chapter 11 how the equilibrium constant of a chemical reaction is modified when the same reaction is "transferred" from an ideal gas to a liquid phase. The modification of the equilibrium constant involves the *solvation* Gibbs energies of all the reactants and products of the chemical reaction.

Finally, solvation quantities are important *tools* to probe the interaction of a molecule with its environment.[12] This is of particular importance in the study of solvation quantities in aqueous solutions, where a host of interesting properties of pure water are revealed by the study of solvation thermodynamics of simple solute in water. We shall mention some of these in chapter 12, when we discuss aqueous solutions of simple solutes, such as argon in water.

10.8.2 *Solvation process and solvation quantities*

Solvation quantities were traditionally studied only for dilute solutions and ideal gases. Let A be a *solute* very dilute in a solvent B, such that the chemical potential can be written as

$$\mu_A^B = \mu_A^{0,B} + RT \ln C_A^B \tag{10.100}$$

where C_A^B is any measure of concentration of A in B and $\mu_A^{0,B}$ is the standard chemical potential defined in section 10.5.2. If the vapor at equilibrium with the solution can be assumed to be an ideal gas, then

$$\mu_A^{gas} = \mu_A^{0,gas} + RT \ln C_A^{gas} \tag{10.101}$$

At equilibrium we must have the equality of the two chemical potentials, hence from (10.100) and (10.101) it follows that

$$\mu_A^{0,B} - \mu_A^{0,gas} = RT \ln(C_A^{gas}/C_A^B)_{eq} \tag{10.102}$$

The quantity on the r.h.s. of (10.102) is a measurable quantity, as it depends on the ratio of the two concentrations of A in the gaseous and the liquid phase (B). The quantity on the l.h.s. was referred to as the "solution free energy".

---

[12]Conventionally, the *solute* is solvated by a *solvent*. However, here we talk of any molecule being surrounded by any environment.

Note carefully that (10.102) is valid only for an ideal gas, hence for only one of the phases in a dilute ideal solution of A in B. This is a serious limitation on the range of applicability of (10.102). However, a more serious problem with this relationship is that thermodynamics does not provide any meaning to the quantity we referred to as the free energy of solution, which is supposed to measure the solvation of A in B.[13] Unfortunately, these difficulties cannot be resolved within thermodynamics.

Because of its central importance in the study of solutions in general, and aqueous solutions in particular, we present here the modern definition of the solvation processes, together with the corresponding thermodynamics, based on one result from statistical mechanics, which will not be proven here. The reader is required to accept it on faith, but can find more details on the statistical mechanical derivation of the solvation quantities, as well as a comparison with older definitions of *solution* quantities, in Ben-Naim (2006b).

As we have seen in section 7.2.3, the chemical potential of an ideal monoatoamic gas (i.e. simple particles having no internal degrees of freedom at very low pressure) is

$$\mu_A = RT \ln(\rho_A \Lambda_A^3) \qquad (10.103)$$

where ρ_A is the density expressed in particles per unit of volume (or number of moles per unit of volume), and Λ_A^3 is the momentum partition function in statistical mechanics. $\Lambda = h/\sqrt{2\pi m k_B T}$ is also referred to as the thermal de Broglie wavelength in quantum mechanics. Here, we refer to the quantity defined in (10.103) as the translational part of the chemical potential.

Note that classical statistical mechanics only holds if

$$\rho_A \Lambda_A^3 \ll 1 \qquad (10.104)$$

which is always satisfied by a gas at very low pressure (hence at very low density; a Maxwell-Boltzmann gas). When the thermal de Broglie wavelength is no more negligible compared to the intermolecular distances, quantum effects cannot be ignored and the correct distribution (Fermi-Dirac or Bose-Einstein) must be used.

---

[13]Furthermore, the situation has been made even more confuse by the fact that different authors used different concentration scales, leading to different definitions of the free energy of solution.

To generalize (10.103) for real molecules in a non-ideal system, we need to introduce correction terms which are easy to understand on qualitative grounds. First, if the molecule A has internal degrees of freedom such as vibration, rotation, etc., we have to add one term μ_A^{int} to keep track of them, and rewrite (10.103) as

$$\mu_A = \mu_A^{int} + RT\ln(\rho_A \Lambda_A^3) \qquad (10.105)$$

Equation (10.105) has also a simple interpretation in terms of an idealized two-step addition of one molecule of A (or one mole of A) to an ideal gas system. First, we place A at some fixed position (no motion allowed: we are discussing classical systems for which it is meaningful to talk about a particle at a specific location with zero velocity). This part involves only the change in the Gibbs energy due to the addition of the internal degrees of freedom of the A molecule. Next, we can release the A-molecule. The resulting change in the Gibbs energy is the second term on the r.h.s. of (10.105), which is the same as the term on the r.h.s. of (10.103).

It is advisable to pause and ponder on the significance of the last step. Releasing the particle from a fixed position causes three changes in the Gibbs energy of the system. First, the molecule which was confined to a fixed point can now access the entire volume of the system. This is a big change in its spatial distribution, and contributes with a term $-RT\ln V$. Second, the molecule being at a fixed position is *distinguishable* from all other particles of the same species in the system, which are identical. Once it is released, the distinguishability of that specific particle is lost, and it assimilates with all other A particles in the system. The assimilation contributes $+RT\ln N_A$ to the chemical potential, and together with the previous term it gives the contribution $RT\ln\rho_A$. Third, a still molecule at a fixed position has no translational degree of freedom. However, once it is released, it acquires a (3-dimensional) distribution of velocities or momenta. This contributes with $RT\ln\Lambda_A^3$ and completes (10.103).[14]

When classical statistical mechanics is applicable, condition (10.104) holds, which implies that (10.103) is negative. This picture is consistent

---

[14]Please note that the product $\rho_A \Lambda_A^3$ is dimensionless, hence it can appear as argument of the logarithm function. On the other hand, writing $\ln V$ implicitly assumes that the volume has been divided by its units of measure (e.g. m^3).

with the behavior of the Gibbs energy: releasing an internal constraint always causes a decrease in the Gibbs energy. The resulting contribution, i.e. (10.103), may be referred to as either the ***translational Gibbs energy*** or the ***liberation Gibbs energy***.

We can further generalize (10.105) by introducing intermolecular interactions, which are important when the mixture is not an ideal gas. In this case we can also split the process of adding one A to a mixture characterized by P, T, N_A, N_B in discrete steps: first, we place the A-molecule at a fixed position, then we add intermolecular interactions, and finally we release it.

As before, the first step corresponds to introducing the internal degrees of freedom of A. We will assume here that they are not affected by the intermolecular interactions that represent the new ingredient here, although this may be not the case in real systems. It is sufficient to keep in mind that, upon "turning on" the interactions, the μ_A^{int} in (10.106) below may be different from the μ_A^{int} in (10.105). There will be some change in the Gibbs energy when we enable the interaction of A at the fixed position with the entire system. We shall refer to this part of the change in free energy as the ***coupling work***, as if the interactions were "switched on" continuously (alternatively, as if the particle were brought to that point from an infinitely long distance). We denote the coupling work by $W(A|P, T, x_A)$ and write the chemical potential of A in the mixture as

$$\begin{aligned}
\mu_A &= W(A|P, T, x_A) + \mu_A^{\text{int}} + RT\ln(\rho_A \Lambda_A^3) \\
&= \mu_A^* + RT\ln(\rho_A \Lambda_A^3)
\end{aligned} \tag{10.106}$$

In (10.106) we have introduced the symbol μ_A^* for what we we shall refer to as the ***pseudo chemical potential***. It is the sum of the coupling work and of the Gibbs energy associated to the internal degrees of freedom.

Thus, the chemical potential is the sum of two terms: the pseudo chemical potential and the liberation Gibbs energy. Note that for classical systems the liberation Gibbs energy in (10.106) is the same as in (10.105) and (10.103).

We now define the ***solvation process*** as the process of transferring a single molecule, say A, from an ideal gas into a solution characterized by (P, T, x). The corresponding ***solvation Gibbs energy*** is defined by

$$\Delta G_A^* \equiv \mu_A^{*,\text{liq}} - \mu_A^{*,\text{i.g.}} \tag{10.107}$$

that is the difference between the pseudo chemical potential in the liquid and gas phases. This is the change in Gibbs energy associated to the solvation process.

It is clear that ΔG_A^* involves the effects of the interactions between all molecules of the system and the newly added A molecule. Although the solvation process is not a process that one can actually perform in laboratory, ΔG_A^* can be determined experimentally. This provides precious information about the local neighborhood of each A molecule in the solution, hence connects macroscopic observations to microscopic properties.

For any two phases of α and β at equilibrium, each containing molecules of type A, we can write the equality

$$\mu_A^\alpha = \mu_A^\beta$$

Using (10.106) for each phase and assuming that Λ_A^3 is the same in the two phases, we obtain

$$\mu_A^{*,\alpha} + RT \ln \rho_A^\alpha = \mu_A^{*,\beta} + RT \ln \rho_A^\beta$$

or equivalently

$$\mu_A^{*,\beta} - \mu_A^{*,\alpha} = RT \ln \left(\frac{\rho_A^\alpha}{\rho_A^\beta} \right)_{eq} \tag{10.108}$$

Thus, measuring the concentration of A in the two phases at equilibrium gives the change in the Gibbs energy for the transfer of one A-molecule from α to a *fixed position* in β.

In particular, when the phase α is an ideal gas and the phase β is a solution, then equation (10.108) provides an experimental method of determining the solvation Gibbs energy:

$$\Delta G_A^{*,liq} = RT \ln \left(\frac{\rho_A^{i.g.}}{\rho_A^{liq}} \right)_{eq} \tag{10.109}$$

(just multiply by the Avogadro number to obtain the change per mole of A, i.e. replace RT with $k_B T$).

An important special case is when the liquid phase is pure A. The solvation Gibbs energy of A in pure liquid A is

$$\Delta G_A^{*,p,liq} \equiv \mu_A^{*,p,liq} - \mu_A^{*,i.g.} = RT \ln \left(\frac{\rho_A^{i.g.}}{\rho_A^{p,liq}} \right)_{eq} \tag{10.110}$$

Although $\Delta G_A^{*,\mathrm{p,liq}}$ is just a particular case of the more general quantity $\Delta G_A^{*,\mathrm{liq}}$, it is important to note that in the traditional approach to solvation both phases had to be ideal (an ideal gas, and A very diluted in the liquid phase). The definition of the solvation process given above, together with the corresponding thermodynamic quantities, does apply to any mixture at any composition, including a "solvent" which is pure A.

Once we have defined the solvation Gibbs energy, we can proceed with the definition of all other thermodynamic quantities of solvation. Of particular interest are the solvation entropy, enthalpy and volume:

$$\Delta S_A^{*,\mathrm{liq}} = \left(\frac{\partial \Delta G_A^{*,\mathrm{liq}}}{\partial T}\right)_{P,N} \tag{10.111}$$

$$\Delta H_A^{*,\mathrm{liq}} = \Delta G_A^{*,\mathrm{liq}} + T\,\Delta S_A^{*,\mathrm{liq}} \tag{10.112}$$

$$\Delta V_A^{*,\mathrm{liq}} = \left(\frac{\partial \Delta G_A^{*,\mathrm{liq}}}{\partial P}\right)_{T,N} \tag{10.113}$$

One can also define the energy and the Helmholtz energy of solvation by

$$\Delta E_A^{*,\mathrm{liq}} = \Delta H_A^{*,\mathrm{liq}} - P\,\Delta V_A^{*,\mathrm{liq}} \tag{10.114}$$

$$\Delta A_A^{*,\mathrm{liq}} = \Delta E_A^{*,\mathrm{liq}} - T\,\Delta S_A^{*,\mathrm{liq}} \tag{10.115}$$

All these solvation quantities reveal a host of information on the local environment of each A-molecule, which is of particular interest in the study of aqueous solutions.

Chapter 11

Chemical Equilibrium

In this chapter, we will discuss the application of thermodynamics to chemical equilibrium. We start out with a simple example of conversion between two isomers A and B. We then generalize to any chemical reaction and we derive a general expression for the equilibrium constant. Next we discuss chemical equilibrium in solutions, and derive expressions for the temperature, pressure and solute dependence of the equilibrium constant.

11.1 The Simple Isomerization Reaction

Consider a simple conversion between two isomers, say cis and trans dichloroethylene (figure 11.1), or helix-coil transition of a polypeptide from a random coil to the so-called α-helix, the regular helicoil structure of proteins. We assume that the conversion between the two isomers can be controlled by either an inhibitor or a catalyst. In the first case we assume that the conversion between A and B is maintained unless we add an *inhibitor*. In the second case we assume that the conversion is possible only in the presence of a *catalyst*.

The system is closed (i.e. no material flow is allowed in or out of the system), and is maintained at constant pressure and temperature. In such a system the fundamental thermodynamic function is the Gibbs energy (if the volume and temperature are constant, the appropriate function is the Helmholtz energy).

When considering chemical reactions, it is more practical to deal with moles rather than particles, hence we write the Gibbs energy function as $G = G(P, T, n_A, n_B)$.

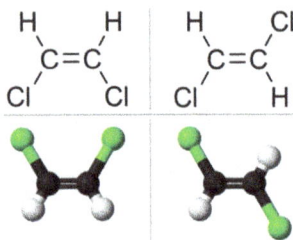

Figure 11.1: Cis- (left) and trans- (right) 1,2-dichloroethylene $C_2H_2Cl_2$.

In the presence of an inhibitor (or in the absence of a catalyst) we can prepare the system with any number of moles of A and B, whose sum $n = n_A + n_B$ is constant, as the system is closed.

We now remove the inhibitor (or introduce the catalyst) for a very short time, so that dn_A moles convert from the isomer A into the isomer B. We write this conversion reaction as:

$$A \to B \tag{11.1}$$

The corresponding change of the Gibbs energy is

$$dG = \left(\frac{\partial G}{\partial n_A}\right)_{P,T} dn_A + \left(\frac{\partial G}{\partial n_B}\right)_{P,T} dn_B$$
$$= \mu_A \, dn_A + \mu_B \, dn_B \tag{11.2}$$

Note that P and T are kept constant, as indicated in the partial derivatives. As the system is closed, the total number n of moles in the system is unchanged. This implies $dn_A = -dn_B$, hence we can rewrite (11.2) as

$$dG = (\mu_A - \mu_B) dn_A \tag{11.3}$$

If the transformation (11.1) is spontaneous, then the Gibbs energy must decrease, i.e.

$$dG < 0 \tag{11.4}$$

Therefore, if $\mu_A < \mu_B$ then $dn_A > 0$. Alternatively, if $\mu_A > \mu_B$ then $dn_A < 0$. This means that matter will "flow" from a higher chemical potential to a lower chemical potential.

Here the flow of matter is from "state" A to "state" B, and occurs under constant pressure and temperature. At equilibrium G must reach a minimum, at which point $dG = 0$.

It is convenient to express the Gibbs energy change in terms of the mole fraction $x_A = n_A/n$, and rewrite (11.3) as

$$dG = n(\mu_A - \mu_B)dx_A \tag{11.5}$$

At equilibrium

$$\left(\frac{\partial G}{\partial x_A}\right)_{P,T,n} = 0 \quad \Rightarrow \quad \mu_A = \mu_B \tag{11.6}$$

The equilibrium condition (11.6) is very general for reactions of the type (11.1). It says that at equilibrium the chemical potentials of the two isomers must be equal.

If the reaction occurs in an ideal gas phase, then we can write

$$\mu_A = \mu_A^p + RT\ln x_A \tag{11.7}$$

$$\mu_B = \mu_B^p + RT\ln x_B \tag{11.8}$$

and the condition of chemical equilibrium (11.6) becomes

$$\left(\frac{x_A}{x_B}\right)_{eq} = \exp\left[\frac{\mu_B^p - \mu_A^p}{RT}\right] \tag{11.9}$$

where μ_A^p and μ_B^p are the chemical potentials of pure A and pure B at the same values of T and P, respectively.

We define the **equilibrium constant** for the reaction (11.1) as

$$K_x \equiv \left(\frac{x_A}{x_B}\right)_{eq} = \exp\left[\frac{\mu_B^p - \mu_A^p}{RT}\right] \tag{11.10}$$

Note that K_x depends only on the properties of *pure* A and *pure* B, and not on the composition of the system. The suffix x reminds us that it has been defined by taking the ratio between mole fractions.

Since $x_B = 1 - x_A$, we can rewrite (11.10) as

$$(x_A)_{eq} = \frac{K_x}{1 + K_x} \tag{11.11}$$

It is sometimes convenient to express the chemical potentials μ_A and μ_B in terms of the partial pressures P_A and P_B of A and B. These are related to the mole fractions by Raoult's law (10.43). At equilibrium, the latter can be combined with Dalton's law of partial pressure (10.27) to give

$$P_A = Px_A \quad \text{and} \quad P_B = Px_B \tag{11.12}$$

where P is the total pressure of the system. Hence, we rewrite (11.7) and (11.8) as

$$\mu_A = \mu_A^p - RT\ln P + RT\ln P_A = \mu_A^0 + RT\ln P_A \qquad (11.13)$$

$$\mu_B = \mu_B^p - RT\ln P + RT\ln P_B = \mu_B^0 + RT\ln P_B \qquad (11.14)$$

In terms of the partial pressures we can write the equilibrium condition as

$$K_P \equiv \left(\frac{P_A}{P_B}\right)_{eq} = \exp\left[\frac{\mu_B^0 - \mu_A^0}{RT}\right] \qquad (11.15)$$

In this particular reaction K_P is the same as K_x, as can be verified by substituting μ_A^0 and μ_B^0 from (11.13) and (11.14) into (11.15). However, in general K_P and K_x are different, as we will see below.

11.2   A General Chemical Reaction

Before we generalize the concept of chemical equilibrium consider again the reaction $A \rightarrow B$. Clearly, the changes in the number of moles of A and B are related by the condition $dn_B = -dn_A$ or, equivalently,

$$\frac{dn_B}{1} = \frac{dn_A}{-1} \qquad (11.16)$$

We define the parameter ξ, in such a way that for an infinitesimal change in the reaction we have

$$d\xi = \frac{dN_B}{1} = \frac{dN_A}{-1} \qquad (11.17)$$

We now consider a more complex chemical reaction, like the formation of ammonia from nitrogen and hydrogen. The balanced reaction is:

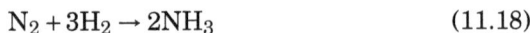

$$N_2 + 3H_2 \rightarrow 2NH_3 \qquad (11.18)$$

The integer numbers in front of each **reactant** (substance at the left hand side) and **product** (substance at the right hand side) are the **stoichiometric coefficients**, and are taken by convention with *negative sign for reactants* and with *positive sign for products*. In (11.18) they are -1, -3, and 2. By the law of conservation of mass, their ratios define the proportions among the numbers of moles of each reactant and product in the reaction.

For an infinitesimal change in the reaction (11.18) we must have

$$d\xi = \frac{dn_{NH_3}}{2} = \frac{dn_{H_2}}{-3} = \frac{dn_{N_2}}{-1} \qquad (11.19)$$

which simply means that, for a small change in the compositions, if the number of product molecules NH_3 *increases*, then the number of the reactant molecules N_2 and H_2 will *decrease* (and vice versa). The extent of change will be proportional to the stoichiometric coefficients.

The change in the Gibbs energy for this reaction is

$$dG = \frac{\partial G}{\partial n_{NH_3}} dn_{NH_3} + \frac{\partial G}{\partial n_{H_2}} dn_{H_2} + \frac{\partial G}{\partial n_{N_2}} dn_{N_2} \qquad (11.20)$$

$$= \mu_{NH_3} dn_{NH_3} + \mu_{H_2} dn_{H_2} + \mu_{N_2} dn_{N_2}$$

or equivalently

$$dG = (2\mu_{NH_3} - 3\mu_{H_2} - \mu_{N_2})d\xi \qquad (11.21)$$

At equilibrium we must have

$$\left(\frac{\partial G}{\partial \xi} \right)_{P,T,n} = 2\mu_{NH_3} - 3\mu_{H_2} - \mu_{N_2} = 0 \qquad (11.22)$$

Note that this linear combination of chemical potentials has the (signed) stoichiometric coefficients as weights.

Now we consider a general chemical reaction, in a closed system at some specified P and T and composition $n_1, \ldots, n_c$, where n_i is the number of moles of the species A_i. We also assume that all these components are involved in a single chemical reaction, which we write as

$$\nu_1 A_1 + \nu_2 A_2 + \cdots \nu_i A_i \rightarrow \nu_{i+1} A_{i+1} + \cdots \nu_c A_c \qquad (11.23)$$

where ν_i are the (unsigned) stoichiometric coefficients.

The Gibbs energy of the system, viewed as a c-component system, is $G(P, T, n_1, \ldots, n_c)$. For an infinitesimal change in the composition of the system, due to the removal of an inhibitor or introduction of a catalyst, the corresponding change in the Gibbs energy is

$$dG = \sum_{i=1}^{c} \mu_i \, dn_i \qquad (11.24)$$

Now the parameter ξ is defined such that for each substance in (11.23) we have

$$dn_i = \nu_i \, d\xi \qquad (11.25)$$

where v_i is taken with the positive sign for products and with negative sign for reactants.

Thus, if A_i is a product, dn_i is positive while the reaction is proceeding, i.e. the quantity of A_i moles increases proportionally to v_i. On the other hand, if A_i is a reactant, then its quantity of moles decreases proportionally to v_i.

From (11.24) and (11.25) we get

$$dG = \sum_{i=1}^{c} v_i \mu_i \, d\xi \tag{11.26}$$

Hence at equilibrium we must have the general condition:

$$\boxed{\left(\frac{\partial G}{\partial \xi}\right)_{P,T,n} = \sum_{i=1}^{c} v_i \mu_i = 0} \tag{11.27}$$

with *signed* stoichiometric coefficients v_i.

Note that in all reactions considered so far, all derivatives are at constant P and T, and the system is closed. All changes in composition are due to the conversion of molecules from the reactant "states" to product "states".

Next, we derive an expression for the chemical constant. If the system is an ideal gas mixture, then the chemical potential of each of the species can be written as

$$\mu_i = \mu_i^{\mathrm{p}} + RT \ln x_1 \tag{11.28}$$

where μ_i^{p} is the chemical potential of the pure component i at the same P, T values. Inserting (11.28) into (11.27) we get

$$\left(\prod_{i=1}^{c} x_i^{v_i}\right)_{\mathrm{eq}} = \exp\left[\frac{-\sum_i v_i \mu_i^{\mathrm{p}}}{RT}\right] \tag{11.29}$$

We define the **equilibrium constant** for the general reaction (11.23) as

$$K_x \equiv \left(\prod_{i=1}^{c} x_i^{v_i}\right)_{\mathrm{eq}} = \exp\left[\frac{-\sum_i v_i \mu_i^{\mathrm{p}}}{RT}\right] \tag{11.30}$$

Exercise 11.1. Express the chemical potentials in terms of the partial pressures P_i and write the corresponding equilibrium constant.

11.3 Chemical Equilibrium in a Solution

In this section we show how the chemical equilibrium is modified when the same reaction is carried out in a solvent. We start with an ideal gas mixture, for which we have

$$\mu_i = \mu_i^{*\,ig} + RT \ln \rho_i \Lambda_i^3 \qquad (11.31)$$

Here $\mu_i^{*\,ig}$ is the pseudo chemical potential of the i-th component in an ideal gas mixture. It depends on the temperature, the pressure and the internal properties of the i-th species. Λ_i is the thermal de Broglie wavelength (see section 7.2.3).

The equilibrium constant in the ideal gas phase is

$$K_\rho^{ig} = \left[\prod_{i=1}^c (\rho_i^g \Lambda_i^3)^{\nu_i} \right]_{eq} = \exp \left[\frac{-\sum_i \nu_i \mu_i^{*\,ig}}{RT} \right] \qquad (11.32)$$

When the same reaction occurs in a liquid phase, the chemical potential of each species is written as[1]

$$\mu_i = \mu_i^{*\,liq} + RT \ln \rho_i \Lambda_i^3 \qquad (11.33)$$

This expression is very general. It applies for any solvent and any concentration of the i-th component. In most applications one uses the limit of this expression when the i-th component is very diluted in the solvent. In this limit, $\mu_i^{*\,liq}$ becomes independent of the density ρ_i.

The corresponding equilibrium constant in the liquid phase is

$$K_\rho^{liq} = \left[\prod_{i=1}^c (\rho_i^{liq} \Lambda_i^3)^{\nu_i} \right]_{eq} = \exp \left[\frac{-\sum_i \nu_i \mu_i^{*\,liq}}{RT} \right] \qquad (11.34)$$

The relation between the equilibrium constants of the same reaction in the two phases is obtained from (11.32) and (11.34):

$$\frac{K_\rho^{liq}}{K_\rho^{ig}} = \exp \left[\frac{-\sum_i \nu_i \Delta G_i^{*\,liq}}{RT} \right] \qquad (11.35)$$

where $\Delta G_i^{*\,liq}$ is the solvation Gibbs energy defined in (10.107).

This is an important relationship. It shows that, if we know the chemical equilibrium in an ideal gas phase and we know the solvation Gibbs

---

[1]Note that Λ_i^3 is assumed to be the same in the two phases.

energies of all the species involved in the reaction, we can calculate the chemical equilibrium of the reaction in the liquid phase.

It should be noted that $\Delta G_i^{*\,\text{liq}}$ is the solvation Gibbs energy of the i-th component in the liquid phase, which in general can have any composition. However, the traditional study of solvation was restricted to dilute ideal solutions, in which case $\Delta G_i^{*\,\text{liq}}$ becomes independent of ρ_i^{liq}.

One can derive in a similar way other equilibrium constants for special systems, like symmetric ideal solutions, or dilute ideal solutions with respect to all the components involved in the reaction.

11.4   The Temperature Dependence of the Equilibrium Constant

As we have seen in the previous sections, one can define the different equilibrium constants for different choices of concentration scales. We shall use the equilibrium constant defined in (11.32), whose temperature dependence is obtained from

$$\left(\frac{\partial \ln K_\rho^{\text{ig}}}{\partial T}\right)_P = \frac{\partial}{\partial T}\left[\frac{-\sum_i \nu_i \mu_i^{*\,\text{ig}}}{RT}\right] = \frac{\sum_i \nu_i H_i^{*\,\text{ig}}}{RT^2} \tag{11.36}$$

where the last term follows from

$$\left(\frac{\partial(\mu_i/T)}{\partial T}\right)_P = \frac{T\frac{\partial \mu_i}{\partial T} - \mu_i}{T^2} = \frac{-T\bar{S}_i - \mu_i}{T^2} = -\frac{\bar{H}_i}{T^2} \tag{11.37}$$

Equation (11.36) is known as **van 't Hoff equation** for the change of the equilibrium constant with the temperature, named after Jacobus Henricus van 't Hoff, Jr. (1852–1911). The quantity $\sum_i \nu_i H_i^{*\,\text{ig}}$ may be referred to as a *standard enthalpy* of the reaction. It is standard in the sense that the reaction occurs for species devoid of translational degrees of freedom.

There are other standard enthalpies of reaction, depending on the choice of the concentration scales. For instance, if all the components form an ideal gas mixture, then for each component we can write

$$\mu_i = \mu_i^{\text{p}} + RT \ln x_i$$

and the corresponding equilibrium constant is given in (11.30). The temperature dependence of K_x is

$$\left(\frac{\partial \ln K_x}{\partial T}\right)_P = \frac{\sum_i \nu_i H_i^{\text{p}}}{RT^2} \tag{11.38}$$

where now $\sum_i \nu_i H_i^{\mathrm{p}}$ is the enthalpy of a **unit reaction**, i.e. it refers to the complete conversion of all reactants into the products. In other words, if we consider the chemical reaction (11.23) and start with exactly ν_j moles of reactant A_j, with $j = 1, \ldots, i$ running over all the reactants, a unit reaction ends up with a state in which no reactant exists any more, because all A_j's have been consumed to produce exactly ν_k moles of product A_k, with $k = i + 1, \ldots, c$ running over all the products.

As a simple example consider the isomerization reaction

$$A \rightleftharpoons B \tag{11.39}$$

In this particular case $\Lambda_A^3 = \Lambda_B^3$, because the thermal de Broglie wavelength Λ_i defined in (7.18) depends only on the mass of the i-th molecule. Hence the equilibrium constant is

$$K_\rho^{\mathrm{ig}} = \left(\frac{\rho_B}{\rho_A}\right)_{\mathrm{eq}} = \exp\left[\frac{\mu_B^{*\,\mathrm{ig}} - \mu_A^{*\,\mathrm{ig}}}{RT}\right] \tag{11.40}$$

and the temperature dependence of the equilibrium constant is

$$\left(\frac{\partial \ln K_\rho^{\mathrm{ig}}}{\partial T}\right)_P = \frac{H_B^{*\,\mathrm{ig}} - H_A^{*\,\mathrm{ig}}}{RT^2} \tag{11.41}$$

where $H_B^{*\,\mathrm{ig}} - H_A^{*\,\mathrm{ig}}$ is the enthalpy change for the conversion of one A molecule into a B molecule. Since the translational energies of the two species are equal in this case, this change in enthalpy is essentially the difference in the internal energies of the two species A and B.

When the same conversion occurs in a solution, we can differentiate (11.34) with respect to the temperature to obtain

$$\left(\frac{\partial \ln K_\rho^{\mathrm{liq}}}{\partial T}\right)_P = \frac{H_B^{*\,\mathrm{ig}} - H_A^{*\,\mathrm{ig}}}{RT^2} + \frac{\Delta H_B^{*\,\mathrm{liq}} - \Delta H_A^{*\,\mathrm{liq}}}{RT^2} \tag{11.42}$$

Thus, in addition to $H_B^{*\,\mathrm{ig}} - H_A^{*\,\mathrm{ig}}$ we need to know also the solvation enthalpies of the two species A and B.

At this stage we derive an identity which will be found useful in Chapter 12 when we discuss the outstanding properties of water.

For the general reaction at equilibrium we rewrite the condition of equilibrium (11.27) as

$$\Delta \mu \equiv \sum_i \nu_i \mu_i = 0 \tag{11.43}$$

We now view $\Delta\mu$ as a function of $(P,T,\boldsymbol{n})$, where $\boldsymbol{n} \equiv (n_1,\ldots,n_c)$ represents the number of moles of each species involved in the chemical reaction. The total differential of $\Delta\mu$ is

$$d(\Delta\mu) = \left(\frac{\partial\Delta\mu}{\partial P}\right)_{T,\boldsymbol{n}} dP + \left(\frac{\partial\Delta\mu}{\partial T}\right)_{P,\boldsymbol{n}} dT + \sum_{j=1}^{c} \frac{\partial\Delta\mu}{\partial n_j} dn_j \qquad (11.44)$$

We use the definition of $d\xi$ in (11.25) and $\Delta\mu$ in (11.43) to rewrite the last term on the r.h.s. of (11.44) as

$$\sum_{j=1}^{c} \frac{\partial\Delta\mu}{\partial n_j} dn_j = \sum_{j=1}^{c}\sum_{i=1}^{c} \nu_i \nu_j \mu_{ij} \, d\xi = \eta \, d\xi \qquad (11.45)$$

where

$$\mu_{ij} \equiv \frac{\partial^2 G}{\partial n_i \partial n_j} = \frac{\partial\mu_i}{\partial n_j} = \frac{\partial\mu_j}{\partial n_i} \qquad (11.46)$$

and

$$\eta \equiv \sum_{j=1}^{c}\sum_{i=1}^{c} \nu_i \nu_j \mu_{ij} \qquad (11.47)$$

Exercise 11.2. Prove that η is always positive.

Next, we take the derivative of $\Delta\mu$ with respect to the temperature while *maintaining* the *equilibrium* condition. In other words, we follow the change of $\Delta\mu$ with temperature along the equilibrium line. At equilibrium $\Delta\mu$ is identically zero, and therefore its derivative must also be zero:

$$0 = \left(\frac{d\Delta\mu}{dT}\right)_{P,\mathrm{eq}} = \left(\frac{\partial\Delta\mu}{\partial T}\right)_{P,\boldsymbol{n}} + \eta\left(\frac{\partial\xi}{\partial T}\right)_{P,\mathrm{eq}} \qquad (11.48)$$

The entropy change for the reaction is

$$\Delta S = \sum_{i=1}^{c} \nu_i \bar{S}_i = -\left(\frac{\partial\Delta\mu}{\partial T}\right)_{P,\boldsymbol{n}} \qquad (11.49)$$

hence (11.48) implies

$$\eta\left(\frac{\partial\xi}{\partial T}\right)_{P,\mathrm{eq}} = \Delta S \qquad (11.50)$$

Since at equilibrium we have (11.43) or equivalently $\Delta G = 0$,

$$\Delta\mu = \Delta H - T\Delta S = 0 \qquad (11.51)$$

and can write (11.50) as

$$\eta\left(\frac{\partial\xi}{\partial T}\right)_{P,\mathrm{eq}} = \frac{\Delta H}{T} \tag{11.52}$$

The last relationship is an example of the **Le Chatelier principle**[2], which states that, if a system is at equilibrium and we make a change in one of the parameters affecting the equilibrium (here the temperature), the system will respond in such a way as to reduce or to minimize the effect of the parameter.

In this particular case, if the reaction is **exothermic**, i.e. if it happens spontaneously with some energy release ($\Delta H > 0$), then an increase in temperature will drive the equilibrium in the direction that absorbs heat, therefore reducing the effect of the increase in temperature.

The special case of the isomerization is a simple example:

$$A \rightarrow B \tag{11.53}$$

for which $v_B = 1$ and $v_A = -1$, and

$$\mathrm{d}n_A = -\mathrm{d}\xi, \quad \mathrm{d}n_B = \mathrm{d}\xi \tag{11.54}$$

Hence, for this reaction (11.52) becomes

$$\eta\left(\frac{\partial n_B}{\partial T}\right)_{P,\mathrm{eq}} = \frac{\Delta H}{T} \tag{11.55}$$

Suppose we know that $\Delta H = \overline{H}_B - \overline{H}_A$ is positive, i.e. (11.53) is exothermic. Since $\eta = \mu_{AA} - 2\mu_{AB} + \mu_{BB} > 0$, (11.55) means that by increasing the temperature, n_B will change in the same direction as the sign of ΔH.

11.5 The Pressure Dependence of the Equilibrium Constants

We repeat the steps as in the previous section, but now we take the pressure derivative of the equilibrium constant K_ρ^{ig}:

$$\left(\frac{\partial \ln K_\rho^{\mathrm{ig}}}{\partial P}\right)_T = -\frac{\partial}{\partial P}\left(\frac{\sum_i v_i \mu_i^{*\,\mathrm{ig}}}{\partial P}\right)_T = 0 \tag{11.56}$$

---

[2]Named after the French chemist Henry Louis Le Châtelier (1850–1936).

Thus, this particular equilibrium constant is independent of pressure. The reason is that $\mu_i^{*\,\text{ig}}$ depends only on the internal properties of the i-molecule, and these do not depend on the pressure.

If we use the mole fractions scale, then we write

$$\mu_i = \mu_i^{\text{p}} + RT\ln x_i \tag{11.57}$$

Here, μ_i^{p} is the chemical potential of a pure i at the same temperature and pressure as in the solution. The corresponding equilibrium condition is

$$\sum_i \nu_i \mu_i = 0$$

or equivalently

$$\left(\prod_i x_i^{\nu_i}\right)_{\text{eq}} = \exp\left[\frac{-\sum_i \nu_i \mu_i^{\text{p}}}{RT}\right] \tag{11.58}$$

The equilibrium constant in this case is (11.30) and its pressure derivative is

$$\left(\frac{\partial \ln K_x}{\partial P}\right)_T = -\frac{\partial}{\partial P}\left[\frac{\sum_i \nu_i \mu_i^{\text{p}}}{RT}\right] = -\frac{\sum_i \nu_i V_i^{\text{p}}}{RT} \tag{11.59}$$

where V_i^{p} is the molar volume of pure i. Relation (11.59) is valid for any ideal solution for which the chemical potential has the form (11.57).

For the most general case, we use the general expression for the chemical potential (11.33), which is valid for any mixture with any composition. We have seen that the equilibrium constant in this case is given by (11.34). The pressure dependence of K_ρ^{liq} is therefore

$$\left(\frac{\partial \ln K_\rho^{\text{liq}}}{\partial P}\right)_T = -\frac{\partial}{\partial P}\left[\frac{\sum_i \nu_i \Delta G_i^{*\,\text{liq}}}{RT}\right] = -\frac{\sum \nu_i \Delta V_i^{*\,\text{liq}}}{RT} \tag{11.60}$$

where $\Delta V_i^{*\,\text{liq}}$ is the solvation volume of the i-th component in the liquid phase at the specified temperature, pressure and composition. We note again that (11.60) is very general and applies to any mixture, not necessarily an ideal one.

Next, we derive an identity similar to (11.50). From the total differential (11.44) we can get the pressure derivative of $\Delta\mu$ along the equilibrium line

$$0 = \left(\frac{\partial\Delta\mu}{\partial P}\right)_{T,eq} = \left(\frac{\partial\Delta\mu}{\partial P}\right)_{T,n} + \sum_{j=1}^{c}\sum_{i=1}^{c} v_i v_j \mu_{ij} \left(\frac{\partial\xi}{\partial P}\right)_{T,eq}$$

or equivalently

$$\eta\left(\frac{\partial\xi}{\partial P}\right)_{T,eq} = -\Delta V = -\sum_i v_i \overline{V}_i \tag{11.61}$$

which is the analog of (11.50).

This is again an example of the Le Chatelier principle. If $\Delta V > 0$, then increasing the pressure $dP > 0$ will cause a change in the reaction in such a way as to "absorb" the increase of pressure.

For the special isomerization reaction (11.39) we have the result

$$\eta\left(\frac{\partial n_B}{\partial P}\right)_{T,eq} = -\Delta V = -(\overline{V}_B - \overline{V}_A) \tag{11.62}$$

11.6 The Dependence of the Equilibrium Constant on the Concentration of a Component which is not Involved in the Reaction

We derive here one more identity, which is another example of the Le Chatelier principle, and which is useful for aqueous solution (to be discussed in chapter 12).

Consider again the simplest example, the isomerization reaction

$$A \rightarrow B \tag{11.63}$$

The condition of chemical equilibrium is

$$\Delta\mu = \mu_B - \mu_A = 0 \tag{11.64}$$

Now assume that, in addition to A and B, there is at least one component C in the system, which is not involved in the reaction (11.63).

The total differential of $\Delta\mu$, viewed as a function of T, P, n_A, n_B and n_C is

$$d\Delta\mu = \frac{\partial\Delta\mu}{\partial T}dT + \frac{\partial\Delta\mu}{\partial P}dP + \frac{\partial\Delta\mu}{\partial n_A}dn_A + \frac{\partial\Delta\mu}{\partial n_B}dn_B + \frac{\partial\Delta\mu}{\partial n_C}dn_C \tag{11.65}$$

Taking the derivative of $\Delta\mu$ with respect to n_C along the equilibrium line, at constant T and P, we get (Ben-Naim, 2006b)

$$0 = \left(\frac{\partial \Delta\mu}{\partial n_C}\right)_{P,T,\text{eq}} = (\mu_{AA} - \mu_{AB})\frac{\partial n_A}{\partial n_C} + (\mu_{AB} - \mu_{BB})\frac{\partial n_B}{\partial n_C} + \frac{\partial \Delta\mu}{\partial n_C} \quad (11.66)$$

Note that the system is closed with respect to A and B, i.e. $n_A + n_B =$ constant. Hence $dn_A = -dn_B$. Re-arranging the terms in (11.66) we get

$$(\mu_{AA} - 2\mu_{AB} + \mu_{BB})\left(\frac{\partial n_B}{\partial n_C}\right)_{P,T,\text{eq}} = -\left(\frac{\partial \Delta\mu}{\partial n_C}\right)_{P,T,n_A,n_B} \quad (11.67)$$

Since $\mu_{AA} - 2\mu_{AB} + \mu_{BB} > 0$, the identity (11.67) means that, if we add C to the system (i.e. $dn_C > 0$), this will affect the reaction (11.63) in such a way that $dn_B > 0$ (if $\Delta\mu < 0$), or $dn_B < 0$ (if $\Delta\mu > 0$).

Chapter 12

Water

This chapter is devoted to the most important substance, water. Water is found everywhere in nature. It is all around us and inside of us. It is a so familiar liquid that it is difficult to believe that it is one of the most unusual liquids. Indeed, the study of water offers a plethora of challenging problems to scientists, from understanding on the molecular level its unusual properties, to solving technological problems such as desalination and purification.

The study of water is presented here as an example to demonstrate the applicability of what we have learned so far in this book. We will use the principles of thermodynamics, the phase rule, the theory of mixtures, and the theory of equilibrium to explain some of the outstanding properties of liquid water.

12.1 Relevance to Biology

There has been a dramatic increase over the interest in water for the past 40–50 years[1], especially since the discovery that water in living systems is not a "passive" liquid that merely dissolves and transports substances from one part of an organism to another. Now we know that water does not only play a major role, but perhaps even a decisive role in biochemical processes, and that it is also vital to the maintenance of life as we know it on our planet.

Nearly all of the water on earth is found in the oceans. The rest, which makes up only a small fraction, is fresh or "sweet". Most of this

---

[1]For review of selected works, see Ben-Naim (2009).

fresh water is stored as ice in the Arctic and Antarctic.

As far as we know, water is the only liquid that supports life. It is believed that the specific anomalous properties of water are intimately related to the origin and evolution of life on earth. Thus, we are motivated to study water both because we are curious about it, and because such knowledge has many practical applications.

Among the most important aspects, we can mention that ice has larger molar volume of liquid water, that the latter has a high surface tension and a high thermal capacity, and a large dielectric constant. The latter is very important, as it implies the water capability of dissolving large variety of molecules, from simple salts to proteins. This is fundamental for every living being, as water is the transport medium of all substances in every organism. Water, as a solvent, helps distribute various chemicals needed in the different parts of the body, and at the same time it helps to discharge other waste chemicals that are not needed, or those that are even harmful.

The surface tension is the primary force that draws fluids up through capillaries. The higher the surface tension of a liquid, the higher the liquid can rise in capillaries. Its high surface tension allows water to rise in the interior of tall trees, even up to the highest leaves where it is needed in photosynthesis and temperature regulation.

But water makes much more than simply transporting substances. It also plays a very important role in the cell chemistry and in the regulation of body temperature. Many of the chemical reactions happening in each living cell are exothermal, i.e. they release energy. If this heat is not absorbed by some mechanism, the temperature will rise to a dangerous level. The large heat capacity of water means that for a given amount of heat absorbed by one unit volume of water, the temperature increase is smaller than it would be in a normal liquid. Thus, most of the heat released in chemical reactions is absorbed by the water, with a minimal increase in its temperature. The large thermal capacity is also the reason why laboratories use water in a thermostat-bath when a constant temperature must be maintained.

Another property of water that helps regulate body temperature is its large heat of vaporization. When the atmospheric temperature is high, both animals and plants evaporate some of their water content. This is the familiar phenomenon of sweating. Because the heat of vaporization of water is large, only a small quantity of water needs to be evaporated to maintain the temperature of the body.[2]

At a much larger scale, it is important to notice that most of the energy that Earth receives from the Sun is absorbed by evaporating ocean water, which mitigates the temperature excursion that would otherwise characterize day-night changes.

The molar volume of ice is larger than that of liquid water. Everyone is familiar with the phenomenon of ice floating on top of water. If the molar volume of ice were not larger than that of water, ice formed on the surface of the ocean would sink to the bottom. Over time, more and more layers of ice would accumulate on the bottom in a process that would gradually freeze nearly the entire ocean.

The fact that ice floats on top of the water helps to maintain the relatively high temperature of liquid water underneath. This in turn helps to maintain life in water, in spite of lower temperatures above the surface. In addition, when the temperature of the atmosphere increases, it is the ice on the surface that melts first, also contributing to the maintenance of the water temperature underneath.

On the other hand, the larger molar volume of ice can also be harmful to living organisms. When a living cell (containing lot of water) freezes, the expanding ice can break the cell membrane, possibly killing the cell.

Water has also a large latent heat of fusion, or better ***enthalpy of fusion***, defined as the energy necessary to convert a given quantity of the substance from a solid to a liquid state at constant pressure. This means that a large amount of heat is absorbed when ice melts. It also means that significant energy is released when water freezes.

If the temperature of pure water or any aqueous solution falls below the freezing point (which is normally lower for aqueous solutions than for pure water), ice begins to form and a large amount of heat is released.

---

[2] Note that the actual amount of sweat visible on the skin in hot weather or after exercising is not an indication of the amount of water that is evaporating. The evaporating sweat is released into the air as a gas and is not conspicuous. Sweat becomes visible only when liquid is released by the body more rapidly than the rate of evaporation. The latter is affected by the relative humidity of the air. When the relative humidity is low, water evaporates quickly from the skin. When it is high, water evaporates slowly from the skin because the air has a lowered capacity to absorb additional water, and temperature regulation by sweating becomes inefficient.

This release of heat decelerates the rate of any further ice formation, preventing or at least delaying the conversion of all the water in the body into ice. This mitigates the damage of freezing cells.

It is apparent that the properties of water are "fine-tuned" to support life. This has been viewed as the evidence of the existence of an Intelligent Designer. This is essentially the same argument as the so-called anthropic principle, invoked by creationists in favor of the existence of an Intelligent Designer: if the Universe is so finely tuned to support life, then an intelligent being has designed the Universe for this purpose. However, from a scientific point of view, it must be emphasized that our observer's role strongly biases the observable sample: clearly we cannot observe an universe in which we do not exist. Thus, it is also possible that, out of many (or even infinite) conceivable universes, we can only observe the few ones (or possibly the only one) supporting intelligent life.

12.2   Hydrogen Bonds

We can compare water against other substances made of *isoelectronic* molecules, all having in total 10 electrons (hence also 10 protons since the molecules themselves are electrically neutral): methane (CH_4), ammonia (NH_3), water (H_2O), hydrofluoric acid (HF), and neon (Ne). Their melting and boiling temperatures are shown in table 12.1 and in figure 12.1, in the left column of each plot. The same figure also shows the melting and boilint temperatures of other substances, whose

Table 12.1: **Properties of Isoelectronic Substances with 10 Electrons**. T_m is the melting point, T_v is the boiling point, and H_v is the molar heat of vaporization.

Substance	T_m (°C)	T_v (°C)	H_v (cal/mol)
CH_4	−184	−161.5	2200
NH_3	−78	−33.4	5550
H_2O	0	+100.0	9750
HF	−92	+19.4	7220
Ne	−249	−246	415

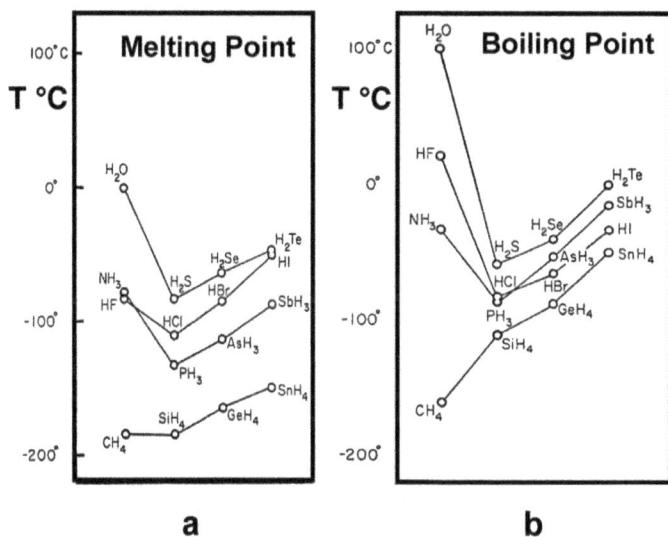

Figure 12.1: Melting and boiling points of different substances, with generic chemical formula XH_n. The abscisa indicates the row in the periodic table from which the element X has been taken.

molecules have the generic form XH_n. It is clear that the values appearing in the first column of each plot are higher than what one would obtain by extrapolating to the left the quite regular behavior of the other 3 points in each *homologous* series XH_n with given n, indicating that the configuration with 10 electrons is peculiar.[3] Furthermore, this difference is maximal for water.

Melting and boiling points of isoelectronic substances with 10 electrons are also shown, together with the molar enthalpy of vaporization, in table 12.1. Among these substances, water achieves maximum values in all cases.

As it has been clearly explained by Linus Carl Pauling[4] (1901–1994) in his classic book "The Nature of the Chemical Bond" (Pauling, 1939), the high values of H_2O and NH_3 are due to the strength of a **hydrogen bond** (HB).

---

[3] The first two electronic shells are fully occupied in this configuration. This and other notions of chemical physics are mentioned in this chapter without full treatment.

[4] Pauling was awarded the Nobel Prize twice: in Chemistry in 1954 and for Peace in 1962.

Chemical bonds may have different nature. A *ionic bond* is formed by the electrostatic force attracting two ions with opposite charges, and is very strong. Also the *covalent bond* is strong, in which two electrons are shared by two atoms. This allows each of them to reach a more stable configuration, characterized by a complete electronic shell. Other types of bonds are weaker, like the dipole-dipole interactions acting between polar molecules and the *Van der Waals forces* acting among neutral atoms and molecules.

A hydrogen bond is formed between the lone pair of an atom and a hydrogen atom that is bonded to either nitrogen, oxygen, or fluorine. A *lone pair* is a pair of *valence electrons* (i.e. electrons that are available to form a chemical bond) that are not shared with another atom. The HB is a sort of strong electrostatic dipole-dipole interaction, also stronger than a van der Waals interaction. It is directional and produces interatomic distances shorter than the sum of van der Waals radii.

A hydrogen atom is made of a single proton and a single electron. When this electron is shared in a covalent bond, the otherwise spherically symmetric electrical distribution around the hydrogen nucleus becomes very asymmetric. This "exposes" somewhat the positive charge of the proton, which can strongly attract the negative charge represented by the lone pair. In addition, as the size of the hydrogen atom involved in a covalent bond is very small, the atom attracted by the HB gets very close to the atom bonded covalently to the H, preventing additional atoms to get also close to the hydrogen.

The greater the *electronegativity* of the atom (i.e. the larger is its capability to attract electrons), the stronger the hydrogen bond. Fluoride forms very strong hydrogen bonds, while oxygen has somewhat weaker hydrogen bonds, nitrogen with even weaker hydrogen bonds, and finally carbon, with almost no detectable hydrogen bonds.

Note that although the strength of a single hydrogen bonds increases from nitrogen to oxygen to fluoride, the effect of HB is more dramatic in water than in hydrofluoric acid. Why? Because in liquid water a water molecule can form up to *four* hydrogen bonds with neighboring molecules! This property is unique to liquid water. As we shall see later, it is also primarily the molecular origin of the anomalous properties of liquid water.

Viewing the phenomenon of hydrogen-bonding as a force of adhesion between water molecules in the liquid state, we can immediately understand the high values of the melting point, boiling point and heat of vaporization. Since the forces that hold water molecules together are strong, a relatively high temperature is needed to melt ice.[5] Similarly, throughout the entire liquid range (0–100°C) the average number of hydrogen bonds decreases as the temperature increases, but even at the boiling point there are many water molecules still engaged in hydrogen bonding. Therefore, a high temperature is required to completely detach water molecules from the liquid and to send them into the gaseous phase.

12.3 Water Molecules

H_2O forms an isosceles triangle with H-O-H angle of 104.523° and equilibrium O-H bond length of 0.9572 Å (figure 12.2). For comparison, the van der Waals radii of the hydrogen and oxygen atoms are 1.2 Å and 1.4 Å. The angstrom unit 1 Å = 10^{-10} m.[6]

Heavy water (D_2O, where deuterium differs from hydrogen only for the neutron bound to the proton in the nucleus) has a similar molecule. D_2O forms an isosceles triangle with D-O-D angle of 104.474° and equilibrium O-D bond length of 0.9575 Å.

The formation energy of H_2O is $\Delta E = -219.337\,\mathrm{kcal\,mol^{-1}}$. The dipole moment of a water molecule has magnitude $\mu = 1.84\,\mathrm{D}$ and is directed from the Oxygen atom toward the line bisecting the H-O-H angle. The debye unit is $1\,\mathrm{D} = 0.2082\,\mathrm{e\AA}$, where e is the proton charge.[7]

Hydrogen bonds give a tetrahedral geometry to the ice lattice, with side of 4.5 Å (figure 12.3). Each O-H "arm" is directed toward the oxygen atom of another water molecule (along one of the two lone-pair directions of the latter), with O-O distance of 2.76 Å, about 2.9 times larger than

---

[5]Note that we are comparing the melting point of pure water with the melting points of only those substances shown in figure 12.1. There are many other substances with much stronger binding forces, hence higher melting points.

[6]Named after the Swedish physicist Anders Jonas Ångström (1814–1874), one of the founders of the field of spectroscopy.

[7]Named after Peter Joseph William Debye (1884–1966), a Dutch-American physicist and physical chemist, who got the Nobel prize in Chemistry for his contributions to the study of molecular structure, with his work on dipole moments and X-ray diffraction.

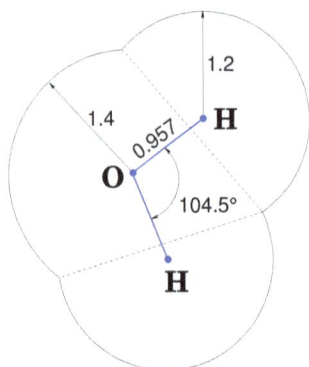

Figure 12.2: Single water molecule. Lengths are in angstroms (Å).

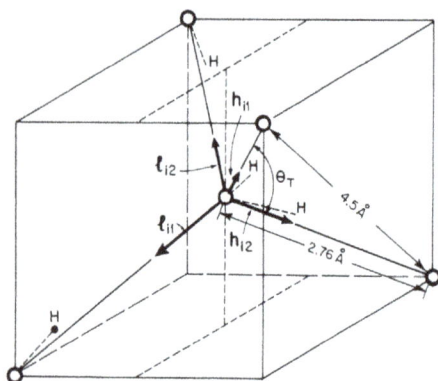

Figure 12.3: Tetrahedral geometry and hydrogen bonds.

the O-H distance.

The interaction energy of the HB (5-6 kcal/mol) is one order of magnitude larger than typical van der Waals interactions between two Neon atoms (also with 10 electrons), but one order of magnitude smaller than a typical covalent bond (e.g. O-H has 102 kcal/mol).

One should notice that, inside the ice structure, each water molecule keeps its own identity, although the tetrahedral angle of 109.46° is larger than the H-O-H angle of a single molecule in the gas phase (104.523°).[8]

The interaction between two water molecules depends on their distance and on the relative orientation of all O-H "arms" (5 angles), which in turn depend on the presence of other molecules in the local surrounding. The interaction energy between two molecules is defined as the difference in energy when moving them from an infinite separation to a distance r. Its analytic form is not known. Most studies are performed taking an effective pair potential with the form

$$U(\boldsymbol{r}_1, \boldsymbol{r}_2) = U_{\mathrm{LJ}} + U_{\mathrm{DD}} + U_{\mathrm{HB}} \tag{12.1}$$

where U_{LJ} is a Lennard-Jones potential modeling the repulsive force when the two molecules come to a very short distance[9], U_{DD} is the dipole-

---

[8]The tetrahedral geometry is formed via hybridization of 2s and 2p orbitals to form sp³ wavefunctions.

[9]The following values can be set in (10.8): $\varepsilon = 5.01 \times 10^{-15}\,\mathrm{erg} = 7.21 \times 10^{-3}\,\mathrm{kcal\,mol^{-1}}$, $\sigma = 2.82$ Å.

dipole interaction accounting for the long-range interaction[10], and U_{HB} is the hydrogen bond component acting at intermediate distances (no analytic expression is known).

12.4 Ice Structure

The tetrahedral structure of ice was first determined by Bragg[11] in 1922 with X-ray crystallography. Each oxygen atom is connected with hydrogen bonds to 4 other O atoms, all at distance of 2.76 Å. The resulting crystal (figure 12.4) is anisotropic: the properties of ice differ when measured along different axes. Figure 12.4 show the hexagonal structure, which is visible along one direction and gives the name of hexagonal ice (I_h) to ordinary ice.

Ordinary ice has a structure obeying two conditions, formulated by Bernal & Fowler (1933):

(1) Each O-O line accomodates exactly one H atom.
(2) Each O has 2 H atoms at about 1 Å and 2 H atoms at 1.76 Å.

When ice is formed from gas or liquid water, each H_2O mantains its own identity (D_2O is very similar), in the sense that H atoms "remember" which O atoms they were bound to. On the other hand, the geometry of the H-O-H molecule is different in ice, with an angle of 109.5° larger than the natural angle of 104.5°. Note that OH^- and H^+ do not conform to the ice conditions, but are being neglected here.

Pauling (1935) computed the approximate number of configurations for the H atoms that are compatible with the two ice conditions. The number of allowed configurations for a crystal with N oxygen atoms is $\Omega = 1.5N$. Assuming that all locations are equally probable (which is not justifiable a priori, but apparently it works well enough), the

---

[10]The dipole-dipole interaction depends as $-\mu^2/d^3$ on the electric dipole moment μ and the distance d, and also depends on the relative orientations of the two molecules. However, the average dipole-dipole interaction energy scales as d^{-6} with the distance.

[11]Sir William Lawrence Bragg (1890–1971), an Australian-born British physicist, is the famous discoverer of Bragg's law of X-ray diffraction, which is basic for the determination of crystal structure. Together with his father, William Henry Bragg, he was awarded the Nobel Prize in Physics in 1915 for the development of X-ray crystallography.

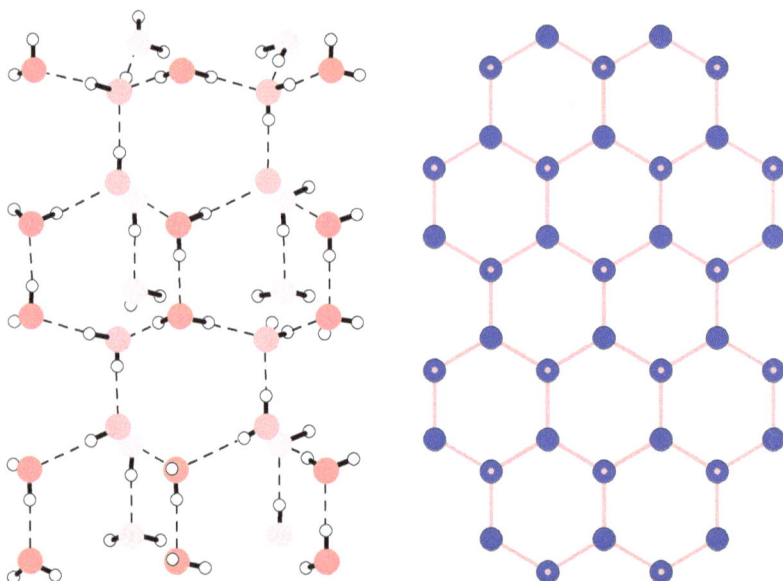

Figure 12.4: Ice I_h structure. Left: darker filled circles represend oxygen atoms in the foreground, and lighter ones are in the background. Right: hexagonal cells are visible when looking from a direction such that oxygen atoms belonging to lower layers are occulted by those in the foreground. Light lines represent hydrogen bonds. In each hexagon, 3 HBs point to the higher layer (they are shown as light dots) and 3 HBs point to the lower layer.

number of allowed configurations corresponds to a residual entropy[12] $S(0) = k_B \ln \Omega = 0.805 \, \mathrm{cal \, mol^{-1} \, K^{-1}}$, computed with Boltzmann's formula (2.13), which is compatible with the value of $0.81 \, \mathrm{cal \, mol^{-1} \, K^{-1}}$ known experimentally. This agreement shows that the distribution of the $2N$ H atoms is not unique.

There are several forms of solid water: at least ten ice phases have been recognized, many shown in figure 12.5. Ice I exists in the exagonal form I_h, with density $0.93 \, \mathrm{g \, cm^{-3}}$, and in a cubic form, with density $0.94 \, \mathrm{g \, cm^{-3}}$ (data from Haynes *et al.* 2015). Both are less dense than liquid water, which has density $1.00 \, \mathrm{g \, cm^{-3}}$. This explains why we see ice floating on the see. The tip of a iceberg contains only few per-

---

[12]The residual entropy is the difference between the measured entropy and the value computed from the partition function of an ideal gas accounting for translation, rotation, and vibration modes. When the residual entropy is not null, it means that the substance can exist in different states when cooled down to very low temperatures (approaching 0 K).

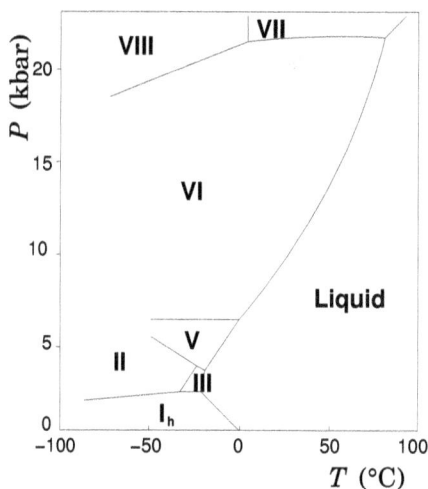

Figure 12.5: Water phase diagram. Ice IV is a metastable phase existing in the region of ice V. Ice IX exists in the region below −100°C and pressures in the range 2–4 kbar. Ice X only exists at pressure above 440 kbar.

cent of its mass. All other forms have larger densities than liquid water, the lightest being the tetragonal forms III ($\rho = 1.15\,\mathrm{g\,cm^{-3}}$) and IX ($\rho = 1.16\,\mathrm{g\,cm^{-3}}$). Triple points exist, in which different ice forms coexist: with I and III at −21.99°C and 209.9 MPa, with III and V at −16.99°C and 350.1 MPa, with V and VI at 0.16°C and 632.4 MPa, and with VI and VII at 82°C and 2216 MPa.[13]

The slope of the liquid-vapor and solid-vapor coexistence curves, given by the Clausius-Clapeyron equation (9.16), is always positive, as these transitions involve both an increase in volume and an increase of entropy (or equivalently of enthalpy). For most substances the solid-liquid curve has also a positive slope, but water is anomalous: the changes in volume and enthalpy have different signs, resulting into a negative slope.

The negative slope is due to the larger molar volume of ice compared to water, induced by the strong directional forces of the hydrogen bonds connecting different H_2O molecules. The more widely spaced structure of ice is energetically favored over configurations in which the molecules

---

[13] 100 MPa = 1 kbar.

are more closely packed. At high pressures, different polymorphs of ice are characterized by higher densities, hence more packed structures. In some case, the number of neighbors is larger than 4 (forms VII, VIII and X have 8 neighbors), but the binding energy is smaller than for ordinary (and less dense) ice.

The relation between local density and binding energy is the most important aspect of the mode of packing of water molecules in the liquid state.

12.5   Mixture-Model Approach to Liquid Water

We have already emphasized the importance of liquid water for life, and listed some key properties. Few properties unique of water (as far as we know) are the negative temperature dependence of the volume, the large negative entropy of solvation of inert solute, and the continuous decrease of the molar volume of water upon increasing the temperature between 0 and 4°C (for heavy water, D_2O, up to 11°C; figure 12.6). They are the effect of the unique correlation between low local density and strong binding energy, at the origin of the tetrahedral structure of ice, which also persists in liquid water.

If one considers liquid water as a mixture of two components with different properties, the apparently surprising behavior may be explained in terms of changes of the proportion between the two components.

One example is the isothermal compressibility of aqueous solutions. The addition of some solutes, such as ethyl ether $(CH_3CH_2)_2O$ or methyl

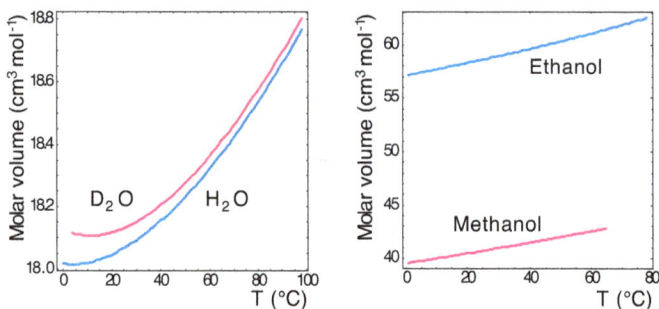

Figure 12.6: Molar volume as a function of the temperature for water, heavy water, ethanol (CH_3CH_2OH) and methanol (CH_3OH).

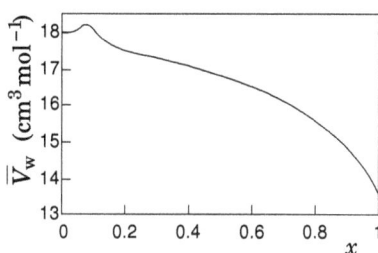

Figure 12.7: Partial molar volume in mixture of water and ethanol as a function of the mole fraction of ethanol.

acetate CH_3COOCH_3 to water causes a *decrease* in the compressibility of the system, in spite of the fact that the compressibility of the pure solute is about three times *larger* than the compressibility of pure water.

This puzzling observation may be explained by viewing water as consisting of two components: one with a relatively smaller compressibility (similar to ice), and a second one with a relatively larger compressibility. In this model, adding ether to pure water causes a shift in the equilibrium concentration of the two components in favor of the ice-like component, giving rise to a net decrease in the compressibility of the system. At this stage, this is only a qualitative explanation of the observed phenomena. A more precise treatment will be shown in section 12.6.5 below.

Similarly, the anomalous behavior of the molar volume of water shown in figure 12.6 may be explained by assuming that one component has larger molar volume than the other, and that the relative proportion between the former (ice-like) and the latter changes with the temperature. More details on section 12.6.3 below.

Another example is the ***partial molar volume*** of water in a mixture of water and ethanol

$$\overline{V}_w \equiv \left(\frac{\partial V}{\partial N_w} \right)_{T,P,N_e} \tag{12.2}$$

which is equivalent to the change of volume upon adding an infinitesimal quantity of water to the mixture, at given T,P and composition $x = N_e/(N_w + N_e)$. As shown in figure 12.7, over most of the domain of composition the partial molar volume of water is *smaller* than that of pure water ($18\,cm^3\,mol^{-1}$). However, in the region of dilute alcohol in water the partial molar volume of water is *larger* than for pure water.

To explain this we invoke once again the same two-structure model for water, with one component ice-like with larger molar volume, and assume as before that the addition of a small amount of alcohol to water causes a shift in the equilibrium concentrations of the two components. Note that "small amount" is important: this assumption is not valid for large amounts of alcohol.

Note that in the mixture model of water, all molecules are identical. The difference between the two components is exclusively the local "structure" around each water molecule. One component has smaller local density and smaller compressibility, like ordinary ice, and the other has higher local density and larger compressibility. However, it is important to emphasize that there are no domains nor microscopic cells made of a single component. All is about the local composition at molecular level.

In addition, the mixture is not ideal (no symmetric ideal, no very dilute solution), as the following argument shows. Following the two-structure model of water by Wada (1961), let i and p indicate the two components of a symmetrical ideal solution (i for "icy" and p for "packed") and consider the "reaction" $i \rightleftharpoons p$. At equilibrium their chemical potentials must be equal: $\mu_i = \mu_p$. For a symmetrical ideal solution, the chemical potential of component j has the form (10.44), which we rewrite as $\mu_j = \mu_j^p + k_B T \ln x_j$ (i.e. we work with particle number $N = n N_{Av}$ rather than with mole number n). The equilibrium constant (11.10) then is

$$K = \frac{x_p}{x_i} = \exp \frac{\mu_i^p - \mu_p^p}{k_B T} \qquad (12.3)$$

Assuming some numerical values for the chemical potentials of pure i and p, one can then get the dependence of the concentration x_i on the temperature.

The molar volume of water is a linear combination of the molar volumes of pure i and p:

$$V_w = x_i V_i^p + x_p V_p^p \qquad (12.4)$$

The temperature dependence of the volume is

$$\left(\frac{\partial V_w}{\partial T} \right)_P = x_i \frac{\partial V_i^p}{\partial T} + x_p \frac{\partial V_p^p}{\partial T} + (V_i^p - V_p^p) \left(\frac{\partial x_i}{\partial T} \right)_P \qquad (12.5)$$

From (12.3) one gets

$$\left(\frac{\partial x_i}{\partial T}\right)_P = \frac{x_i \, x_p}{k_B T^2} \Delta H \qquad (12.6)$$

which means that the concentration of component i increases (or decreases) with the temperature when $\Delta H \equiv H_i^{\mathrm{p}} - H_p^{\mathrm{p}}$ is positive (or negative). Combining the two previous equations we finally get

$$\left(\frac{\partial V_w}{\partial T}\right)_P = \left[x_i \frac{\partial V_i^{\mathrm{p}}}{\partial T} + x_p \frac{\partial V_p^{\mathrm{p}}}{\partial T}\right] + \frac{x_i \, x_p}{k_B T^2} \Delta H \, \Delta V \qquad (12.7)$$

where $\Delta V \equiv V_i^{\mathrm{p}} - V_p^{\mathrm{p}}$.

One assumes that the pure components i and p behave normally, i.e. that they both expand upon increasing the temperature. Therefore, the first term in the square brackets in (12.7) is expected to be positive. The second term depends on the product of the mole fractions $x_i x_p$.[14] This term is positive for normal liquids, characterized by changes in enthalpy and volume with the same sign. However for water the situation is different. By our choice of the components $\Delta V > 0$, hence ΔH must be negative. This means that the interaction energy in the icy component is more negative than in the packed structure. This is the essential difference between liquid water and other "normal" liquids.

As we increase the temperature, both components expand. In addition one component, the one with more negative interaction energy (or enthalpy), melts (i.e. $i \to p$) upon heating. Since this component has also larger molar volume, the second term in (12.7) is negative. In liquid water, at low temperatures (close to $0°C$) this term is the dominant one, which requires both large values of ΔH and ΔV and that $x_i x_p$ be not negligible. As we continue to increase the temperature beyond $4°C$, there are two effects that cause a decrease of the second term on the right-hand side of (12.7). First, the dependence on T^{-2} rapidly damps this term. Second, as the temperature increases, the component i "melts" reducing x_i according to (12.6), and at some point x_i becomes so small that the product $x_i x_p$ becomes small too.

Let's consider now the molar enthalpy of water, and write it as a linear combination of the molar volumes of pure i and p:

$$H_w = x_i H_i^{\mathrm{p}} + x_p H_p^{\mathrm{p}} \qquad (12.8)$$

---

[14]This specific form results from assuming an ideal solution. However, the qualitative argument below is independent of this assumption.

Its temperature dependence is given by the molar heat capacity at constant pressure, which can be written as

$$C_{P,\mathrm{w}} \equiv \left(\frac{\partial H_\mathrm{w}}{\partial T} \right)_P = \left[x_i \frac{\partial H_i^\mathrm{p}}{\partial T} + x_p \frac{\partial H_p^\mathrm{p}}{\partial T} \right] + \frac{x_i\, x_p}{k_\mathrm{B} T^2} \Delta H^2 \qquad (12.9)$$

The last term is always positive (independent of the sign of ΔH), and is responsible for the outstandingly large value of the heat capacity of water.

The explanation provided by this model of the temperature dependence of the molar volume and of the heat capacity is valid, independently of the assumption of ideality of the solution. At the same time, (12.7) and (12.9) clearly show that the product $x_i\, x_p$ cannot be very small, i.e. the mixture is not very diluted, over the entire range of temperatures in which liquid water exists.

12.6 Exact Mixture-Model Approach to the Theory of Liquids

In this section an exact mixture-model approach to the theory of liquids is presented, which can be also applied to a one-component system like liquid water. What is needed is some property which distinguishes the molecules and allows us to group them in different categories. The only requirement for the rule of classification of molecules into various groups is that it must be unique and exhaustive. Here we illustrate the approach with two categories, using the coordination number as molecular property.

12.6.1 *Coordination number*

Let's consider a fixed distance R_C and focus on a sphere of radius R_C centered on any given particle in the system (taken far enough from boundaries). The ***coordination number*** is defined as the average number of particles contained within such a sphere, and is computed with the help of the radial distribution function $g(r)$ defined in (10.12) (also known as pair correlation function)[15]:

---

[15]Radial distribution functions of water and ice at various temperature and pressure values are shown by Soper (2000).

$$CN(R_C) = \rho \int_0^{R_C} g(r) 4\pi r^2 \, dr \tag{12.10}$$

Thus the coordination number may be viewed as a property conveying the local density around the particles. The larger the CN of a specific particle, the larger its local density.

As an example, if we choose R_C as the distance at which $g(r)$ has the first minimum following its first maximum, we obtain the **first coordination number** n_{CN}. For liquid argon at 84.25 K and 0.7 bar $n_{CN} \approx 10$, tyipical of close-packed liquids (see table 12.2 for a comparison against different substances). On the other hand, for water at 4°C and 1 atm $n_{CN} = 4.4$, an unusual value for a liquid and close to the coordination number of the solid hexagonal ice (4). This suggests that the "local structure" of water is very similar to the local structure of ice. No other simple liquid has this property. Furthermore, the first coordination number of simple liquids decrease with increasing temperature, whereas for water n_{CN} initially increases but above 50–60°C it decreases with increasing temperature (Ben-Naim, 2009).

In order to compare the radial distribution function of different substances, one considers the dimensionless distance $r^* = r/\sigma$ obtained by measuring the radial distance in units of the Lennard-Jones parameter σ (see figure 10.3). For argon $g(r^*)$ has a the second peak for $r^* \approx 2$, which is what one would expect from simple spherical particles. For water the second peak is at $r^* \approx 1.6$, which corresponds to about 4.5 Å, the

Table 12.2: First Coordination Number in Solids and Liquids

Substance	n_{CN}^{sol}	n_{CN}^{liq}
Li	14	9.8
Na	14	9.3
K	14	8
Ar	12	10.2–10.9
Xe	12	8.5
Ga	7	7.8
Ge	4	8

second-nearest neighbor distance between oxygen atoms in ordinary ice. This means that the local structure around a water molecule is close to the local structure of ice, a unique property.

Coming back to the mixture-model approach, we take the coordination number for a distance $\sigma \le R_C \le 1.5\sigma$ close to the molecule diameter. The property to be considered here is the coordination number of the specific particle i, given a configuration $\boldsymbol{R}^N$ of the system specifying the locations of all N molecules. We are interested into the *distribution* of the coordination number, i.e. into the number $n_{CN}(K|\boldsymbol{R}^N)$ of particles for which $CN(R_C)$ is equal to K, where $K = 0, 1, 2, \ldots$, given $\boldsymbol{R}^N$. The mole fraction of the molecules with coordination number equal to K is then

$$x_{CN}(K) = \frac{n_{CN}(K|\boldsymbol{R}^N)}{N} \qquad (12.11)$$

The "composition" of the system is then viewed as the vector

$$\boldsymbol{x}_{CN} \equiv (x_{CN}(0), x_{CN}(1), x_{CN}(2), \ldots) \qquad (12.12)$$

Of course, $\sum_K x_{CN}(K) = 1$ and clearly only the first few elements of the vector (12.12) are significantly different from zero, as for increasing K it becomes more and more difficult to pack the molecules tightly enough to fit K of them into a sphere of radius R_C. This means that the average coordination number of the particles in the system

$$\langle K \rangle = \sum_{k=0}^{\infty} x_{CN}(K) \qquad (12.13)$$

can be computed by summing over few terms (say the first dozen terms, depending on the choice of R_C), although the sum in (12.13) formally runs up to infinity.

We can define the local density around any given particle for a given configuration $\boldsymbol{R}^N$ by dividing the average coordination number by the volume of the sphere in which it is computed:

$$D(\boldsymbol{R}^N) \equiv \frac{\langle K \rangle}{\frac{4}{3}\pi R_C^3} \qquad (12.14)$$

12.6.2 *Two-component mixture*

Now let's focus on the mixture of the following species: the particles with coordination number smaller or equal than K^*, and those with coordination number greater than K^*, where K^* is chosen in some intermediate range, such that the mole fractions of the species with low and high local density

$$x_L = \sum_{k=0}^{K^*} x_{CN}(K) \quad \text{and} \quad x_H = \sum_{k=K^*+1}^{\infty} x_{CN}(K)$$

are both sizable. The molecules are all identical. Hence this mixture of "quasi-components" is different from a mixture of real components. First, the quasi-components do not differ in their chemical composition: they are characterized by the local environment in the immediate surroundings of each molecule. Second, and very important, a system of quasi-components cannot be "prepared" in any desired composition, i.e. the components of the vector (12.12) cannot be chosen at will. One consequence of this restriction is that quasi-components do not exist in the pure state.

Note also that the same molecule can "belong" to different species at different times. This is quite different from a regular mixture having a fixed composition but is similar to a mixture of inter-converting molecules at equilibrium. Once again, it's important to remember that the choice of the coordination number is not mandatory. Other local properties may be used equally well (Ben-Naim, 2009).

Our mixture of quasi-components can be now described by the composition vector (x_L, x_H). For water we might choose for example $K^* = 4$, obtaining $x_L = x_{CN}(4)$ and $x_H = 1 - x_L$. Each component behaves as a normal liquid, and the outstanding properties of water depend on the redistribution of molecules into the two species L and H, with equilibrium numbers of molecules N_L and $N_H = N - N_L$, where N is the total number of water molecules.

12.6.3   *Dependence of the molar volume on the temperature*

Denoting as $\overline{V}_L$ and $\overline{V}_H$ the partial molar (or molecular) volumes of the two quasi-components, the total volume is

$$V = N_L\overline{V}_L + N_H\overline{V}_H$$

and its temperature dependence along the equilibrium line with the respect to the reaction $L \rightleftharpoons H$ is given by the partial derivative with respect to T keeping P and N constant (the latter implies $dN_H = -dN_L$):

$$\left(\frac{\partial V}{\partial T}\right)_{P,N,\text{eq}} = \left(\frac{\partial V}{\partial T}\right)_{P,N_L,N_H} + (\overline{V}_L - \overline{V}_H)\left(\frac{\partial N_L}{\partial T}\right)_{P,N,\text{eq}} \tag{12.15}$$

where

$$\left(\frac{\partial V}{\partial T}\right)_{P,N_L,N_H} = N_L\left(\frac{\partial \overline{V}_L}{\partial T}\right)_{P,N_L,N_H} + N_H\left(\frac{\partial \overline{V}_H}{\partial T}\right)_{P,N_L,N_H} \tag{12.16}$$

may be referred as the "frozen-in" part of the total temperature dependence of the volume, whereas the last term in (12.15) may be referred as the "relaxation" part. It is this latter term that contains the contribution associated with the "structural changes" in the system, i.e. with the redistribution of the molecules between the two species.

It can be shown (Ben-Naim, 2009) that

$$\left(\frac{\partial N_L}{\partial T}\right)_{P,N,\text{eq}} = (\mu_{LL} - 2\mu_{LH} + \mu_{HH})^{-1}\frac{\overline{H}_L - \overline{H}_H}{T} \tag{12.17}$$

where $\mu_{ij} = \partial^2 G/\partial N_i \partial N_j$, is always positive. The identity (12.17) allows us to write the temperature dependence of the molar volume within the exact mixture model as

$$\left(\frac{\partial V}{\partial T}\right)_{P,N,\text{eq}} = \left(\frac{\partial V}{\partial T}\right)_{P,N_L,N_H} + \frac{(\overline{V}_L - \overline{V}_H)(\overline{H}_L - \overline{H}_H)}{T(\mu_{LL} - 2\mu_{LH} + \mu_{HH})} \tag{12.18}$$

From the Kirkwood-Buff theory of solutions, one gets

$$\mu_{LL} - 2\mu_{LH} + \mu_{HH} = \frac{k_B T}{x_L x_H V \eta} \tag{12.19}$$

where

$$\eta \equiv \rho_L + \rho_H + \rho_L \rho_H (G_{LL} - 2G_{LH} + G_{HH}) \tag{12.20}$$

features the Kirkwood-Buff integrals G_{ij} defined in section 10.2.3. For a symmetrical ideal solution $G_{LL} - 2G_{LH} + G_{HH} = 0$. As $(\rho_L + \rho_H)V = N$, (12.19) becomes

$$\mu_{LL} - 2\mu_{LH} + \mu_{HH} = \frac{k_B T}{x_L x_H N} \tag{12.21}$$

When (12.21) is inserted into (12.18), one reproduces the result (12.7) obtained for a symmetrical ideal solution.

In order to explain the peculiar volume dependence of water with (12.18), one must choose a classification procedure such that the frozen-in term (12.16) behaves normally, with both derivatives positive, and that the last term in (12.18) (i.e. the relaxation term) is large and negative, overcoming the positive contribution from the frozen-in term in the temperature range 0–4°C. The relaxation term is negative when the changes in molar volume and enthalpy have opposite signs.

In order for the relaxation term to be large (in absolute value), two more conditions need to be fulfilled. First, the L and H components must be very different, since otherwise the differences $(\overline{V}_L - \overline{V}_H)$ and $(\overline{H}_L - \overline{H}_H)$ would be small. Second, neither of mole fractions x_L and x_H can be very small, otherwise their product would make the "relaxation" term too small. As mentioned above, this means that the mixture is not symmetrical ideal nor very diluted.

In water, the strong directional forces related to the hydrogen bonds are responsible for mantaining a low local density (higher molar volume), because this is the energetically favored state (high binding energy, or high enthalpy). This makes the product $(\overline{V}_L - \overline{V}_H)(\overline{H}_L - \overline{H}_H)$ negative. To explain the peculiar dependence of the molar volume on the temperature, it is also necessary that the product $x_L x_H$ is not too small. This means that liquid water mantains a large degree of "structure".

Exercise 12.1. Examine the conditions for the minimum of the molar volume (figure 12.6) within this model.

12.6.4 *Dependence of the heat capacity on the temperature*

Figure 12.8 shows the heat capacity at constant pressure as a function of the water temperature at different pressures. At ambient pressure C_P has a minimum at about 35°C. For increasing pressure, this minimum moves to lower temperatures and then disappears, giving water a normal behavior.

The total enthalpy of the system can be written as

$$H = N_L \overline{H}_L + N_H \overline{H}_H \tag{12.22}$$

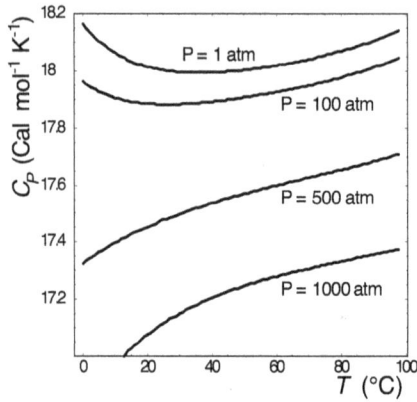

Figure 12.8: Heat capacity at constant pressure as a function of the water temperature at different pressures.

The heat capacity is given by

$$C_P = \left(\frac{\partial H}{\partial T}\right)_{P,N,\text{eq}} = \left(\frac{\partial H}{\partial T}\right)_{P,N_L,N_H} + (\bar{H}_L - \bar{H}_H)\left(\frac{\partial N_L}{\partial T}\right)_{P,N,\text{eq}} \qquad (12.23)$$

where

$$\left(\frac{\partial H}{\partial T}\right)_{P,N_L,N_H} = N_L\left(\frac{\partial \bar{H}_L}{\partial T}\right)_{P,N_L,N_H} + N_H\left(\frac{\partial \bar{H}_H}{\partial T}\right)_{P,N_L,N_H} \qquad (12.24)$$

may be referred as the frozen-in part

The last term in (12.23) is the relaxation part, and can be written as (Ben-Naim, 2009)

$$(\bar{H}_L - \bar{H}_H)\left(\frac{\partial N_L}{\partial T}\right)_{P,N,\text{eq}} = \frac{(\bar{H}_L - \bar{H}_H)^2}{T(\mu_{LL} - 2\mu_{LH} + \mu_{HH})}$$

$$= \frac{x_L x_H V \eta}{k_B T^2}(\bar{H}_L - \bar{H}_H)^2 \qquad (12.25)$$

where η is defined in (12.20). The relaxation part (12.25) is always positive.

One can assume that the frozen-in takes a "normal" value and that the excess heat capacity of liquid water originates from the relaxation term. The difference in binding energy of the two components must be sizable (which rules out symmetrical ideal solutions) and neither of the mole fractions can be very small (i.e. the solution is not very diluted). These are the same conclusions reached above, when speaking about the molar volume. The only difference here is that there is no sign dependence: the relaxation term is always positive.

Since the heat capacity is always positive, addition of heat to any system at equilibrium will shift the distribution of species towards those that have higher energy. In other words, the "structural changes" in the system will always happen in the direction that leads part of the heat to be absorbed by the system. This means that the average coordination number will increase with the temperature, shifting the system toward less negative binding energy.

Exercise 12.2. Examine the conditions for the minimum of the heat capacity C_P within this model.

12.6.5 *Isothermal compressibility*

Normally, as the temperature of a liquid increases the average intermolecular distance between the particles of the liquid increases. This makes it easier to compress a liquid at a higher temperature. Thus, we expect the compressibility to increase as we increase the temperature. Water is anomalous, because it has a region below approximately 45°C, where κ_T actually decreases as the temperature increases (figure 12.9).

The isothermal compressibility has also two contributions, from a frozen-in term and a relaxation term:

$$
\begin{aligned}
\kappa_T &= -\frac{1}{\overline{V}} \left(\frac{\partial V}{\partial P} \right)_{T,N,\text{eq}} \\
&= -\frac{1}{\overline{V}} \left[N_L \left(\frac{\partial \overline{V}_L}{\partial P} \right)_{T,N_L,N_H} + N_H \left(\frac{\partial \overline{V}_H}{\partial P} \right)_{T,N_L,N_H} \right] \\
&\quad - \frac{1}{\overline{V}} (\overline{V}_L - \overline{V}_H) \left(\frac{\partial N_L}{\partial P} \right)_{T,N,\text{eq}}
\end{aligned}
\tag{12.26}
$$

The relaxation part can be written as (Ben-Naim, 2009)

$$
\begin{aligned}
-\frac{1}{\overline{V}} (\overline{V}_L - \overline{V}_H) \left(\frac{\partial N_L}{\partial P} \right)_{T,N,\text{eq}} &= -\frac{(\overline{V}_L - \overline{V}_H)^2}{V(\mu_{LL} - 2\mu_{LH} + \mu_{HH})} \\
&= -\frac{x_L x_H \eta}{k_B T^2} (\overline{V}_L - \overline{V}_H)^2
\end{aligned}
\tag{12.27}
$$

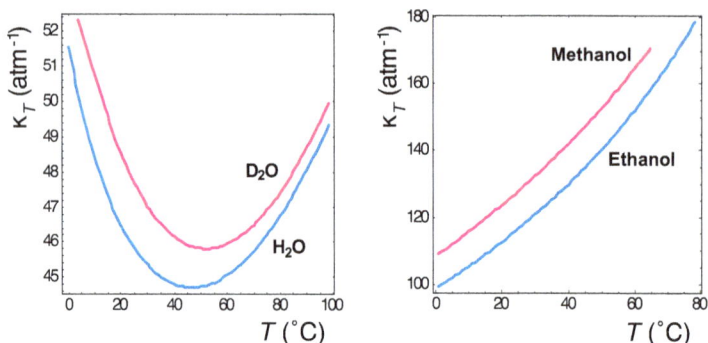

Figure 12.9: Isothermal compressibility of water and heave water (left), compared to methanol and ethanol (right).

and is always negative, independent of the particular definition of the L and H components. The relaxation term only depends on the difference in molar volume between the two components. It gives a significant contribution to the isothermal compressibility whenever the difference in molar volume is large (ruling out symmetrical ideal solutions) and both components have sizable mole fraction (excluding very diluted solutions).

It is worthwhile to consider once more the condition of similarity $G_{LL} - 2G_{LH} + G_{HH} = 0$ which holds for symmetrical ideal solutions, giving $\eta V = N$. We have seen that, in order to reproduce the peculiar properties of liquid water with the two-component model, we must classify molecules in such a way that the two quasi-components have sizable differences in molar volume and enthalpy. But in this model the two components have *identical* chemical composition. Therefore, the differences must arise from the local environment around each molecule. In other words, the water molecules see different degrees of structure.

Exercise 12.3. Examine the condition for the minimum of the compressibility within this model.

Exercise 12.4. Write the expression for the pressure dependence of the volume using the assumption of ideality. What can we conclude from the relaxation term?

Appendix A

Solutions to Exercises

Here we provide solutions for a subset of the exercises given in the text.

Solution of exercise 4.1 on page 80:
The function is $H = -p\log_2 p - (1-p)\log_2(1-p)$, shown in figure 4.1 on page 79.

Solution of exercise 4.2 on page 80:
Let X, Y be two random variables with possible outcomes $\{1,2,3,4,5,6\}$ and consider their sum $S = X + Y$, which ranges from 2 to 12. The corresponding probabilities are shown in the first plot of figure 4.2 and the SMI associated with this distribution is $H = \sum_{n=2}^{12} p_n \log_2 p_n = 3.2744$.

Solution of exercise 4.3 on page 80:
The solutions for $1,\dots,6$ dice are shown in figure 4.2 on page 81. The SMI values are $H_1 = 2.58$, $H_2 = 3.27$, $H_3 = 3.60$, $H_4 = 3.81$, $H_5 = 3.98$, $H_6 = 4.11$.

Solution of exercise 4.4 on page 90:
The total SMI for the distribution of letters in the English language is $-\sum_{i=1}^{27} p_i \log_2 p_i = 4.138$ bits. This means that in the "best strategy" we need to ask on the average 4.138 questions. However, since we cannot divide the events into two groups with exactly equal probabilities at each step, the best we can do is to divide into two groups of about equal probabilities. For example, in the first step choose the first 12 letters with probability 0.504 and the remaining letters with probability 0.496, and

so on in the next steps. This way, you can get the answer in five questions. Clearly, this is much better than asking: "is the letter A?", "is it
B?", and so on.

Solution of exercise 5.1 on page 110:

The first specific configuration of the two marbles in the four boxes is
$(B,R,0,0)$ where B,R denote one black or red marble, respectively. Because there are $4^2 = 16$ possible ways of placing the two marbles, this specific configuration has probability $\frac{1}{16}$. The corresponding generic configuration has state-distribution $(1,1,0,0)$, to which we associate the probability distribution $(\frac{1}{2},\frac{1}{2},0,0)$. As there are two ways of obtaining this
generic configuration, the probability of this generic state is $2/4^2 = \frac{1}{8}$.

Solution of exercise 7.1 on page 168:

The initial SMI is $H_1 = N\log_2 M$ and the final SMI is

$$H_2 = N\log_2(2M) = N + N\log_2 M$$

Thus $\Delta H = H_2 - H_1 = N$.

Solution of exercise 7.2 on page 168:

The initial SMI is $H_1 = N\log_2 M$ and the final SMI is $H_2 = N\log_2(5M)$,
hence $\Delta H = N\log_2 5$. The entropy change is obtained by changing to
thermodynamic units: $\Delta S = k_B\Delta H/\log_2 e = Nk_B\ln 5$.

Solution of exercise 7.3 on page 168:

The initial SMI is $H_1 = N_A\log_2 M + N_B\log_2 M$ and the final SMI is
$H_2 = N_B\log_2(2M) + N_B\log_2(2M)$, hence $\Delta H = N_A + N_B$ and $\Delta S = (N_A + N_B)k_B\ln 2$. Each gas expands from a volume V to a volume $2V$, hence
the uncertainty on the location increases. More comments in section 7.4.

Solution of exercise 7.4 on page 171:

Hint: the final temperature is a weighted average of the initial ones.

Solution of exercise 8.1 on page 201:

The isothermal steps in figure 8.8 on page 201 are similar to the Carnot
cycle. Since no PV-work is done in the isochoric process, the efficiency

of this engine is similar to the efficiency of a Carnot cycle: $\eta = |W_{\mathrm{I}} + W_{\mathrm{III}}|/Q_{\mathrm{I}} = 1 - T_{\mathrm{III}}/T_{\mathrm{I}}$.

Solution of exercise 8.2 on page 205:

The system is shown in figure 8.10 on page 205. Initially, the pressures in the two subsystems are

$$P_{\mathrm{i}}^{(1)} = \frac{Nk_{\mathrm{B}}T}{V} \qquad P_{\mathrm{i}}^{(2)} = \frac{4Nk_{\mathrm{B}}T}{9V}$$

Clearly, since $P_{\mathrm{i}}^{(1)} > P_{\mathrm{i}}^{(2)}$ the partition will move in such a way to expand the first subsystem and compress the second. There is no change in the particle numbers. At equilibrium the pressure is balanced

$$P_{\mathrm{f}}^{(1)} = \frac{Nk_{\mathrm{B}}T}{V_{\mathrm{f}}^{(1)}} = \frac{4Nk_{\mathrm{B}}T}{V_{\mathrm{f}}^{(2)}} = P_{\mathrm{f}}^{(2)}$$

hence $V_{\mathrm{f}}^{(2)}/V_{\mathrm{f}}^{(1)} = 4$. Since the total volume of the system is $10V$, this relation implies $V_{\mathrm{f}}^{(1)} = 2V$ and $V_{\mathrm{f}}^{(2)} = 8V$. The final densities are $\rho_{\mathrm{f}}^{(1)} = \frac{1}{2}N/V$ and $\rho_{\mathrm{f}}^{(2)} = \frac{1}{2}N/V$, a quite natural result.

Solution of exercise 8.3 on page 205:

Initially, the chemical potential from eq. (7.19) in each subsystems is

$$\mu_{\mathrm{i}}^{(1)} = k_{\mathrm{B}}T\ln(\tfrac{N}{V}\Lambda^3) \qquad \mu_{\mathrm{i}}^{(2)} = k_{\mathrm{B}}T\ln(\tfrac{4N}{9V}\Lambda^3)$$

As $\mu_{\mathrm{i}}^{(1)} > \mu_{\mathrm{i}}^{(2)}$, there will be a net flow of particles from the first to the second subsystem, until the two chemical potentials are equal. Setting $\mu_{\mathrm{f}}^{(1)} = \mu_{\mathrm{f}}^{(2)}$ or

$$k_{\mathrm{B}}T\ln(\tfrac{N_{\mathrm{f}}^{(1)}}{V}\Lambda^3) = k_{\mathrm{B}}T\ln(\tfrac{N_{\mathrm{f}}^{(2)}}{9V}\Lambda^3)$$

one obtains $N_{\mathrm{f}}^{(1)}/N_{\mathrm{f}}^{(2)} = 1/9$. Since there are in total $5N$ moles, one obtains

$$N_{\mathrm{f}}^{(1)} = N/2 \qquad N_{\mathrm{f}}^{(2)} = 4.5N$$

Again, the final densities are equal: $\rho_{\mathrm{f}}^{(1)} = \rho_{\mathrm{f}}^{(1)} = \frac{1}{2}N/V$. One obtains a uniform density in the entire system.

Solution of exercise 8.4 on page 221:
Solution 1: Using the function $S(E,V,N)$ in the form (7.1) for an ideal gas expanding from V to $2V$, we get $\Delta S = S(E,2V,N) - S(E,V,N) = Nk_B \ln(2V/V) = Nk_B \ln 2$. Alternatively, we can consider the two halves as separate subsystems, and integrate (8.81) for each subsystem:

$$\Delta S_{ex}^{(1)} = \int_V^{2V} \frac{P}{T}\,dV = \int_V^{2V} \frac{Nk_B}{V}\,dV = Nk_B \ln 2 \qquad \Delta S_{ex}^{(2)} = 0$$

Solution 2: check section 7.3.1.

Solution of exercise 8.5 on page 221:
Hint: check section 7.4.

Solution of exercise 8.6 on page 221:
Hint: check section 7.4.

Solution of exercise 8.7 on page 221:
Hint: check section 7.3.2.

Solution of exercise 8.8 on page 221:
Hint: postulate the extension of the validity of the differential (8.81).

Solution of exercise 9.1 on page 255:
Integrate equation (9.16), assuming that ΔH and ΔV are independent of the temperature, to obtain

$$P_2 - P_1 \approx \frac{\Delta H}{\Delta V} \ln \frac{T_2}{T_1}$$

Since $P_2 - P_1 = 399$ atm, and $T_1 = 273$, we can calculate T_2.

Solution of exercise 9.2 on page 262:
The phase diagram of phosphorous is shown in figure 9.9. For the liquid-vapour coexisting line of white phosphorous we write eq. (9.20) as

$$\ln P = \frac{-\Delta H_{liq}}{RT} + c_1 \qquad (A.1)$$

and determine the value of the constant c_1 by imposing that the point A is a solution: $c_1 = \ln P_A + \Delta H_{liq}/(RT_A)$.

For the red solid-vapour coexisting line, we write

$$\ln P = \frac{-\Delta H_r}{RT} + c_2 \qquad (A.2)$$

and fix the constant c_2 by imposing that point D is a solution: $c_2 = \ln P_D + \Delta H_r/(RT_D)$.

The triple point B is the intersection of (A.1) and (A.2).

Solution of exercise 9.3 on page 269:

The conservation of the total material implies $N^{(\alpha)} + N^{(\beta)} = N_{\text{tot}}$. The conservation of the phenol reads then $N^{(\alpha)}x_p^{(\alpha)} + N^{(\beta)}x_p^{(\beta)} = N_{\text{tot}}x_p^{(\text{tot})} = N^{(\alpha)}x_p^{(\text{tot})} + N^{(\beta)}x_p^{(\text{tot})}$ from which we get eq. (9.29).

Solution of exercise 9.4 on page 275:

At point a, we have a liquid solution of A and B with composition x_0. As we cool down the system at constant pressure, we move in a region with two degrees of freedom until we reach the point b, at which pure solid A precipitates. As a result of this removal of pure A from the solution, the composition of the liquid solution becomes richer in B. However, note that the overall composition of the system (solid plus liquid) remains constant at x_0. If we cool down the system further, say at point c, we have two phases, solid A and liquid mixture L with composition x' (but the overall system has still composition x_0)[16]. The liquid fraction decreases until we reach point d, at which we have two phases at equilibrium: pure A and a solution with composition solution with composition x_e: we reached the eutectic point. Further removal of heat from the system won't change the temperature until the conversion from liquid solution to pure solid A and pure solid B is complete. When only two solid phases remain, we gain one more degree of freedom and the temperature can decrease along the line d-e.

Solution of exercise 10.1 on page 296:

At equilibrium we have (10.34), that is

$$\mu_A^{\text{liq}} = \mu_A^{\text{gas}} = \mu_A^{0,\text{gas}} + RT\ln P_A$$

---

[16]More details can be found e.g. on https://www.southampton.ac.uk/~pasr1/tielines.htm.

Using Henry's law (10.33), valid when $x_A \approx 0$, one gets

$$\mu_A^{\text{liq}} = \mu_A^{0,\text{gas}} + RT \ln k_{\text{H,A}} x_A = \mu_A^{0,\text{liq}} + RT \ln x_A$$

Here we have defined a new standard chemical potential in the liquid phase

$$\mu_A^{0,\text{liq}} \equiv \mu_A^{0,\text{gas}} + RT \ln k_{\text{H,A}}$$

Using the Gibbs-Duhem equation (8.102) at constant T,P one gets

$$x_A \, d\mu_A + x_B \, d\mu_B = 0$$

Dividing by dx_A and making use of the derivative of (10.36) with respect to x_A one gets

$$x_A \frac{d\mu_A}{dx_A} + x_B \frac{d\mu_B}{dx_A} = RT + x_B \frac{d\mu_B}{dx_A} = 0$$

As $x_B = 1 - x_A$ and $dx_A = -dx_B$, the previous result give the differential equation

$$d\mu_B = \frac{RT}{x_B} \, dx_B$$

whose solution is

$$\mu_B = RT \ln x_B + \text{constant}$$

The integration constant is found by taking the limit $x_B = 1$, as the result is valid when the concentration of the solute is very low. This way one obtains (10.37).

Solution of exercise 10.4 on page 308:
See Appendix P of (Ben-Naim, 2006b).

Appendix B

Mathematics

B.1 Proof of $\log x \leq x - 1$

First note that the equality $\log x = x - 1$ holds for $x = 1$. For $x > 0$ and $x \neq 1$ the inequality $\log x < x - 1$ can be proven with the help of the auxiliary function $g(x) = \log x - (x - 1)$, which is always negative. The derivative of $g(x)$ is $g'(x) = 1/x - 1 = (1 - x)/x$.

In the open interval $0 < x < 1$ the derivative $g'(x)$ is always positive. This means that $g(x)$ is a monotonic increasing function of x in this region. Since we have $g(1) = 0$, it follows that $g(x) < 0$ for $0 < x < 1$ or, equivalently, $\log x < x - 1$ in this interval.

For $x > 1$ the derivative $g'(x)$ is negative, which means that $g(x)$ is a monotonic decreasing function of x in this region. Since $g(1) = 0$, it follows that $g(x) < 0$ for any $x > 1$ or, equivalently, $\log x < x - 1$ in this region.

The geometrical meaning of the inequality $\log x \leq x - 1$ is the following. If we draw the two functions $y_1(x) = \log x$ and the straight line $y_2(x) = x - 1$, we see that $y_2(x)$ is tangential to $y_1(x)$ at the point $x = 1$. At any other value $x > 0$ the straight line $y_2(x)$ is above the line $y_1(x)$.

B.2 Euler's Theorem

A *homogeneous function* of degree n is a function satisfying the scaling property

$$f(\lambda x_1, \lambda x_2, \ldots, \lambda x_k) = \lambda^n f(x_1, x_2, \ldots, x_k) \tag{B.1}$$

For example, all extensive thermodynamic quantities are homogeneous functions of degree 1.

The derivative of (B.1) with respect to the scaling parameter λ is

$$n\lambda^{n-1} f(x_1,x_2,\ldots,x_k) = \frac{d}{d\lambda}[\lambda^n f(x_1,x_2,\ldots,x_k)]$$

$$= \frac{d}{d\lambda} f(\lambda x_1, \lambda x_2, \ldots, \lambda x_k) \qquad \text{(B.2)}$$

$$= \sum_i x_i \frac{\partial f}{\partial(\lambda x_i)}$$

For $\lambda = 1$ one obtains **Euler's theorem**:

$$\sum_i x_i \frac{\partial f}{\partial x_i} = nf(x_1,x_2,\ldots,x_k) \qquad \text{(B.3)}$$

For example, the volume $V(T,P,N_1,\ldots,N_c)$ is a homogeneous function of $N_1,\ldots,N_c$, hence Euler's theorem gives the important identity

$$V(T,P,N_1,\ldots,N_c) = \sum_{i=1}^{c} N_i \overline{V}_i \qquad \text{(B.4)}$$

where $\overline{V}_i$ is the partial molar volume.

Similarly, for the Gibbs function $G(T,P,N_1,\ldots,N_c)$ one obtains

$$G(T,P,N_1,\ldots,N_c) = \sum_{i=1}^{c} N_i \overline{G}_i = \sum_{i=1}^{c} N_i \mu_i \qquad \text{(B.5)}$$

where $\mu_i = \partial G/\partial N_i$ is the partial molar Gibbs energy. Taking the total differential

$$dG = \sum_{i=1}^{c} N_i \, d\mu_i + \sum_{i=1}^{c} \mu_i \, dN_i$$

and equating it to (8.94)

$$dG = -S \, dT + V \, dP + \sum_{i=1}^{c} \mu_i \, dN_i$$

we get the Gibbs-Duhem equation (8.102)

$$\sum_{i=1}^{c} N_i \, d\mu_i = -S \, dT + V \, dP$$

Bibliography

ATLAS Collaboration. 2012. Observation of a new particle in the search for the Standard Model Higgs boson with the ATLAS detector at the LHC. *Phys. Lett. B*, **716**, 1. doi: 10.1016/j.physletb.2012.08.020.

Ben-Naim, Arieh. 1987. Is mixing a thermodynamic process? *American Journal of Physics*, **55**, 725–733. doi: 10.1119/1.15064.

Ben-Naim, Arieh. 2006a. The entropy of mixing and assimilation: An information-theoretical perspective. *American Journal of Physics*, **74**, 1126. doi: 10.1119/1.2338545.

Ben-Naim, Arieh. 2006b. *Molecular theory of solutions*. Oxford University Press.

Ben-Naim, Arieh. 2008. *A Farewell to Entropy: Statistical Thermodynamics Based on Information*. World Scientific.

Ben-Naim, Arieh. 2009. *Molecular theory of water and aqueous solutions*. World Scientific.

Ben-Naim, Arieh. 2010. *Discover Entropy and the Second Law of Thermodynamics: A Playful Way of Discovering a Law of Nature*. World Scientific. http://ariehbennaim.com/books/discover.html.

Bernal, J.D., & Fowler, R.H. 1933. A Theory of Water and Ionic Solution, with Particular Reference to Hydrogen and Hydroxyl Ions. *J. Chem. Phys.*, 515. doi: 10.1063/1.1749327.

Bernoulli, Daniel. 1738. *Hydrodynamica*. Johannes Reinhold Dulsecker, Basilea. http://www.scribd.com/doc/73264185/Daniel-Bernoulli-Hydrodynamica.

Boltzmann, Ludwig. 1995. *Lectures on Gas Theory*. Reprinted by Dover Publications.

Boyle, Robert. 1682. *New Experiments Physico-Mechanical: Touching the Spring of the Air and its Effects*. Miles Flesher, London. doi: 10.3931/e-rara-16019.

Bundy, F.P. 1989. Pressure-temperature phase diagram of elemental carbon. *Physica A*, **156**, 169.

CMS Collaboration. 2012. Observation of a new boson at a mass of 125 GeV with the CMS experiment at the LHC. *Phys. Lett. B*, **716**, 30. doi: 10.1016/j.physletb.2012.08.021.

Cooper, Leon N. 1969. *An Introduction to the Meaning and Structure of Physics*. Harper & Row, New York.

Dalton, John. 1808, 1810, 1827. *A New System of Chemical Philosophy.* Vol. 3 vols. Bickerstaff, Strand, London. http://webserver.lemoyne.edu/faculty/giunta/dalton.html.

de Finetti, Bruno. 1974, 1975. *Theory of Probability.* Vol. 1 & 2. John Wiley & Sons.

Denbigh, K.G. 1981. *The Principles of Chemical Equilibrium With Applications in Chemistry and Chemical Engineering.* Cambridge University Press. doi: 10.1002/aic.690300227.

Feynman, Richard, Leighton, Robert, & Sands, Matthew. 2010. *The Feynman Lectures on Physics.* New millennium edition edn. California Institute of Technology. http://www.feynmanlectures.info.

Galilei, Galileo. 1612. *Discorso intorno alle cose che stanno in su l'acqua.* Cosimo Giunti, Firenze. online: http://www.liberliber.it/mediateca/libri/g/galilei/discorso_intorno_alle_cose/html/index.htm; English translation: http://www.gutenberg.org/ebooks/37729.

Galilei, Galileo. 1623. *Il Saggiatore.* Giacomo Mascardi, Roma. online: http://it.wikisource.org/wiki/Il_Saggiatore; English translation (selection): http://www.princeton.edu/~hos/h291/assayer.htm.

Gibbs, Josiah Willard. 1874. *On the equilibrium of heterogeneous substances.* Transactions of the Connecticut Academy of Arts and Sciences, Vol. III, Part I. http://archive.org/details/OnequilibriumheOOGibb.

Gibbs, Josiah Willard. 1906. *The scientific papers of J. Willard Gibbs.* Longmans-Green, London. https://archive.org/stream/scientificpapers01gibbuoft.

Goh, & Chia. 1983. Teaching phase diagrams of sulphur and phosphorus. *Teaching and Learning,* **4**, 18–25.

Greene, Brian. 2004. *The fabric of the cosmos. Space, time, and the texture of reality.* Random House, Inc., New York.

Haynes, W.M., Bruno, T.J., & Lide, D.R. (eds). 2015. *Handbook of Chemistry and Physics.* 85-th edn. CRC Press. http://www.hbcpnetbase.com/.

Jaynes, Edwin T. 2003. *Probability Theory: The Logic of Science.* Cambridge University Press.

Khinchin, A.I. 1957. *Mathematical Foundation of Information Theory.* Dover, New York.

Kirkwood, J.G., & Buff, F.P. 1951. The Statistical Mechanical Theory of Solutions. I. *J. Chem. Phys.,* **19**, 774. doi: 10.1063/1.1748352.

McNaught, A.D., & Wilkinson, A. 1997. *IUPAC Compendium of Chemical Terminology.* 2-nd edn. Blackwell Science. ISBN 0-86542-6848. Internet edition: http://old.iupac.org/publications/compendium/index.html.

Pauling, Linus. 1935. The Structure and Entropy of Ice and of Other Crystals with Some Randomness of Atomic Arrangement. *J. Am. Chem. Soc.,* 2680–2684. doi: 10.1021/ja01315a102.

Pauling, Linus. 1939. *The Nature of the Chemical Bond and the Structure of Molecules and Crystals: An Introduction to Modern Structural Chemistry.* Cornell University Press. (3rd edition, 1960).

Ploetz, E.A., Bentenitis, N., & Smith, P.E. 2010. "Kirkwood-Buff integrals for ideal solutions". *Journal of Chemical Physics,* **132**, 164501.

Prigogine, Ilya, & Defay, Raymond. 1954. *Treatise on Thermodynamics: Chemical thermodynamics.* Longmans, Green.

Shannon, C.E. 1948. A Mathematical Theory of Communication. *The Bell System Technical Journal,* **27**, 379.

Sheehan, D.P., & Gross, D.H.E. 2006. Extensivity and the thermodynamic limit: Why size really does matter. *Physica A,* **370**, 461–482. doi: 10.1016/j.physa.2006.07.020.

Soper, A.K. 2000. The radial distribution functions of water and ice from 220 to 673 K and at pressures up to 400 MPa. *Chemical Physics,* 121–137.

Tribus, Myron, & McIrvine, Edward C. 1971. Energy and Information. *Scientific American,* **225**(3), 179–188.

Wada, Goro. 1961. A Simplified Model for the Structure of Water. *Bulletin of the Chemical Society of Japan,* **34**(7), 955–962. doi: 10.1246/bcsj.34.955.

West, John B. 2005. Robert Boyle's landmark book of 1660 with the first experiments on rarified air. *Journal of Applied Physiology,* **98**(1), 31–39. doi: 10.1152/japplphysiol.00759.2004.

Index

absolute zero, 32
activity coefficient, 304, 305, 307
adiabatic, 184
adiabatic process, 193
allotropes, 242
allotropy, 259
Amontons' law of
 pressure-temperature, 31
anthropic principle, 338
athermal, 184
atom, 40
atomic mass unit, 161
atomism, 40
average of the function, 62
average outcome, 60
average value, 61
Avogadro constant, 33
Avogadro's law, 33
azeotrope
 negative azeotrope, 279
 positive azeotrope, 279
azeotrope, 279

barometer, 28
Bernoulli's distribution, 66
Bernoulli's theorem, 54
binodal curve, 270
binomial coefficient, 50
binomial distribution, 66
binomial theorem, 50
bits, 85
boiling, 243

boiling point, 243
Boltzmann constant, 33
Boltzmann distribution, 119
Bose-Einstein condensate, 244
Bose-Einstein statistics, 244
bosons, 244
Boyle-Mariotte law, 29
Brownian motion, 43
bulk density, 284

canonical ensemble, 46
carbon nanotubes, 264
central moment, 64
certain event, 50
Charles's law, 31
chemical equilibrium, 248
chemical potential, 164, 282
Clapeyron equations, 254
Clausius-Clapeyron equation, 252
closed system, 183, 217
coefficient of thermal expansion, 196
coexistence curve, 270
complementary event, 51
complete, 76
compound event, 49
condensation, 243
conditional density, 284
conditional information, 95
conditional probability, 58
conditional SMI, 94
congruent composition, 276
congruent freezing point, 276